Human-Robot Interaction

Edited by

Daisuke Chugo

Human-Robot Interaction

Edited by Daisuke Chugo

Republished by InTech

Janeza Trdine 9, 51000 Rijeka, Croatia

Technical Editor Teodora Smiljanic

Additional hard copies can be obtained from orders@intechopen.com

Human-Robot Interaction, Edited by Daisuke Chugo
p. cm.
ISBN 978-953-307-051-3

Meet the editor

Daisuke Chugo, Associate Professor, School of Science and Technology, Department of Human System Interaction. Born in 1976. Graduated from the Department of Mechanical Engineering in the Faculty of Science and Technology at Tokyo University of Science. Completed the Master's Course in Mechanical Engineering from the Graduate School of Science and Technology in the Department of Mechanical Engineering at Tokyo University of Science. Completed the Doctoral Course in Mathematics, Electronics and Informatics from the Graduate School of Science and Engineering at Saitama University. Received his Ph.D. in engineering. In 2009, became an Assistant Professor in the School of Science and Technology at Kwansei Gakuin University after serving in such positions as researcher in a Tokyo University research facility and Assistant Professor in the Graduate School of Information Systems at The University of Electro-Communications. Assumed his current position in 2013. Has received numerous awards from, and presented numerous papers at, international conferences.

Preface

Robot's performance is increased greatly in recent years and their applications are not limited for industries and manufacturing. Robots are becoming a necessary part of our life and in near future, this trend will be increasing. In our future society, robots may nurse elderly, may look after children and may assist our household work. Furthermore, robots may work together with us in factories, offices and our homes. Robots may become our good friends. For realizing such a wonderful futures, there are still many hard problems both in technically and socially. As good friends know their characteristics each other, robots should have enough performance to know and understand the human.

Human-robot interaction (HRI) is the study of interactions between people (users) and robots. HRI is multidisciplinary with contributions from the fields of human-computer interaction, artificial intelligence, robotics, speech recognition, and social science (psychology, cognitive science, anthropology, and human factors). There has been a great deal of work done in the area of human-robot interaction to understand how a human interacts with a computer. However, there has been very little work done in understanding how people interact with robots. For robots becoming our friends, these studies will be required more and more.

Therefore, the aim of this book is to provide an overview of the state-of-art, to present new ideas, original results and practical experiences. The content of this book has been structured into 5 technical research sections with 18 chapters written by well-recognized researchers world-side. I hope the readers of this book enjoy its reading and this book helps their understanding on HRI.

Editor

Daisuke CHUGO
Kwansei Gakuin University, Hyogo
Japan

Contents

Human-Robot Communication

1

Understanding Activities and Intentions for Human-Robot Interaction

Richard Kelley, Alireza Tavakkoli, Christopher King,
Monica Nicolescu and Mircea Nicolescu
University of Nevada, Reno
United States of America

1. Introduction

As robots move from the factory and into the daily lives of men, women, and children around the world, it is becoming increasingly clear that the skills they will require are vastly different from the majority of skills with which they were programmed in the 20th century. In fact, it would appear that many of these skills will center on the challenge of interacting with humans, rather than with machine parts or other robots. To this end, modern-day roboticists are actively studying the problem of human-robot interaction - how best to create robots that can interact with humans, usually in a social setting. Among the many problems of human robot interaction, one of the most interesting is the problem of *intent recognition*: the problem of predicting the intentions of a person, usually just by observing that person. If we understand intentions to be non-observable goal-directed mental activities, then we may (quite understandably) view the intent recognition problem for robots as one of *reading peoples' minds.*

As grandiose as this claim may sound, we believe that this understanding of intent recognition is quite reasonable; it is this interpretation that we seek to justify in the following pages.

Every day, humans observe one another and on the basis of their observations "read people's minds," correctly inferring the intentions of others. Moreover, this ability is regarded not as remarkable, but as entirely ordinary and effortless. If we hope to build robots that are similarly capable of successfully interacting with people in a social setting, we must endow our robots with an ability to understand humans' intentions.

In this paper, we review the intent recognition problem, and provide as an example a system we have been developing to recognize human intentions. Our approach is ultimately based on psychological and neuroscientific evidence for a theory of mind (Premack & Woodruff, 1978), which suggests that the ease with which humans recognize the intentions of others is the result of an innate mechanism for representing, interpreting, and predicting other's actions. The mechanism relies on taking the perspective of others (Gopnick & Moore, 1994), which allows humans to correctly infer intentions.

Although this process is innate to humans, it does not take place in a vacuum. Intuitively, it would seem that our understanding of others' intentions depend heavily on the contexts in which we find ourselves and those we observe. This intuition is supported by

neuroscientific results (Iacobini et al., 2005), which suggest that the context of an activity plays an important and sometimes decisive role in correctly inferring underlying intentions.
Before considering this process in detail, we first look at some of the related work on the problem of intent recognition. After that, we reconsider the problem of intent recognition, looking at it from a new perspective that will shed light on how the process is accomplished. After looking at this re-framing of the problem, we consider some more general questions related to intent recognition, before moving on to describe a specific example system. We describe the architecture of our system, as well as experimental results we have obtained during validation of our system. We move on to describe some of the challenges facing future intent recognition systems, including planning based on recognized intentions, complexity of recognition, and the incorporation of novel sources of information for intent recognition systems. We then conclude with a summary of the central issues in the field of intent recognition.

2. Related work

Whenever one wants to perform statistical classification in a system that is evolving over time, hidden Markov models may be appropriate (Duda et al., 2000). Such models have been very successfully used in problems involving speech recognition (Rabiner, 1989). Recently, there has been some indication that hidden Markov models may be just as useful in modelling activities and intentions. For example, HMMs have been used by robots to perform a number of manipulation tasks (Pook and Ballard, 93), (Hovland et al., 96), (Ogawara et al., 2002). These approaches all have the crucial problem that they only allow the robot to detect that a goal has been achieved *after* the activity has been performed; to the extent that intent recognition is about prediction, these systems do not use HMMs in a way that facilitates the recognition of intentions. Moreover, there are reasons to believe (see Sec. 3) that without considering the disambiguation component of intent recognition, there will be unavoidable limitations on a system, regardless of whether it uses HMMs or any other classification approach.
The use of HMMs in intent recognition (emphasizing the prediction element of the intent recognition problem) was first suggested in (Tavakkoli et al., 2007). That paper also elaborates on the connection between the HMM approach and theory of mind. However, the system proposed there has shortcomings that the present work seeks to overcome.
The problem of intent recognition is also of great interest to researchers in neuroscience. Recent research in that field informs us that the mirror neuron system may play a role in intent recognition, and that contextual information is employed by the brain when ascribing intentions to others (Iacobini et al., 2005).

3. Reconsidering the intent recognition problem

Although some researchers consider the problems of activity recognition and intent recognition to be essentially the same, a much more common claim is that intent recognition differs from activity recognition in that intent recognition has a predictive component: by determining an agent's intentions, we are in effect making a judgment about what we believe are the likely actions of the agent in the immediate or near future. Emphasizing the predictive component of intent recognition is important, but may not reveal all of the significant facets of the problem.

In contrast with the more traditional view of intent recognition, we contend that *disambiguation* is an essential task that any completely functional intent recognition system must be capable of performing. In emphasizing the disambiguation component of an intent recognition system, we recognize that there are some pairs of actions that may appear identical in all respects *except* for their underlying intentions. To understand such pairs of activities, our system must be able to recognize intentions even when making intent-based predictions is not necessary.

For an example of intent recognition as disambiguation, consider an agent playing chess. When the agent reaches for a chess piece, we can observe that activity and ascribe to the agent any number of possible intentions. Before the game, an agent reaching for a chess piece may putting the piece into its initial position; during the game, the agent may be making a move using that piece; and after the game, the agent may be cleaning up and putting the piece away. In each of these cases, it is entirely possible (if not likely) that the activity of reaching for the piece will appear identical to the other cases. It is only the intentional component of each action that distinguishes it from the others. Moreover, this component is determined by the context of agent's activity: before, during, or after the game. Notice that we need to infer the agent's intention in this example even when we are not interested in making any predictions. Disambiguation in such circumstances is essential to even a basic understanding of the agent's actions.

4. Vision-based capabilities

We provide a set of vision-based perceptual capabilities for our robotic system that facilitate the modelling and recognition of actions carried out by other agents. As the appearance of these agents is generally not known a priori, the only visual cue that can be used for detecting and tracking them is image motion. Although it is possible to perform segmentation from an image sequence that contains global motion, such approaches -- typically based on optical flow estimation (Efros et al., 2003) -- are not very robust and are time consuming. Therefore, our approach uses more efficient and reliable techniques from real-time surveillance, based on background modelling and segmentation:

- During the *activity modelling* stage, the robot is moving while performing various activities. The appearance models of other mobile agents, necessary for tracking, are built in a separate, prior process where the static robot observes each agent that will be used for action learning. The robot uses an enhanced mean-shift tracking method to track the foreground object.
- During the *intent recognition* stage, the static robot observes the actions carried out by other agents. This allows the use of a foreground-background segmentation technique to build appearance models on-line, and to improve the speed and robustness of the tracker. The robot is stationary for efficiency reasons. If the robot moves during intent recognition we can use the approach from the modelling stage.

Fig. 1 shows the block diagram of the proposed object tracking frameworks.

4.1 Intent recognition visual tracking module

We propose an efficient Spatio-Spectral Tracking module (SST) to detect objects of interest and track them in the video sequence. The major assumption is that the observer robot is static. However, we do not make any further restrictions on the background composition, thus allowing for local changes in the background such as fluctuating lights, water fountains, waving tree branches, etc.

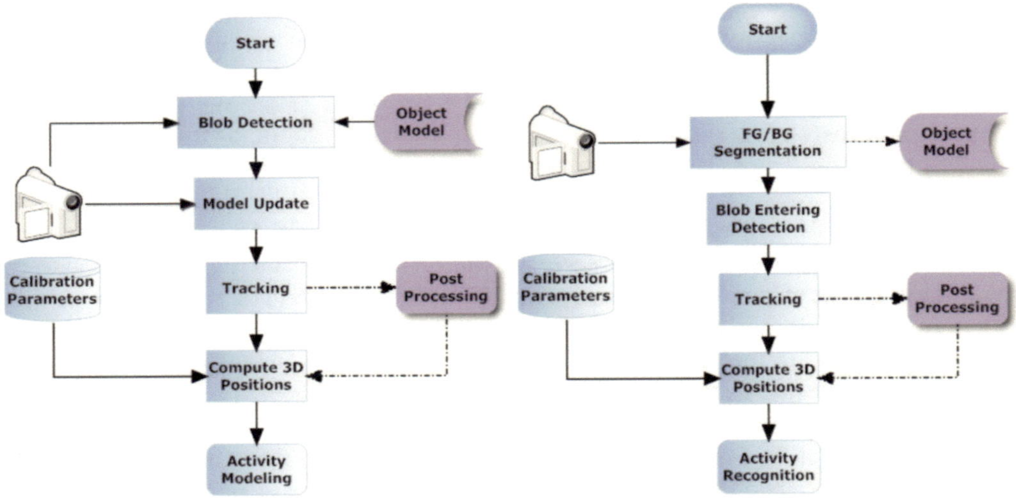

Fig. 1. The two object tracking frameworks for (a) *activity modelling* using a modified mean-shift tracker and (b) *intent recognition* using a spatio-spectral tracker.

The proposed system models the background pixel changes using an Incremental Support Vector Data Description module. The background model is then used to detect foreground regions in new frames. The foreground regions are processed further by employing a connected component processing in conjunction with a blob detection module to find objects of interest. These objects are tracked by their corresponding statistical models that are built from the objects' spectral (color) information. A laser-based range finder is used to extract the objects' trajectories and relative angles from their 2-D tracking trajectories and their depth in the scene. However, the spatio-spectral coherency of tracked objects may be violated in cases when two or more objects occlude each other.

A collision resolution mechanism is devised to address the issue of occlusion of objects of interest. This mechanism uses the spatial object properties such as their size, the relative location of their center of mass, and their relative orientations to predict the occlusion (collision).

4.2 Incremental support vector data description

Background modelling is one of the most effective and widely used techniques to detect moving objects in videos with a quasi-stationary background. In these scenarios, despite the presence of a static camera, the background is not completely stationary due to inherent changes, such as water fountains, waving flags, etc. Statistical modelling approaches estimate the probability density function of the background pixel values. If the data is not drawn from a mixture of normal distributions the parametric density estimation techniques may not be useful. As an alternative, non-parametric density estimation approaches can be used to estimate the probability of a given sample belonging to the same distribution function as the data set (Tavakkoli et al., 2006). However, the memory requirements of the non-parametric approach and its computational costs are high since they require the evaluation of a kernel function for all data samples.

Support Vector Data Description (SVDD) is a technique that uses support vectors in order to model a data set (Tax & Duin, 2004). The SVDD represents one class of known data samples in such a way that for a given test sample it can be recognized as known, or rejected as novel. Training of SVDDs is a quadratic programming optimization problem. This optimization converges by optimizing only on two data points with a specific condition (Platt, 1998) which requires at least one of the data points to violate the KKT conditions – the conditions by which the classification requirements are satisfied (Osuna et al., 1997). Our experimental results show that our SVDD training achieves higher speed and require less memory than the online and the canonical training (Tax & Duin, 2004).

4.3 Blob detection and object localization

In the blob detection module, the system uses a spatial connected component processing to label foreground regions from the previous stage. However, to label objects of interest a blob refinement framework is used to compensate for inaccuracies in physical appearance of the detected blobs due to unintended region split and merge, inaccurate foreground detection, and small foreground regions. A list of objects of interest corresponding to each detected blob is created and maintained to further process and track each object individually. This raw list of blobs corresponding to objects of interest is called the spatial connected component list.

Spatial properties about each blob such as its center and size are kept in the spatial connected component list. The list does not incorporate individual objects' appearances and thus is not solely useful for tracking purposes. The process of tracking individual objects based on their appearance in conjunction with their corresponding spatial features is carried out in the spatio-spectral tracking mechanism.

4.4 Spatio-spectral tracking mechanism

A system that can track moving objects (i.e. humans) requires a model for individual objects. These appearance models are employed to search for correspondences among the pool of objects detected in new frames. Once the target for each individual has been found in the new frame they are assigned a unique ID. In the update stage the new location, geometric and photometric information for each visible individual are updated. This helps recognize the objects and recover their new location in future frames.

Our proposed appearance modelling module represents an object with two sets of histograms, for the lower and upper half of the body. In the spatio-spectral tracking module a list of known objects of interest is maintained. This list represents each individual object and its corresponding spatial and color information along with its unique ID. During the tracking process the system uses the raw spatial connected component list as the list of observed objects and uses a statistical correspondence matching to maintain the ordered objects list and track each object individually. The tracking module is composed of three components:

- **Appearance modelling.** For each object in the raw connected component list a model is generated which contains the object center of mass, its height and width, the upper and lower section foreground masks, and the multivariate Gaussian distribution models of its upper and lower section pixels.
- **Correspondence matching.** The pixels in the upper and lower sections of each object in the raw list are used against each model in the ordered list of tracked objects. The winner model's ID then is used to represent the object.
- **Model update.** Once the tracking is performed the models will be updated. Any unseen object in the raw list is then assigned a new ID and their models are updated accordingly.

4.5 Collision resolution

In order for the system to be robust to collisions -- when individuals get too close so that one occludes the other-- the models for the occluded individual may not reliable for tracking purposes. Our method uses the distance of detected objects and uses that as a means of detecting a collision. After a collision is detected we match each of the individual models with their corresponding representatives. The one with the smallest matching score is considered to be occluded. The occluded object's model will not be updated but its new position is predicted by a Kalman filter. The position of the occluding agent is updated and tracked by a well-known mean-shift algorithm. After the collision is over the spatio-spectral tracker resumes its normal process for these objects.

5. Recognition system

5.1 Low-level recognition via hidden Markov models

As mentioned above, our system uses HMMs to model activities that consist of a number of parts that have intentional significance. Recall that a hidden Markov model consists of a set of *hidden states*, a set of *visible states*, a probability distribution that describes the probability of transitioning from one hidden state to another, and a probability distribution that describes the probability of observing a particular visible state given that the model is in a particular hidden state. To apply HMMs, one must give an interpretation to both the hidden states and the visible states of the model, as well as an interpretation for the model as a whole. In our case, each model represents a single well-defined activity. The hidden states of represent the intentions underlying the parts of the activity, and the visible symbols represent changes in measurable parameters that are relevant to the activity. Notice in particular that our visible states correspond to *dynamic* properties of the activity, so that our system can perform recognition as the observed agents are interacting.

As an example, consider the activity of *meeting* another person. To a first approximation, the act of meeting someone consists of approaching the person up to a point, interacting with the stationary person in some way (talking, exchanging something, etc.), and then parting. In our framework, we would model meeting using a single HMM. The hidden states would correspond to *approach, halt,* and *part,* since these correspond with the short-term intermediate goals of the meeting activity. When observing two people meeting, the two parameters of interest that we can use to characterize the activity are the distance and the angle between the two agents we're observing; in a meeting activity, we would expect that both the distance and the angle between two agents should decrease as the agents approach and face one another. With this in mind, we make the visible states represent changes in the distance and angle between two agents. Since each of these parameters is a real number, it can either be positive, negative, or (approximately) zero. There are then nine possibilities for a pair representing "change in distance" and "change in angle," and each of these nine possibilities represents a single visible state that our system can observe.

We train our HMMs by having our robot perform the activity that it later will recognize. As it performs the activity, it records the changes in the parameters of interest for the activity, and uses those to generate sequences of observable states representing the activity. These are then used with the Baum-Welch algorithm (Rabiner, 1989) to train the models, whose topologies have been determined by a human operator in advance.

During recognition, the stationary robot observes a number of individuals interacting with one another and with stationary objects. It tracks those individuals using the visual capabilities described above, and takes the perspective of the agents it is observing. Based on its perspective-taking and its prior understanding of the activities it has been trained to understand, the robot infers the intention of each agent in the scene. It does this using maximum likelihood estimation, calculating the most probable intention given the observation sequence that it has recorded up to the current time for each pair of interacting agents.

5.2 Context modeling

To use contextual information to perform intent recognition, we must decide how we want to model the relationship between intentions and contexts. This requires that we describe what intentions and contexts *are*, and that we specify how they are *related*. There are at least two plausible ways to deal with the latter consideration: we could choose to make intentions "aware" of contexts, or we might make contexts "aware" of intentions. In the first possibility, each intention knows all of the contexts in which it can occur. This would imply that we know in advance all contexts that are possible in our environment. Such an assumption may or may not be appropriate, given a particular application. On the other hand, we might make contexts aware of intentions. This would require that each context know, either deterministically or probabilistically, what intentions are possible in it. The corresponding assumption is that we know in advance all of the possible (or at least likely) intentions of the agents we may observe. Either of these approaches is possible, and may be appropriate for a particular application. In the present work, we adopt the latter approach by making each context aware of its possible intentions. This awareness is achieved by specifying the content of *intention models* and *context models*.

An intention model consists of two parts: first, an activity model, which is given by a particular HMM, and secondly a name. This is the minimal amount of information necessary to allow a robot to perform disambiguation. If necessary or desirable, intentions could be augmented with additional information that a robot could use to support interaction. As an example we might augment an intention model to specify an action to take in response to detecting a particular sequence of hidden states from the activity model.

A context model, at a minimum, must consist of a name or other identifier to distinguish it from other possible contexts in the system, as well as some method for discriminating between intentions. This method might take the form of a set of deterministic rules, or it might be a discrete probability distribution defined over the intentions about which the context is aware. In general, a context model can contain as many or as few features as are necessary to distinguish the intentions of interest. Moreover, the context can be either *static* or *dynamic*.

A static context consists of a name for the context and a probability distribution over all possible intentions. This is the simplest approach to context-based intent recognition in our framework, and is useful for modelling context that depends on unchanging location of an observer robot (as we would see in the case of a guard or service robot that only works in a single room or building), or on time or the date.

A dynamic context consists of features that are inferred by the observer. This could include objects that are being manipulated by the observed agents, visually detected features of the agents, or aspects of the environment that vary in hard-to-predict ways. In general, a dynamic context consists of a name and a probability distribution over *feature values* given the context. While being obviously more general than static context, a dynamic-context approach depends on good algorithms outside of the intent recognition domain, and can be (very) computationally expensive. However, the flexibility of the approach may justify the cost in a large number of potential applications.

Suppose that we have an activity model (*i.e.* an HMM) denoted by w. Let s denote an intention, let c denote a context, and let v denote a sequence of visible states from the activity model w. If we are given a context and a sequence of observation, we would like to find the intention that is maximally likely. Mathematically, we would like to find the s that maximizes $p(s \mid v, c)$, where the probability structure is determined by the activity model w.

We can further simplify matters by noting that the denominator is independent of our choice of s. Moreover, because the context is simply a distribution over intention names, the observable symbols are independent of the current context. Based on these observations, we can say that $p(s \mid v,c)$ is approximately equal to $p(v \mid s)p(s \mid c)$.

This approximation suggests an algorithm for determining the most likely intention given a series of observations and a context: for each possible intention s for which $p(s \mid c) > 0$, we compute the probability $p(v \mid s)p(s \mid c)$ and choose as our intention that s whose probability is greatest. Because we assume a static context, the probability $p(s \mid c)$ is available by assumption, and if the HMM w represents the activity model associated with intention s, then we assume that $p(v \mid s) = p(v \mid w)$. In our case this assumption is justified since our intention models contain only a name and an activity model, so that our assumption only amounts to assuming that observation sequences are independent of intention names.

5.3 Intention-based control

In robotics applications, simply determining an observed agent's intentions may not be enough. Once a robot knows what another's intentions are, the robot should be able to act on its knowledge to achieve a goal. With this in mind, we developed a simple method to allow a robot to dispatch a behavior based on its intent recognition capabilities. The robot first infers the global intentions of all the agents it is tracking, and for the activity corresponding to the inferred global intention determines the most likely local intention. If the robot determines over multiple time steps that a certain local intention has the largest probability, it can dispatch a behavior in response to the situation it believes is taking place.

For example, consider the activity of stealing an object. The local intentions for this activity might include "approaching the object," "picking up the object," and "walking off with the object." If the robot knows that in its current context the local intention "picking up the object" is not acceptable and it infers that an agent is in fact picking up the object, it can execute a behavior, for example stopping the thief or warning another person or robot of the theft.

6. Experimental validation

6.1 Setup

To validate our approach, we performed a set of experiments using a Pioneer 3DX mobile robot, with an on-board computer, a laser rangefinder, and a Sony PTZ camera. We trained our robot to understand three basic activities: *following*, in which one agent trails behind another; *meeting*, in which two agents approach one another directly; and *passing*, in which two agents move past each other without otherwise directly interacting.

We placed our trained robot in an indoor environment and had it observe the interactions of multiple human agents with each other, and with multiple static objects. In our experiments, we considered both the case where the robot acts as a passive observer and the case where the robot executes an action on the basis of the intentions it infers in the agents under its watch.

We were particularly interested in the performance of the system in two cases. In the first case, we wanted to determine the performance of the system when a single activity could have different underlying intentions based on the current context (so that, returning to our example in Sec. 3, the activity of "moving one's hand toward a chess piece" could be interpreted as "making a move" during a game but as "cleaning up" after the game is over). This case deals directly with the problem that in some situations, two apparently identical activities may in fact be very different, although the difference may lie entirely in contextually determined intentional component of the activity.

In our second case of interest, we sought to determine the performance of the system in disambiguating two activities that were in fact different, but due to environmental conditions appeared superficially very similar. This situation represents one of the larger stumbling blocks of systems that do not incorporate contextual awareness.

In the first set of experiments, the same visual data was given to the system several times, each with different a context, to determine whether the system could use the context alone to disambiguate agents' intentions. We considered three pairs of scenarios, which provided the context we gave to our system: leaving the building on a normal day/evacuating the building, getting a drink from a vending machine/repairing a vending machine, and going to a movie during the day/going to clean the theater at night. We would expect our intent recognition system to correctly disambiguate between each of these pairs using its knowledge of its current context.

The second set of experiments was performed in a lobby, and had agents meeting each other and passing each other both with and without contextual information about which of these two activities is more likely in the context of the lobby. To the extent that meeting and passing appear to be similar, we would expect that the use of context would help to disambiguate the activities.

Lastly, to test our intention-based control, we set up two scenarios. In the first scenario (the "theft" scenario), a human enters his office carrying a bag. As he enters, he sets his bag down by the entrance. Another human enters the room, takes the bag and leaves. Our robot was set up to observe these actions and send a signal to a "patrol robot" in the hall that a theft had occurred. The patrol robot is then supposed to follow the thief as long as possible.

In the second scenario, our robot is waiting in the hall, and observes a human leaving the bag in the hallway. The robot is supposed to recognize this as a suspicious activity and follow the human who dropped the bag for as long as possible.

6.2 Results

In all of the scenarios considered, our robot was able to effectively observe the agents within its field of view and correctly infer the intentions of the agents that it observed.

To provide a quantitative evaluation of intent recognition performance, we use two measures:

- *Accuracy rate* = the ratio of the number of observation sequences, of which the winning intentional state matches the ground truth, to the total number of test sequences.
- *Correct Duration* = C/T, where C is the total time during which the intentional state with the highest probability matches the ground truth and T is the number of observations.

The accuracy rate of our system is 100%: the system ultimately chose the correct intention in all of the scenarios in which it was tested. We consider the correct duration measure in more detail for each of the cases in which we were interested.

6.3 One activity, many intentions

Table 1 indicates the system's disambiguation performance. For example, we see that in the case of the scenario *Leave Building,* the intentions *normal* and *evacuation* are correctly inferred 96.2 and 96.4 percent of the time, respectively. We obtain similar results in two other scenarios where the only difference between the two activities in question is the intentional information represented by the robot's current context. We thus see that the system is able to use this contextual information to correctly disambiguate intentions.

Scenario (With Context)	Correct Duration [%]
Leave Building (Normal)	96.2
Leave Building (Evacuation)	96.4
Theater (Cleanup)	87.9
Theater (Movie)	90.9
Vending (Getting a Drink)	91.1
Vending (Repair)	91.4

Table 1. Quantitative Evaluation.

6.4 Similar-looking activities

As we can see from Table 2, the system performs substantially better when using context than it does without contextual information. Because *meeting* and *passing* can, depending on the position of the observer, appear very similar, without context it may be hard to decide what two agents are trying to do. With the proper contextual information, though, it becomes much easier to determine the intentions of the agents in the scene.

Meet (No Context) – Agent 1	65.8
Meet (No Context) – Agent 2	74.2
Meet (Context) - Agent 1	97.8
Meet (Context) – Agent 2	100.0

Table 2. Quantitative Evaluation.

6.5 Intention-based control

In both the scenarios we developed to test our intention-based control, our robot correctly inferred the ground-truth intention, and correctly responded the inferred intention. In the theft scenario, the robot correctly recognized the theft and reported it to the patrol robot in the hallway, which was able to track the thief (Figure 2). In the bag drop scenario, the robot correctly recognized that dropping a bag off in a hallway is a suspicious activity, and was able to follow the suspicious agent through the hall. Both examples indicate that intention-based control using context and hidden Markov models is a feasible approach.

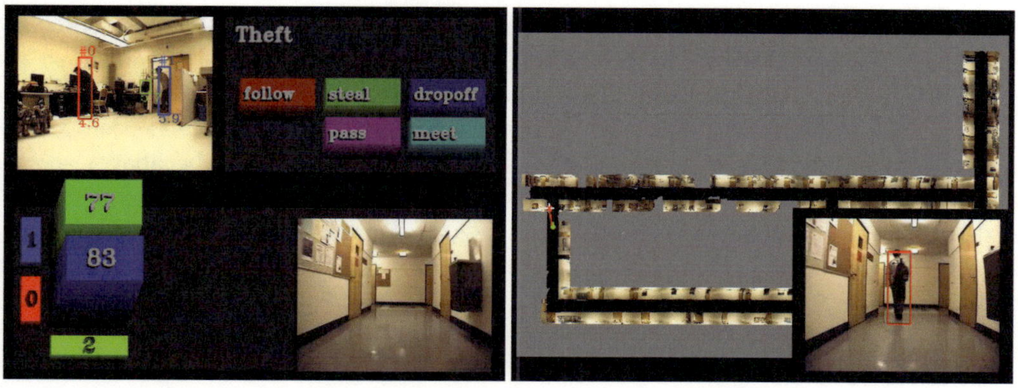

Fig. 2. An observer robot catches a human stealing a bag (left). The top left view shows the robot equipped with our system. The bottom right is the view of a patrol robot. The next frame (right) shows the patrol robot using vision and a map to track the thief.

6.6 Complexity of recognition

In real-world applications, the number of possible intentions that a robot has to be prepared to deal with may be very large. Without effective heuristics, efficiently performing maximum likelihood estimation in such large spaces is likely to be difficult if not impossible. In each of the above scenarios, the number of possible intentions the system had to consider was reduced through the use of contextual information. In general, such information may be used as an effective heuristic for reducing the size of the space the robot has to search to classify agents' intentions. As systems are deployed in increasingly complex situations, it is likely that heuristics of this sort will become important for the proper functioning of social robots.

7. Discussion

7.1 Strengths

In addition to the improved performance of a context-aware system over a context-agnostic one that we see in the experimental results above, the proposed approach has a few other advantages worth mentioning. First, our approach recognizes the importance of context in recognizing intentions and activities, and can successfully operate in situations that previous intent recognition systems have had trouble with.

Most importantly, though, from a design perspective it makes sense to separately perform inference for activities and for contexts. By "factoring" our solution in this way, we increase modularity and create the potential for improving the system by improving its individual parts. For example, it may turn out that another classifier works better than HMMs to model activities. We could then use that superior classifier in place of HMMs, along with an unmodified context module, to obtain a better-performing system.

7.2 Shortcomings

Our particular implementation has some shortcomings that are worth noting. First, the use of static context is inflexible. In some applications, such as surveillance using a set of stationary cameras, the use of static context may make sense. However, in the case of robots, the use of static context means that it is unlikely that the system will be able to take much advantage of one of the chief benefits of robots, namely their mobility.

Along similar lines, the current design of the intention-based control mechanism is probably not flexible enough to work "in the field." Inherent stochasticity, sensor limitations, and approximation error make it likely that a system that dispatches behaviors based only on a running count of certain HMM states is likely to run into problems with false positives and false negatives. In many situations (such as the theft scenario describe above), even a relatively small number of such errors may not be acceptable.

In short, then, the system we propose faces a few substantial challenges, all centering on a lack of flexibility or robustness in the face of highly uncertain or unpredictable environments.

8. Extensions

To deal with the problems of flexibility and scalability, we extend the system just described in two directions. First, we introduce a new source for contextual information, the lexical digraph. These data structures provide the system with contextual knowledge from linguistic sources, and have proved thus far to be highly general and flexible.

To deal with the problem of scalability, we introduce the *interaction space*, which abstracts the notion that people who are interacting are "closer" to each other than people who aren't, we are careful about how we talk about "closeness." In what follows, we outline these extensions, discussing how they improve upon the system described thus far.

9. Lexical digraphs

As mentioned above, our system relies on contextual information to perform intent recognition. While there are many sources of contextual information that may be useful to infer intentions, we chose to focus primarily on the information provided by object affordances, which indicate the actions that one can perform with an object. The problem, once this choice is made, is one of training and representation: given that we wish the system to infer intentions from contextual information provided by knowledge of object affordances, how do we learn and represent those affordances? We would like, for each object our system may encounter, to build a representation that contains the likelihood of all actions that can be performed on that object.

Although there are many possible approaches to constructing such a representation, we chose to use a representation that is based heavily on a graph-theoretic approach to natural language -- in particular, English. Specifically, we construct a graph in which the vertices are words and a labeled, weighted edge exists between two vertices if and only if the words corresponding to the vertices exist in some kind of grammatical relationship. The label indicates the nature of the relationship, and the edge weight is proportional to the frequency with which the pair of words exists in that particular relationship. For example, we may have vertices *drink* and *water*, along with the edge *((drink, water), direct_object, 4)*, indicating that the word "water" appears as a direct object of the verb "drink" four times in the experience of the system. From this graph, we compute probabilities that provide the necessary context to interpret an activity.

There are a number of justifications for and consequences of the decision to take such an approach.

9.1 Using language for context

The use of a linguistic approach is well motivated by human experience. Natural language is a highly effective vehicle for expressing facts about the world, including object affordances. Moreover, it is often the case that such affordances can be easily inferred directly from grammatical relationships, as in the example above.

From a computational perspective, we would prefer models that are time and space efficient, both to build and to use. If the graph we construct to represent our affordances is sufficiently sparse, then it should be space efficient. As we discuss below, the graph we use has a number of edges that is linear in the number of vertices, which is in turn linear in the number of sentences that the system "reads." We thus attain space efficiency. Moreover, we can efficiently access the neighbors of any vertex using standard graph algorithms.

In practical terms, the wide availability of texts that discuss or describe human activities and object affordances means that an approach to modelling affordances based on language can scale well beyond a system that uses another means for acquiring affordance models. The act of "reading" about the world can, with the right model, replace direct experience for the robot in many situations.

Note that the above discussion makes an important assumption that, although convenient, may not be accurate in all situations. Namely, we assume that for any given action-object pair, the likelihood of the edge representing that pair in the graph is at least approximately equal to the likelihood that the action takes place in the world. Or in other words, we assume that linguistic frequency well approximates action frequency. Such an assumption is intuitively reasonable. We are more likely to read a book than we are to throw a book; as it happens, this fact is represented in our graph. We are currently exploring the extent to which this assumption is valid and may be safely relied upon; at this point, though, it appears that the assumption is valid for a wide enough range of situations to allow for practical use in the field.

9.2 Dependency parsing and graph representation

To obtain our pairwise relations between words, we use the Stanford labeled dependency parser (Marneffe et al., 2006). The parser takes as input a sentence and produces the set of all pairs of words that are grammatically related in the sentence, along with a label for each pair, as in the "water" example above.

Using the parser, we construct a graph $G = (V,E)$, where E is the set of all labeled pairs of words returned by the parser for all sentences, and each edge is given an integer weight equal to the number of times the edge appears in the text parsed by the system. V then consists of the words that appear in the corpus processed by the system.

9.3 Graph construction and complexity

One of the greatest strengths of the dependency-grammar approach is its space efficiency: the output of the parser is either a *tree* on the words of the input sentence, or a graph made of a tree plus a (small) constant number of additional edges. This means that the number of edges in our graph is a linear function of the number of nodes in the graph, which (assuming a bounded number of words per sentence in our corpus) is linear in the number of sentences the system processes. In our experience, the digraphs our system has produced have had statistics confirming this analysis, as can be seen by considering the graph used in our recognition experiments. For our corpus, we used two sources: first, the simplified-

English Wikipedia, which contains many of the same articles as the standard Wikipedia, except with a smaller vocabulary and simpler grammatical structure, and second, a collection of childrens' stories about the objects in which we were interested. In Figure 3, we show the number of edges in the Wikipedia graph as a function of the number of vertices at various points during the growth of the graph. The scales on both axes are identical, and the graph shows that the number of edges for this graph does depend linearly on the number of vertices.

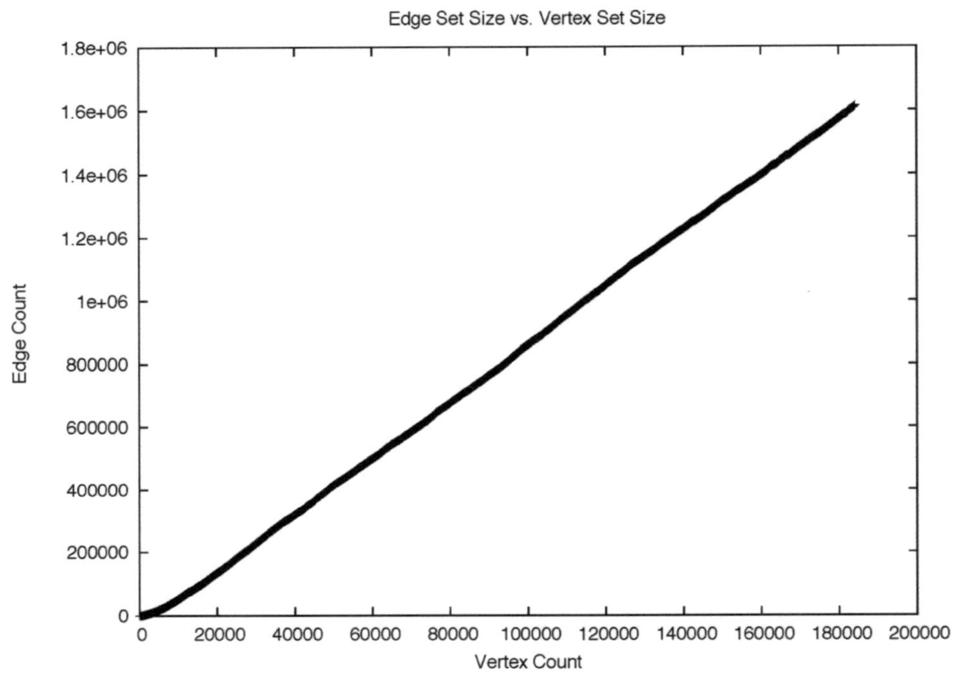

Fig. 3. The number of edges in the Wikipedia graph as a function of the number of vertices during the process of graph growth.

The final Wikipedia graph we used in our experiments consists of 244,267 vertices and 2,074,578 edges. The childrens' story graph is much smaller, being built from just a few hundred sentences: it consists of 1754 vertices and 3873 edges. This graph was built to fill in gaps in the information contained in the Wikipedia graph. The graphs were merged to create the final graph we used by taking the union of the vertex and edge sets of the graphs, adding the edge weights of any edges that appeared in both graphs.

9.4 Experimental validation and results

To test the lexical-digraph-based system, we had the robot observe an individual as he performed a number of activities involving various objects. These included books, glasses of soda, computers, bags of candy, and a fire extinguisher.

To test the lexically informed system, we considered three different scenarios. In the first, the robot observed a human during a meal, eating and drinking. In the second, the human was doing homework, reading a book and taking notes on a computer. In the last scenario, the robot observed a person sitting on a couch, eating candy. A trashcan in the scene then catches on fire, and the robot observes the human using a fire extinguisher to put the fire out.

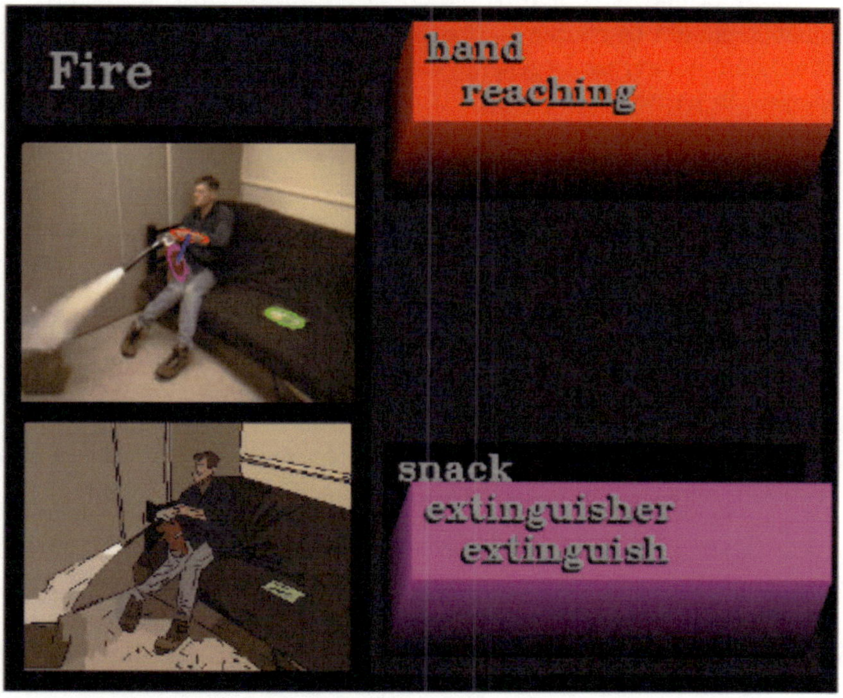

Fig. 4. The robot observer watches as a human uses a fire extinguisher to put out a trashcan fire.

Defining a ground truth for these scenarios is slightly more difficult than in the previous scenarios, since in these scenarios the observed agent performs multiple activities and the boundaries between activities in sequence are not clearly defined. However, we can still make the interesting observation that, except on the boundary between two activities, the correct duration of the system is 100%. Performance on the boundary is more variable, but it isn't clear that this is an avoidable phenomenon. We are currently working on carefully ground-truthed videos to allow us to better compute the accuracy rate and the correct duration for these sorts of scenarios. However, the results we have thus far obtained are encouraging.

10. Identifying interactions

The first step in the recognition process is deciding what to recognize. In general, a scene may consist of many agents, interacting with each other and with objects in the environment. If the scene is sufficiently complex, approaches that don't first narrow down the likely interactions before using time-intensive classifiers are likely to suffer, both in terms of performance and accuracy. To avoid this problem, we introduce the *interaction space* abstraction: for each identified object or agent in the scene, we represent the agent or object as a point in a space with a weak notion of distance defined on it. In this space, the points ideally (and in our particular models) have a relatively simple internal structure to permit efficient access and computation. We then calculate the distance between all pairs of points in this space, and identify as interacting all those pairs of entities for which the distance is less than some threshold. The goal in designing an interaction space model is that the distance function should be chosen so that the probability of interaction is decreasing in distance. We should not expect, in general, that the distance function will be a metric in the sense of analysis. In particular, there is no reason to expect that the triangle inequality will hold for all useful functions. Also, it is unlikely that the function will satisfy a symmetry condition: Alice may intend to interact with Bob (perhaps by secretly following him everywhere) even if Bob knows nothing about Alice's stalking habits. At a minimum, we only require nonnegativity and the trivial condition that the distance between any entity and itself is always zero. Such functions are sometimes known as premetrics.

For our current system, we considered four factors that we identified as particularly relevant to identifying interaction: distance in physical space, the angle of an entity from the center of an agent's field of view, velocity, and acceleration. Other factors that may be important that we chose not to model include sensed communication between two agents (this would be strongly indicative of interaction between two agents), time spent in and out of an agent's field of view, and others. We classify agents as interacting whenever a weighted sum of these distances is less than a human-set threshold.

10.1 Experimental validation and results

To test the interaction space model, we wished to use a large number of interacting agents behaving in a predictable fashion, and compare the results of an intent recognition system that used interaction spaces against the results of a system that did not. Given these requirements, we decided that the best approach was to simulate a large number of agents interacting in pre-programmed ways. This satisfied our requirements and gave us a well-defined ground truth to compare against.

The scenario we used for these experiments was very simple. The scenario consisted of $2n$ simulated agents. These agents were randomly paired with one another, and tasked with approaching each other or engaging in a wander/follow activity. We looked at collections of eight and thirty-two agents. We then executed the simulation, recording the performance of the two test recognition systems. The reasoning behind such a simple scenario is that if a substantial difference in performance exists between the systems in this case, then regardless of the absolute performance of the systems for more complex scenarios, it is likely that the interaction-space method will outperform the baseline system.

The results of the simulation experiments show that as the number of entities to be classified increases, the system that uses interaction spaces outperforms a system that does not. As we can see in Table 3, for a relatively small number of agents, the two systems have somewhat comparable performance in terms of correct duration. However, when we increase the number of agents to be classified, we see that the interaction-space approach *substantially* outperforms the baseline approach.

	8 Agents	32 Agents
System with Interaction Spaces	96%	94%
Baseline System	79%	6%

Table 3. Simulation results – correct duration.

11. Future work in intent recognition

There is substantial room for future work in intent recognition. Generally speaking, the task moving forward will be to increase the flexibility and generality of intent recognition systems. There are a number of ways in which this can be done. First, further work should address the problem of a non-stationary robot. One might have noticed that our work assumes a robot that is not moving. While this is largely for reasons of simplicity, further work is necessary to ensure that an intent recognition system works fluidly in a highly dynamic environment.

More importantly, further work should be done on context awareness for robots to understand people. We contend that a linguistically based system, perhaps evolved from the one described here, could provide the basis for a system that can understand behavior and intentions in a wide variety of situations.

Lastly, beyond extending robots' *understanding* of activities and intentions, further work is necessary to extend robots' ability to *act* on their understanding. A more general framework for intention-based control would, when combined with a system for recognition in dynamic environments, allow robots to work in human environments as genuine partners, rather than mere tools.

12. Conclusion

In this chapter, we proposed an approach to intent recognition that combines visual tracking and recognition with contextual awareness in a mobile robot. Understanding intentions in context is an essential human activity, and with high likelihood will be just as essential in any robot that must function in social domains. Our approach is based on the view that to be effective, an intent recognition system should process information from the system's sensors, as well as relevant social information. To encode that information, we introduced the lexical digraph data structure, and showed how such a structure can be built and used. We demonstrated the effectiveness of separating interaction identification from interaction classification for building scalable systems. We discussed the visual capabilities necessary to implement our framework, and validated our approach in simulation and on a physical robot.

When we view robots as autonomous agents that increasingly must exist in challenging and unpredictable human social environments, it becomes clear that robots must be able to understand and predict human behaviors. While the work discussed here is hardly the final say in the matter of how to endow robots with such capabilities, it reveals many of the challenges and suggests some of the strategies necessary to make socially intelligent machines a reality.

13. References

Duda, R.; Hart, P. & Stork, D. (2000). *Pattern Classification*, Wiley-Interscience

Efros, J.; Berg, A.; Morri, G. & Malik, J. (2003). "Recognizing action at a distance," *Intl. Conference on Computer Vision*.

Gopnick, A. & Moore, A. (1994). "Changing your views: How understanding visual perception can lead to a new theory of mind," in *Children's Early Understanding of Mind*, eds. C. Lewis and P. Mitchell, 157-181. Lawrence Erlbaum

Hovland, G.; Sikka, P. & McCarragher, B. (1996). "Skill acquisition from human demonstration using a hidden Markov model," *Int. Conf. Robotics and Automation* (1996), pp. 2706-2711.

Iacobini, M.; Molnar-Szakacs, I.; Gallese, V.; Buccino, G.; Mazziotta, J. & Rizzolatti, G. (2005). ``Grasping the Intentions of Others with One's Own Mirror Neuron System,'' *PLoS Biol* 3(3):e79

Marneffe, M.; MacCartney, B.; & Manning, C. (2006). "Generating Typed Dependency Parses from Phrase Structure Parses," *LREC*.

Ogawara, K.; Takamatsu, J.; Kimura, H. & Ikeuchi, K. (2002). "Modeling manipulation interactions by hidden Markov models," Int. Conf. Intelligent Robots and Systems (2002), pp. 1096-1101.

Osuna, E.; Freund, R.; Girosi, F. (1997) "Improved Training Algorithm for Support Vector Machines," *Proc. Neural Networks in Signal Processing*

Platt, J. (1998). "Fast Training of Support Vector Machines using Sequential Minimal Optimization," Advances in Kernel Methods - Support Vector Learning. MIT Press 185--208.

Pook, P. & Ballard, D. "Recognizing teleoperating manipulations," *Int. Conf. Robotics and Automation* (1993), pp. 578-585.

Premack D. & Woodruff, G. (1978). ``Does the chimpanzee have a theory of mind?'' *Behav. Brain Sci.* 1(4) 515-526

L. R. Rabiner, (1989). "A tutorial on hidden-Markov models and selected applications in speech recognition," in *Proc. IEEE* 77(2)

Tavakkoli, A., Nicolescu, M., Bebis, G. (2006). "Automatic Statistical Object Detection for Visual Surveillance." *Proceedings of IEEE Southwest Symposium on Image Analysis and Interpretation* 144--148

Tavakkoli, A.; Kelley, R.; King, C.; Nicolescu, M.; Nicolescu, M. & Bebis, G. (2007). "A Vision-Based Architecture for Intent Recognition," *Proc. of the International Symposium on Visual Computing*, pp. 173-182

Tax, D., Duin, R. (2004). "Support Vector Data Description." *Machine Learning* 54. pp. 45-66.

2

Interaction between a Human and an Anthropomorphized Object

Hirotaka Osawa and Michita Imai
Keio University
Japan

1. Introduction

Present-day home appliances have more functions, are more complicated, and are expected to process information together as more home networks and protocols are developed. This situation makes many users feel uneasy as they need to understand more complex information. They cannot intuitively understand what functions objects have and it has become more difficult to accept information from them in these situations. Therefore engineers are faced with a massive challenge to improve their interfaces and design products that facilitate easier use.

However, it is difficult to improve the designs and interfaces of all objects. Instead of improving the designs or the interfaces of objects, we preferred to provide information via anthropomorphic and communicative agents such as though a humanoid robot (Kanda et al., 2003) or a virtual agent (Mukawa et al., 2003), which seemed to be more useful and user friendly.

We propose a "display robot" as one agent system. It transforms an object into an by using anthropomorphization, which makes the interaction between humans and the object more intuitive. Users can understand the functions of objects more intuitively using the display robot and can accept information from them. We also think that the display robot can solve problems with impediments where users accept the agents themselves as "obstacles" to acquisition (Fukayama et al., 2003) (Fig. 1 top). The display robot does not use additional agents that are not related to an object, but it makes the object as additional agent that interacts with users (Fig. 1 bottom). As this situation does not create any additional agents in the field of interaction, users are not encumbered by additional information. It is also possible to identify the object's segments such as its "head" or "stomach" if it is anthropomorphized and has an imaginary body image. It can also use metaphorical and intuitive expressions for functions, such as "Something is wrong with my stomach" using the virtual body image.

We have already conducted an experiment to evaluate the anthropomorphization of an object (Osawa et al. 2006) and its virtual body image (Osawa et al. 2007). We used three anthropomorphized refrigerators in these experiments, the first was anthropomorphized by eye-like parts attached to its top, the second was anthropomorphized by the parts attached to its bottom, and the third was anthropomorphized by voice only. The study found that users can detect requests by an object more easily if it is anthropomorphized using the eye-like parts than if it is just the object itself. This indicated that the eye-like appearance reinforced the "body image of the stomach" in the situation where the Iris-board was attached to the top of the object, and users could recognize its top segment as the "head" and interact with it as such.

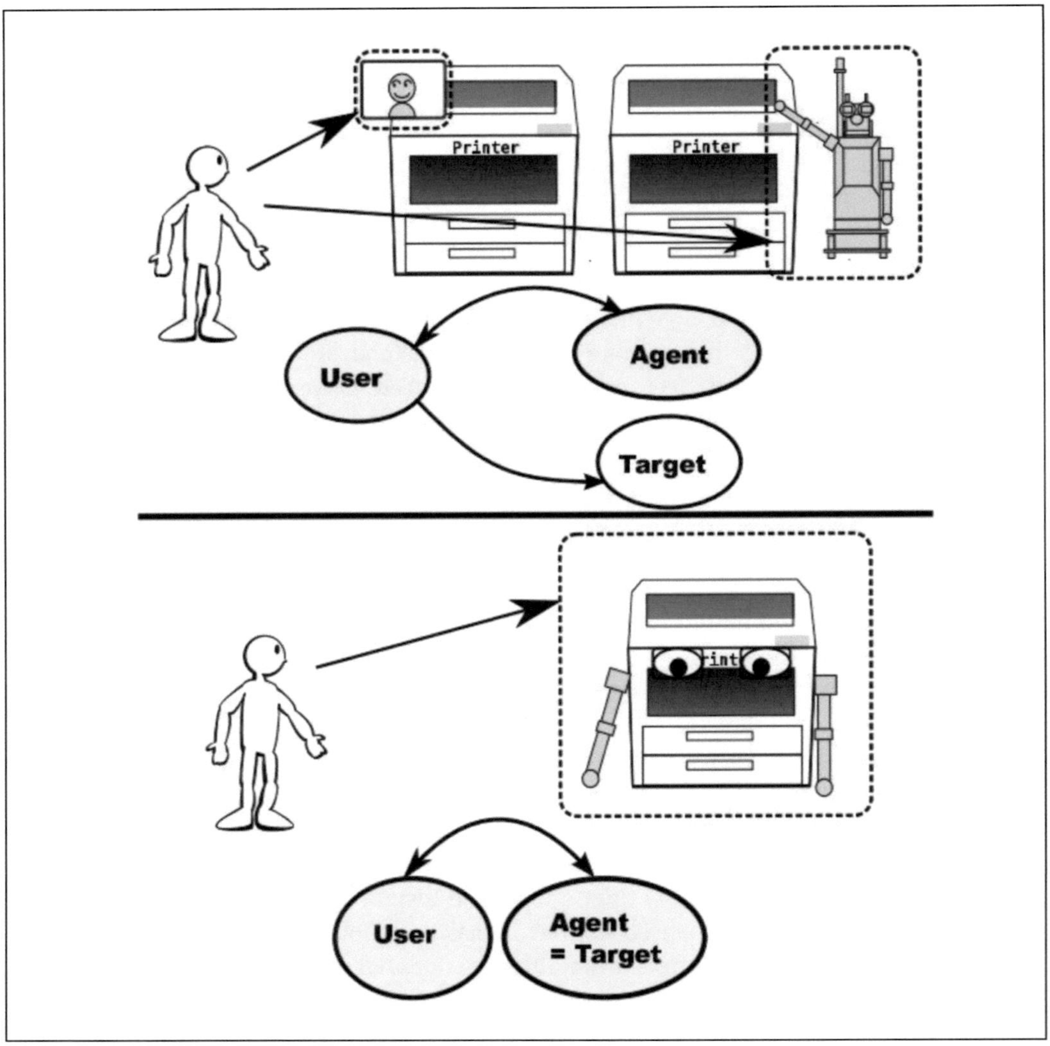

Fig. 1. Difference between anthropomorphic agent and display robot

However, these experiments were conducted with a limited category, i.e., university students. Therefore we needed to find what sorts of people (gender and age) accept anthropomorphized objects.

We developed eye-like parts and arm-like parts for this study, and we did on-the-spot research on human-object interaction by using these. Our result indicates that anthropomorphization by the display robot was accepted mostly by female participants and accepted by everyone except for those aged 10 to 19.

2. Design

2.1 Theoretical background

Reeves noted in Media Equation (Reeves & Nass, 1996) that people can accept objects as communicative subjects and act as if they had a "virtual" body under some circumstances. Their study revealed that we have the tendency to regard non-communicative objects as communicative agents.

Bateson et al. demonstrated the effect of anthropomorphization in an experiment using an honesty box (Bateson et al. 2006). They attached a picture of an eye to the top of a menu and participants gazed 2.76 times more at this than the picture of a flower that had also be attached to its top. Their study revealed that attaching human-like parts to a menu affects human actions.

The display robot extends this "virtual" body of an object that participants basically accept because human-like moving body parts have been attached to it to extend its subjectivity. For example, if washing machines are anthropomorphized, users can accept their door as being "mouths" (Fig. 2). Anthropomorphic agent on the machine is considered by C-Roids (Green et al. 2001). However, a user can accept machine's "virtual body" by attached display robot. So this kind of robot extends expression of machines more than C-Roids.

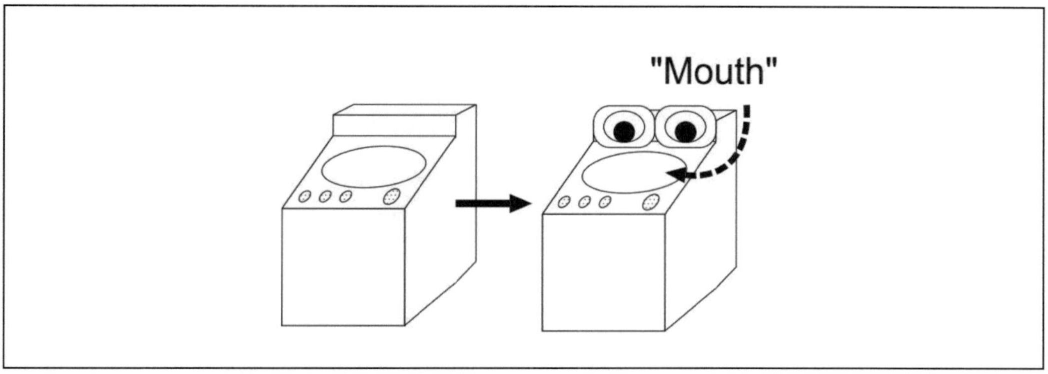

Fig. 2. Difference between anthropomorphic agent and display robot

We can convert instructions from the object using these virtual body images. For example, an anthropomorphized washing machine using a display robot can use intuitive expressions like "please throw it in my mouth" instead of "please throw it through the door." We think that these expressions are intuitive to users and they increase his or her intimacy with the object.

2.2 System construction

Figure 3 outlines the system construction for the display robot.

The display robot first calculates the scale of its virtual body image and determines its basic motions and voices for interaction. The main process runs on the scenario server (Fig. 3 center), which selects an appropriate scenario and generates speech and eye and arm motions according to the selected scenario. The eye and arms motions are affected by the scale and position of the virtual body image constructed according to the location of the user's face and locations of eye-like parts and arm-like parts.

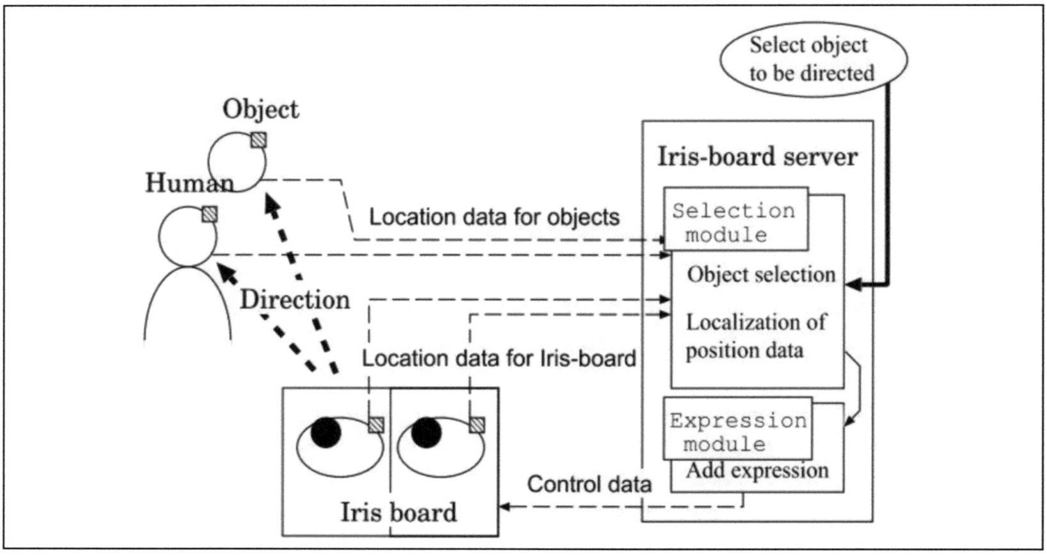

Fig. 3. System construction

2.3 Eye-like parts

The eye-like parts imitated human eyes.

The human eye (1) enables vision and (2) indicates what a person is looking at (Kobayashi & Kohshima, 2001). We focused on objects being looked at and hence used a positioning algorithm design.

The eye-like module that simulates the human eye (Fig. 4) uses an "iris" that represents the human iris and pupil together. The open elliptical region on the right in Fig. 4 represents the sclera and the closed circle, the iris and pupil. Here, the eye-like parts looking at a cup consist of a pair of displays to simulate the eyes. The locations of the irises are calculated with respect to the location of the object, which is acquired by a position sensor.

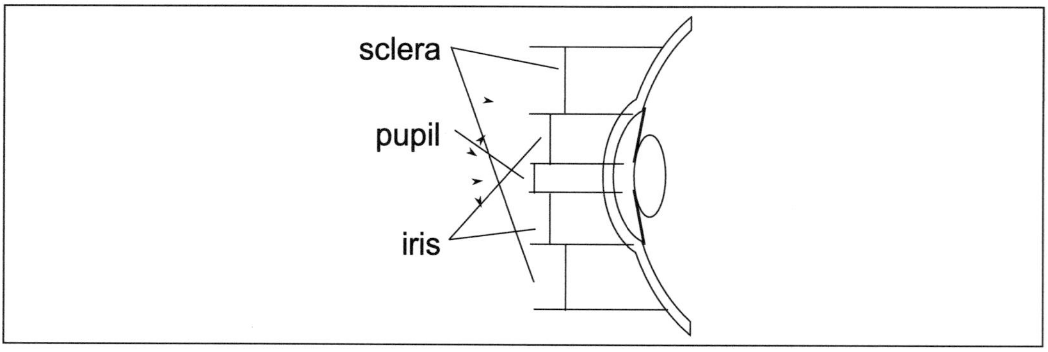

Fig. 4. Human Eye

First, it calculates each iris position as shown below. Each board has an "imaginary eyeball" and it calculates the point of intersection, $\boldsymbol{p}$, of a vector from the object, $\boldsymbol{i}$, to the center of the eyeball, $\boldsymbol{c}$, and board plane $\mathbf{A}$. Based on this point of intersection, the eye-like parts convert the global coordinates of $\boldsymbol{p}$ into display coordinates, $\boldsymbol{i}$; these processes are performed in both eye-like panels (Fig. 5).

Second, it calculates the orientation of the front of anthropomorphized target by the directions of two eye boards as shown below.
While calculating the normal vector ***a*** in certain cases, for example, if the eye-like parts are based on one panel, some additional sensors need to be used, e.g., gyros, to calculate the orientation of panel **A**.

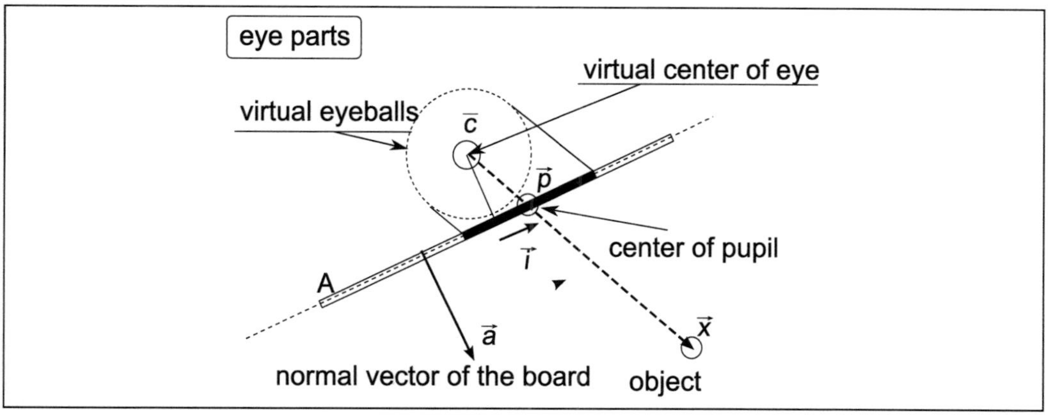

Fig. 5. Positioning of iris on each board

Since the eye-like parts use two panels, ***a*** is calculated from the vector ***r*** between the position sensors in the right and left panels. Restrictions exist when the two panels are symmetrically oriented with plane in the middle of the two boards, when the panels are placed vertically (i.e., their pitch angles are 90 degree), and when the tilt angle is known. Under these restrictions, the eye-like parts calculate the iris positions even if one of the two panels moves.

2.4 Arm-like parts

The arm-like parts of the robot imitated a human arm in all respects except in terms of manipulating objects.
When the arm-like parts pointed at the outside of an attached common object, we used the vector from the root of the limb to the tip of the hand as the pointing vector, as shown on the left side of Fig. 6 according to Sugiyama's study on pointing gestures of a communication robot (Sugiyama et al., 2006). However, when the arm-like parts pointed at the inside of an attached common object, we used the vector from the root of the hand to the tip of the hand as the pointing vector, as shown on the right side of Fig. 6.

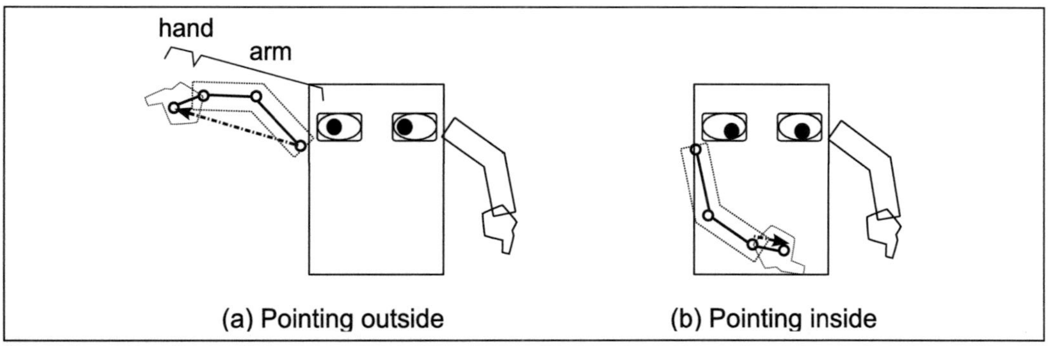

Fig. 6. Pointing vector

2.5 Implementation

The display robot did not need to manipulate other objects. Because the target already has its own task, and our devices are used for just expressionism. Instead of manipulation, these devices must be simple and light so they can be easily attached. We developed human-like robotic devices and attached them to our target by using hook and loop fasteners.
The eye-like parts are consisted of a TFT LC Panel. They were used to determine the positions of the pupils and irises using the 3-D coordinate of the places they were attached to and their direction vectors. The eye-like parts were 2-cm wide. They were thin and could be attached anywhere. They can be used to gaze in any directions as if the implemented eye of the object were watching.
The arm-like parts are consisted of six servo motors. Its hand had three motors and it could express delicate gestures with its fingers. The hands looked like long gloves, were covered with cloth, and concealed the implementation required for intuitive interaction.
The parts' locations are obtained from ultrasonic 3D tags (Nishida et al., 2003) on the parts. They send ultrasonic waves to implemented ultrasonic receivers, which calculate 3D axis of the tags. Humanoid parts search for "anthropomorphize-able" objects according to the locations of the parts.
Specifications of parts for an experiment are presented in Tables 1 and 2, and the parts are depicted in Fig. 7.

Scale	120mm x 160mm x 50mm
Weight	180g
TFT Controller	ITC-2432-035
Wireless module	ZEAL-Z1(19200bps)
Microcontroller	Renesas H8/3694
Connection method	Velcro tape
Cover	Sponge sheet, Plastic board

Table 1. Specification of eye parts

Scale	250mm x 40mm x 40mm
Weight	250g
Motor	Micro-MG x 3, GWS-pico x 3
Wireless module	ZEAL-Z1(9600bps)
Microcontroller	Renesas H8/3694
Connection method	Velcro tape
Cover	Aluminum, sponge, rubber, gloves

Table 2. Specification of arm parts

3. Research

We conducted research to attach the display robot to home appliances to evaluate it. Subjects were given an "invitation task" for interaction where an anthropomorphized home appliance directly invited users with its eyes and arms to interact.
We conducted research in a booth at a university laboratory. The research was conducted over two days. We did experiments for five hours on the first day and seven hours on the second day.

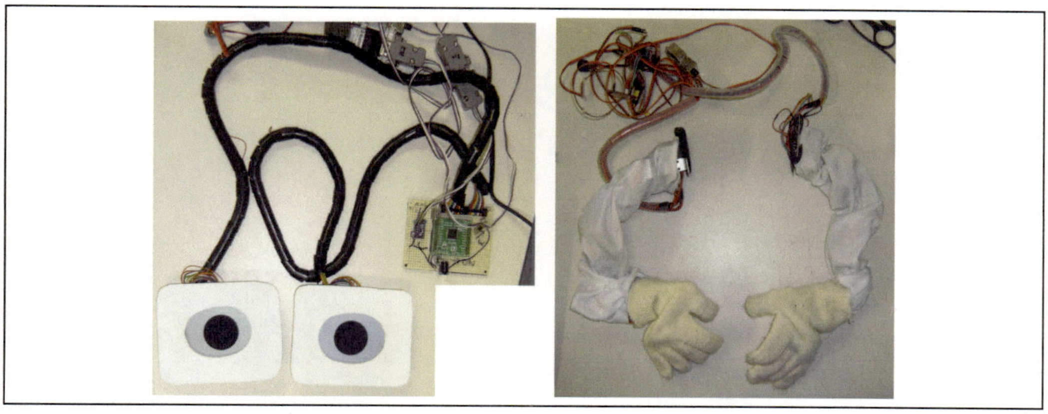

Fig. 7. Humanoid parts

The flow for the interaction between the display robot and users is mapped in Fig. 8. We first attached eye-like parts, arm-like parts, camera and speaker to the object and initialized the coordinates of all the devices. After they had been set up, the display robot detected the user's face with the camera and calculated its position. After it had detected the face, the display robot gazed at it by showing pupil and the iris on eye-like parts and directed him or her with the arm-like parts. If detection lasted 4 s, the display robot randomly chose voices from four alternatives ("Hello!", "Welcome!", "Hey!", and "Yeah!") and said one of these and beckoned to the user. The display robot with the devices attached invited users to a booth at the laboratory according to the flow in Fig. 8.

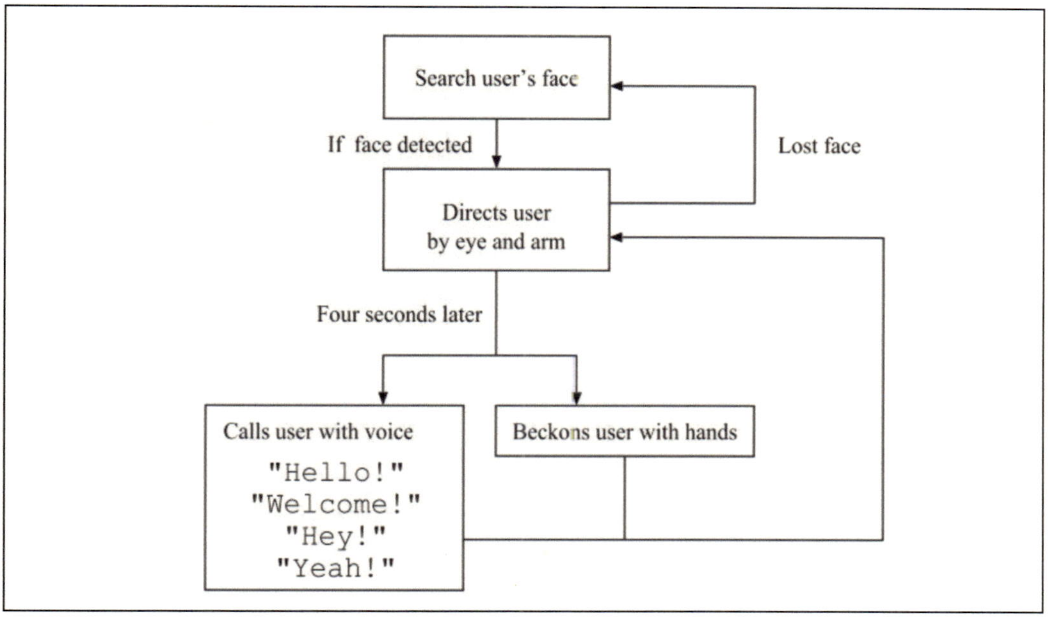

Fig. 8. Flow of interaction between display robot and users

We attached the display robot to a small trash box on a desk on the first day (Fig. 9 left), and attached it to an exercise bike on the second day (Fig. 9 right). We manually input the positions of all devices.

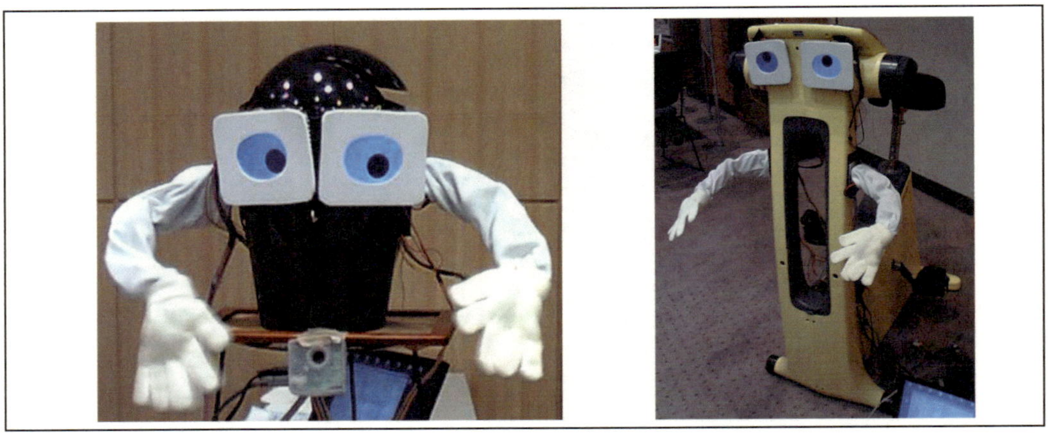

Fig. 9. Anthropomorphized trash box and exercise bike

3.1 Method of evaluation

We sent participants a questionnaire after the interactions. The questionnaire consisted of two parts, and participants answered it voluntarily. The first question consisted of a paired-adjective test (7-level evaluations of the 17 paired-adjective phrases in Table 3) and a free description of their impressions in watching and being called by the display robot.

Formal	Informal
Flexible	Inflexible
New	Old
Horrible	Gentle
Uninteresting	Interesting
Cold	Hot
Intimate	Not intimate
Unpleasant	Pleasant
Lively	Gloomy
Foolish	Wise
Plain	Showy
Slow	Fast
Selfish	Unselfish
Simple	Complex
Difficult to understand	Understandable
Weak	Strong
Cool	Queer

Table 3. Paired-adjective phrases

4. Result

There were 52 valid replies to the questionnaire (17 on the first day and 35 on the second). There were 31 male and 16 female participants (five did not identify their gender). Only 46 participants gave their age. The age of the participants ranged from under ten to over fifty years old. Most participants did not interact with the robots until the experiment started and then all the participants interacted with them.

4.1 Sociability value extracted using basic method of analysis

We could not evaluate the results obtained from the questionnaire (17 values from -3 to 3) by simply using the paired-adjective-test results, because participants were not obliged to complete the questionnaire. We applied a principal component analysis to the results of the paired-adjective-test to find hidden trends. We found six axes where the estimated values exceeded one. The results are listed in Table 4.

PC1: Sociability value (28.8%)	
Hot Cold	0.793
Flexible Inflexible	0.680
Fast Slow	0.657
Showy Plain	0.613
Wise Foolish	0.598
PC2: Uniqueness value (11.26%)	
Cool Weird	0.611
New Old	0.600
Plain Showy	0.526
Flexible Inflexible	0.451
PC3: Intuitiveness value (8.30%)	
Cool Weird	0.458
Understandable Difficult to understand	0.443
Horrible Gentle	0.438
PC4: Simplicity value (7.96%)	
Understandable Difficult to understand	0.490
Simple Complex	0.475
Lively Gloomy	0.391
PC5: Freshness value (7.06%)	
Cool Weird	0.480
Gentle Horrible	0.422
Flexible Inflexible	0.404
PC6: Intimateness value (6.40%)	
Intimate Not intimate	0.679
Selfish Unselfish	0.353
Plain Showy	0.321

Table 4. Categories using basic method of analysis

The most effective axis for evaluating the display robot was PC1 (sociability value) which affected results by approximately 30%. We calculated the sociability values of participants according to gender and age categories. As a result, the average value for male participants was -0.378 and the average value for female participants was 0.434 (Fig. 10). The average values by age are in Fig. 11. We also categorized participants who thought interaction was positive and those who thought interaction was negative according to situations involving watching and calling. The results are listed in Tables 5 and 6.

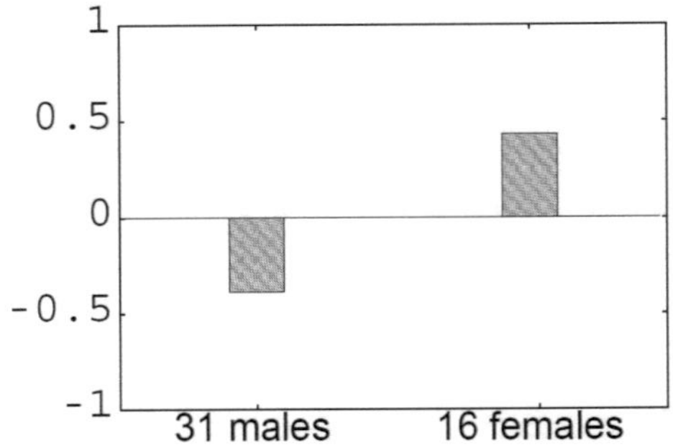

Fig. 10. Distribution of ``sociability value'' by gender

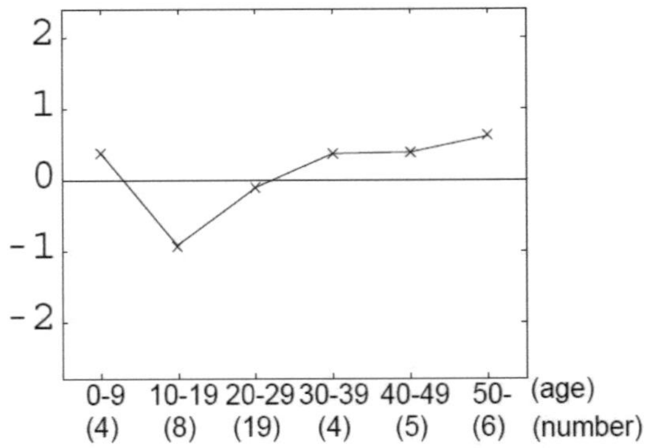

Fig. 11. Distribution of "sociability value" by age

Positive	Joyful! Surprised Cute It is very strange. It has eyeglasses. Very interesting.
Negative	Suspicious Horrible Terrible It was sure it watched me. I do not know whether it gazed at me or not. I did not understand it. I was terrified to think about what it would do. Its upward glance was unnatural.

Table 5. Impressions for watching action

Positive	Wonderful. I woke up Friendly Surprised! I felt good. I felt relieved if it be...
Negative	Surprised. Vague. All thing I could say was "Yes." Its timing was astonishing Machinelike. Confused. What did we must to do? I could not hear its voice. I was amazed. I was surprised because I did not think it could talk.

Table 6. Impressions for calling action

5. Discussion

5.1 Difference between genders

The sociability values between genders are plotted in Fig. 10. These results indicate that female participants had a more favorable impression of the display robot than the males. One female participant said that she felt the display robot was "cute and unique" in an uncoerced answer to the questionnaire. Other descriptions by female participants indicated that they saw the display robot intuitively and the object as a unified agent. Some female participants seemed surprised after the researcher had explained that the display robot and the object were separate devices.
The reason for the difference may have been because the female participants accepted the display robot and object as one unified character (agent) and felt good about it, but male participants accepted the display robot and object as separate devices. Male participants also only paid attention to the display robot's functions and found deficiencies in the devices. They felt the display robot was weirder than the female participants did.
We not only need to improve the accuracy of the display robot's devices but also to design a natural scenario for male users to increase their favorable impressions.

5.2 Differences between age groups

The sociability values for the six different age groups are plotted in Fig. 11. We can see that the values decrease for those under 10 years old an gradually increase for those over 10 years old.

The reasons for this phenomenon may be as follows. If participants are under 10 years old, they freely admit the object has eyes and arms. However, if they are 10 to 19, they think it is embarrassing to interact with anthropomorphized objects and their sociability values are decrease as a result. The experimenters found in observing the participants that those under 10 years of age acted aggressively with the display robot, pulling its arms or pushing its eyes, but those between 10 years of age to university-age students watched the display robot from a distance.

We also found that those who were more than 30 years old had greater sociability values than younger participants. This may have been because they could objectively interact with the anthropomorphized object, and felt less embarrassed because they were older.

These results indicate that 10 to 19 years olds had a tendency to find interaction with anthropomorphized objects to be embarrassing. We need to design a more attractive scenario where the 10 to 19 year old age group can interact with objects without being embarrassed.

5.3 Impressions for watching action

The results are listed in Tables 5 and 6.

Table 5 shows that participants who felt watching were negative said that they felt the object was horrible because it could not gaze at them accurately. It also shows that participants who felt watching was positive said that they felt the Iris-board itself was beneficial. We need to improve its gaze so that it is more precise by developing better accurate facial recognition and capturing a wider area with the camera to improve participants' impressions of the display robot.

5.4 Impressions for calling action

Table 6 shows that participants who felt calling was negative said that they could not understand the intentions of anthropomorphized objects and they could not respond to them. We expected that the invitation by an object using its eye gaze and beckoning would attract participants toward the object. The research results indicate that participants could not understand the "invitation by the object" because the trash box and exercise bike basically had no functions and there was no need to invite people. We found that we needed to design scenarios that extended the "intention of the object." For example, if the trash box is anthropomorphized, it needs to interact in the situation where "it needs to collect garbage" and if the exercise bike is anthropomorphized, it needs to interact in the situation where "it needs participants to exercise." However, participants who felt calling was positive says that they felt it was not only "cute or cool" but also "safe". This indicates that anthropomorphization increased the subjectivity of the objects and participants felt more glances from them.

6. Conclusion

This chapter proposed a display robot that acts as an agent to anthropomorphize objects by changing them, using devices that are like human body parts. We did research on the interaction between users and anthropomorphized objects using the displaying robot. As a result, we found that anthropomorphization by the display robot was mostly appreciated by female participants and accepted by people of all ages except for those aged 10 to 19.

However, we need to clarify how the virtual body image is created in the future and what interaction is possible by conducting more experiments and researches.

7. Acknowledgements

The first author was supported in part by the JSPS Research Fellowships for Young Scientists. This work was supported in part by Grant in Aid for the Global Center of Excellence Program for "Center for Education and Research of Symbiotic, Safe and Secure System Design from the Ministry of Education, Culture, Sport, and Technology in Japan."

8. References

Bateson, M.; Nettle, D. & Roberts, G. (2006). Cues of being watched enhance cooperation in a real-world setting'. *Biology Letters*, Vol.2, 2006, pp.412-414

Fukayama, A.; Pham, V. & Ohno, T. (2003). Acquisition of Body Image by Anthropomorphization Framework. *Proceedings of Joint 3rd International Conference on Soft Computing and Intelligent Systems and 7th International Symposium on Advanced Intelligent Systems*, Vol.103, No.743, pp. 53-58

Green, A. (2001). C-Roids: Life-like Characters for Situated Natural Language User Interface. *Proceedings of 15th International Symposium on Robot and Human Interactive Communication*, Vol.10, pp. 140-145, Bordeaux-Paris, France

Kobayashi, H.& Kohshima, S. (2001). Unique morphology of the human eye and its adaptive meaning: comparative studies on external morphology of the primate eye. *Journal of human evolution*, Vol. 40, No. 5, pp. 419—435

Mukawa, N.; Fukayama, A.; Ohno, T.; Sawaki, N. & Hagita, N. (2001). Gaze Communication between Human and Anthropomorphic Agent. *Proceedings of 10th International Symposium on Robot and Human Interactive Communication*, Vol.10, pp. 366-370, Bordeaux-Paris, France

Nishida, Y.; Aizawa, H.; Hori, T.; Hoffman, NH.; Kanade, T, & Kakikura, M. (2003). 3D ultrasonic tagging system for observing human activity. *Proceedings of International Conference on Intelligent Robots and Systems*, Vol.1, pp. 785-791

Osawa, H.; Mukai, J. & Imai, M. (2006). Anthropomorphization of an Object by Displaying Robot. *Proceedings of 15th International Symposium on Robot and Human Interactive Communication*, Vol.15, pp. 763-768, Hatfield, United Kingdom

Osawa, H.; Mukai, J. & Imai, M. (2007). Anthropomorphization Framework for Human-Object Communication. *Journal of Advanced Computational Intelligence and Intelligent Informatics*, Vol.11 No.8 , pp.1007-1014

Reeves, B. & Nass, C. (1996). The Media Equation: How People Treat Computers, Television, and New Media Like Real People and Places, *Univ. of Chicago Press*.

Sugiyama, O.; Kanda, T.; Imai, M.; Ishiguro, H. & Hagita, N. (2006). Three-layer model for generation and recognition of attention-drawing behavior. *Proceedings of International Conference on Intelligent Robots and Systems,* pp. 5843--5850, IEEE/RSJ

3

Probo, an Intelligent Huggable Robot for HRI Studies with Children

Kristof Goris, Jelle Saldien, Bram Vanderborght and Dirk Lefeber
Vrije Universiteit Brussel
Belgium

1. Introduction

Nowadays, robots are mostly known for their work in factories in industries such as automotive, electrical and electronics, chemical, rubber and plastics and many others. Robots are used for a wide variety of tasks like handling of materials and processes, welding and soldering, etc. The next generation of robots will be used in close collaboration with people in a wide spectrum of applications. For example, service robots will help elderly and assist disabled people. Household robots will be used in our homes and offices. Our children will play with entertainment robots. Medical robots assist in surgery and robotic prostheses replace limbs for amputees. Orthoses and exoskeletons can facilitate the rehabilitation process to regain mobility or manipulation skills. They also enlarge human strengths to carry heavy objects, for instance, nurses lifting a patient in and out of a bed. In lots of applications human and robots will work in a more close collaboration. For instance, NASA is developing a Robonaut to work together with astronauts in space.

Statistics from the International Federation of Robotics (IFR) show worldwide an increasing request of innovative robots, especially in non-automotive sectors. A growth can be noticed in both traditional and new markets, ranging from the industrial field to the service robotics, both for professional use and for domestic applications. However, European society has a different relationship towards robots than the Japanese or US society. The Japanese are more accepting of technological change. For instance, robots have always been a source of comics and amusement in Japan, making it easier to introduce robots into a personal environment. The Japanese Robot Association (JARA) predicts that the personal robot industry will be worth more than $50 billion dollars a year worldwide by 2025, compared with about $5 billion today. The technology of current industrial robots is insufficient to respond to this issue. This means that human-robot interaction using social robots must be studied in the different regions all over the world to address the different needs. Reasons why personal robotics is emerging now are the fact that actuators and sensors can be made very small and cheap, and the required computational power for computing real time all the software components is still increasing. And, last but not least, the markets for such robots are coming. Especially the aging population in Japan, Europe and United States faces a number of daunting societal problems. Also its dwindling work force and the increased cost of care demand out-of-the-box thinking. Companion and service robots are one part of the solution. However, the shift from industrial robots towards these service, personal and domestic robots leads to specific design criteria. For instance, an industrial robots can carry heavy loads with high accelerations to be able to work with high precision at high speed. Safety is established by putting them in cages, away from humans. This is possible because a factory is a well know environment. While the goal of the next generation of robots is to work in close collaboration with humans in daily life circumstances. These environments are often unknown and very dynamic. Tracking and precision performances become less stringent, while safety, cognition aspects, energy efficiency, etc. become the challenges to conquer. For acceptable human robot interaction, a good communication between the human and the robot is needed. According to Mehrabian (1968) its 7%-38%-55%-rule, most of our communication goes over non-verbal means, like facial expression and gestures. In order to communicate properly, the robot must be able to have those capabilities as well. That way the robot becomes a social robot. Idea of this approach is to adapt the communication with human-centered design instead of adapting to the technology of machines, which is now the case with computers and mobile devices.

2. The Probo project

The entire Probo project focuses on physical and cognitive human-robot interaction (HRI) especially with hospitalized children. A hospitalization can have serious physical and mental influences, especially on children. It confronts them with situations that are completely different from those at home. These children need to be distracted from the, in their eyes, scary and unfortunate hospital life, for instance, by getting in contact with their family and friends. Furthermore, they require moral support and they have specific needs for relevant information about their illness, the hospital environment, medical investigations, etc. Several projects already exist that aim to use Information and Communication Technologies (ICT) like internet and webcams to allow hospitalized children to stay in contact with their parents, to virtually attend lectures at school and to provide information as described by Fels et al. (2003). However, these ICT applications are usually computer animations displayed on PC, television screens or laptops. Breazeal (2002) shows the importance of embodied creatures during interaction with the environment and with others.

Animals could be such an embodied creature. In medical applications there exists animal assisted therapy (AAT) and animal-assisted activities (AAA). AAT and AAA are expected to have useful psychological, physiological and social effects. Some psychological studies, by Burch (1991), Allen et al. (1991), Ballarini (2003), have already shown that animals can be used to reduce heart and respiratory rate, lower levels of stress, progress mood elevation and social facilitation. Nonetheless animals are difficult to control, they always have a certain unpredictability, and they are possible carriers of disease and allergies. Therefore, the use of robots instead of animals has more advantages and has a better chance to be allowed in hospitals. There is existing early research on using robots in care settings for the elderly or mentally challenged, but the majority of these studies use wizard-of-oz methodologies to explore the patients' attitudes towards the robots. Recently, social pet robots are utilized just for these purposes, termed robot-assisted therapy (RAT). For example, the seal robot Paro by Shibata et al. (2001a) and Shibata et al. (2001) is used for pediatric therapy at university hospitals. Currently, Sony's dog robot AIBO by Tamura et al. (2004), Philips' cat robot iCat by van Breemen (2005) and Omron's cat robot Necoro (in Libin & Libin (2004)) are also being tested for RAT. However there is few research into using robots in therapeutic settings for young patients. Some research in this area is done by Dautenhahn (1999) and her co-workers studies into the interaction between autistic children and robots.

The development and construction of the social robot Probo, with the main ideas described in the former section in mind, is part of the entire project. Probo will serve as a multi-disciplinary research platform for similar studies where not only the cognitive HRI aspects are important, but also the physical HRI aspects such as touch and hug. The next section will show some remarkable robotic platforms used for research on cognitive human robot interaction and some of their features will be compared with those of the Probo platform in the section after it.

3. Remarkable social robots

In recent decades, research labs and companies all over the world are developing social robots. Social robots could be defined as robots that people anthropomorphize in order to interact with them. Pioneer robot is MIT's robot Kismet by Breazeal (2002). Kismet is an expressive anthropomorphic robotic head with twenty-one degrees of freedom (DOF). Three DOF are used to direct the robot's gaze, another three DOF control the orientation of its head, and the remaining fifteen DOF move its facial features such as eyelids, eyebrows, lips, and ears. To visually perceive the person who interacts with it, Kismet is equipped with a total of 4 color CCD cameras and a lavalier microphone is used to process vocalizations.
Kismets' successor Leonardo is developed in collaboration with the Stan Winston Studio. It combines the studio's artistry and expertise in creating compelling animatronics characters with state of the art research in socially intelligent robots. Leonardo has 69 degrees of freedom. With 32 of those in the face alone, Leonardo is capable of near-human facial expression. Moreover, Leonardo can gesture and is able to manipulate objects in simple ways. Leonardo is about 2.5 feet tall. Unlike the vast majority of autonomous robots today, Leonardo has an organic appearance. It is a fanciful creature, clearly not trying to mimic any living creature today. A camera in Leonardo's right eye captures images and a real-time face recognition system can be trained via simple social interaction with the robot. The interaction allows people to introduce themselves and others to Leonardo, who tries to memorize their faces for use in subsequent interactions.
The Huggable is another type of robotic companion being developed at the MIT Media Lab. It is being used for healthcare, education, and social communication applications (Stiehl et al. (2005)). It has a full body sensitive skin with over thousendfivehundred sensors, quiet back-drivable actuators, video cameras in the eyes, microphones in the ears, an inertial measurement unit, a speaker, and an embedded PC with 802.11g wireless networking. An important design goal of the Huggable is to make the technology invisible to the user. The movements, gestures and expressions of the bear convey a personality-rich character, not a robotic artefact. A soft silicone-based skin covers the entire bear to give it a more lifelike feel and heft, so you do not feel the technology underneath.
Nexi is being developed as a team member of four small mobile humanoid robots that possess a novel combination of mobility, moderate dexterity, and human-centric communication and interaction abilities (Breazeal et al. (2008)). The purpose of this platform is to support research and education goals in HRI, teaming, and social learning. MIT's collaborative partners in this project are UMASS Amherst, Meka Inc. and Xitome Design. Nexi has an expressive head with fifteen DOF in the face to support a diverse range of facial expressions including gaze, eyebrows, eyelids and an articulate mandible for expressive posturing. A four DOF neck mechanism support a lower bending at the base of the neck as well as pan-tilt-yaw of the head. Perceptual inputs include a colour CCD camera in each eye,

an indoor Active 3D IR camera in the head, four microphones to support sound localization and a wearable microphone for speech. The five DOF lower arm has forearm roll and wrist flexion. Each hand has three fingers and an opposable thumb. The thumb and index finger are controlled independently and the remaining two fingers are coupled. The fingers compliantly close around an object when flexed, allowing for simple gripping and hand gestures.

Keepon is a small creature-like robot designed to interact with children by directing attention and expressing emotion. It is developed by BeatBots LLC. The company's core design philosophy centres around cuteness, personality, simplicity, and rhythmic interaction. Keepon's minimal design makes its behaviors easy to understand, resulting in interactions that are enjoyable and comfortable, particularly important in the research on human social development. It has soft rubber skin, cameras in its eyes, and a microphone in its nose. Keepon has 4 degrees of freedom. Attention is directed by turning and nodding, while emotion is expressed by rocking side-to-side and bobbing up. It has been used since 2003 in research on social development and communication. Behaviors such as eye-contact, joint attention, touching, emotion, and imitation between Keepon and children of different ages and levels of social development have been studied. In the case of children with autism and other developmental disorders, one have had encouraging results with the use of Keepon as a tool for therapists, pediatricians, and parents to observe, study, and facilitate social interactions (Kozima et al. (2009)).

TOFU is a project that introduces a robotic platform for enabling new opportunities in robot based learning with emphasis on storytelling and artistic expression. This project introduces a socially expressive robot character designed to mimic the expressive abilities of animated characters by leveraging techniques that have been used in 2d animation for decades. Disney Animation Studios pioneered animation tools such as squash and stretch and secondary motion in the 50's. Such techniques have since been used widely by animators, but are not commonly used to design robots. TOFU can also squash and stretch. Clever use of compliant materials and elastic coupling, provide an actuation method that is vibrant yet robust. Instead of using eyes actuated by motors, TOFU uses inexpensive OLED displays, which offer highly dynamic and lifelike motion (Wistort & Breazeal (2009)).

Philips' robot cat iCat (van Breemen (2005)) is a plug & play desktop user-interface robot that is capable of mechanically rendering facial expressions ideal for studying human-robot interaction. The robot has been made available by Philips Research to stimulate research in this area further and in particular to stimulate research topics such as social robotics, humanrobot collaboration, joint-attention, gaming, and ambient intelligence. For facial expressions and body control iCat has eleven RC servos and two DC motors. Four multi-colour RGBLEDs and capacitive touch sensors are located in the feet and ears. The RGBLEDs can be used, for instance, to communicate iCat's mode of operation (e.g. sleeping, awake, busy, and listening). Besides the iCat itself, the iCat Research Community has been set. The goal of the community is to exchange experiences with the iCat Research Platform, brainstorm on new iCat projects or modifications, track bugs, and benchmark applications.

Another robot cat is NeCoRo developed by Omron. NeCoRo realizes natural human robot communication by its ability to react to human movement and express its own emotions. People pour their affection into this robot and feel attached to it as they would to a pet. NeCoRo has a synthetic fur giving it a feline appearance. Via internal sensors of touch, sound, sight, and orientation, it can perceive human action and thoughts. NeCoRo has fifteen actuators inside the body. NeCoRo has been used as a therapeutic tool for persons with dementia by Libin & Cohen-Mansfield (2002).

Sony's robot dog AIBO was the first commercially available robotic pet. Besides entertaining the user with its behaviours it can also read out web pages and emails and can therefore be considered as a robotic user interface. It is highly autonomous and with the additional "AIBO Life" program it also develops its own character and behaviours. Its interaction with humans is highly reactive. The user can initiate the interaction by giving a voice command or touching the robot, to which AIBO will react with a set of behaviours and expressions by LED display in its head.
Paro is a robotic user interface based on a baby of harp seal. It is developed by the National Institute of Advanced Industrial Science and Technology (AIST). It has a fur coat and is equipped with several sensors, like ubiquitous surface contact sensor, whisker sensor, stereoscopic optical sensor, a microphone for voice recognition and 3D source orientation, temperature sensor to control body temperature, and a posture sensor. Paro has 9 DOF for movement of eyelids, upper body, front paw and hind-limb. It responses to various stimuli like, daily rhythm (morning-midday-night-time) and it shows animal-mimic. Paro has been used as a mental commit robot in AAT by Shibata et al. (2001b).
Since 1997 a platform named ROBOTA dolls exists, it is a family of mini humanoid robots based on a doll. They can engage in complex interaction with humans, involving speech, vision and body imitation. The Robota robots have been applied as assistive technologies in behavioral studies with low-functioning children with autism Dautenhahn (1999).

4. The huggable robot Probo

4.1 A story about Probo

One of the unique features of Probo, compared to other similar projects, is that this character has its own identity, which is of major importance for communication and emotional interaction with children. Classical animators are masters at conveying intentionality through characters. In the "Illusion of Life" ,Thomas & Johnston (1981) stress the importance of emotive expression for making animated characters believable. They argue that it is how characters express themselves that conveys apparent beliefs, intents, and desires to the human observer. In order for Probo to become a believable character, the identity of Probo includes a name, a family and a history. By developing an imaginary creature MIT's philosophy (Breazeal (2003)) is followed. They believe that robots are not and will never be dogs, cats, humans, etc. so there is no need to make them look as such. Rather, robots will be their own kind of creature and should be accepted, measured, and valued on those terms.
The name Probo is derived from the word Proboscidea. Proboscidea is an order that now contains only one family of living animals, Elephantidae or "the elephants", with three species (African Bush Elephant, African Forest Elephant, and Asian Elephant)Wilson & Reeder (2005) (see Figure 1). In the name Probo we can also see the word "ROBO" which emphasizes the robotic nature of Probo. Also the word "PRO" is recognized to underline the positive effects on research aspects on one side and education and welfare of children on the other side. The history of Probo starts in the Ice Age where he lived among other similar species such as the elephant-like mammoths and mastodons. About 12.000 years ago, warmer, wetter weather began to take hold. The Ice Age was ebbing. As their habitats disappeared most of the Ice Age creatures became extinct. Probo managed to migrate north and was frozen underneath the ice-cap at the North Pole. Due to recent global warming the polar caps started to melt and create large floating chunks of ice drifting into open sea. Probo escaped inside such a chunk of ice and finally arrived at mainland Europe. His quest

here is to help children overcome their difficulties and diseases and to bring more joy into their lives.

Fig. 1. Origin of Probo.

4.2 Design of Probo

The first prototype of the robot Probo has a fully actuated head and trunk, giving a total of twenty DOF. By moving its head (3 DOF), eyes (3 DOF), eyelids (2 DOF), eyebrows (4 DOF), ears (2 DOF), trunk (3 DOF) and mouth (3 DOF) the robot is able to express its emotions Goris et al. (2009). Probo is about 80cm tall and it feels like a stuffed animal. In contrast with other robotic heads, a special body part, namely the trunk, is added to intensify certain emotional expressions and to increase interactivity. Due to its actuated head, Probo is, in contrast with other comparable companion robots such as Paro, Huggable, AIBO and Necoro, capable of expressing a wide variety of facial expressions as shown in Figure 2. Philip's iCat has also the ability to render mechanically facial expression with emotions, but lacks the huggable appearance and warm touch that attracts children. A section view of Probo is shown in Figure 2.

To build safety aspects intrinsically in the robot's hardware all the actuators have a flexible components in series, this kind of actuation is referred to as soft or compliant actuation. In case of a collision the robot will be elastic and will not harm the child who's interacting with it. A triple layered construction also contributes to the safe interactions and soft touch for the user. The layered construction (Figure 2) consists of hard ABS covers mounted on the aluminium frame of the robot. The first covers shield the internals and protects the internal mechatronics. These covers are encapsulated in a PUR foam layer, which act as the second layer. The third layer is a removable fur-jacket. The fur-jacket can be washed and disinfected. The use of the soft actuation principle together with well-thought designs concerning the robot's filling and huggable fur, are both essential to create Probo's soft touch feeling and ensure safe interaction. Furthermore, Probo is equipped with a wide range of sensory input devices, such as a digital camera, microphones and force sensing resistor (FSR) touch sensors under the fur. These sensors give the robot the ability to capture the stimuli from its environment. With various inputs from these sensors, and outputs like emotional facial expressions and a affective nonsense speech, Probo becomes a true robotic user interface. Another special and unique feature on the Probo platform is the touch screen in the belly of the robot. This creates a window to the outside world and through the use of wireless internet it opens up a way to implement new and/or existing computer applications such as social networking, teleconferencing, etc Saldien et al. (2008).

Fig. 2. Left: A section view of Probo. Right: 6 basic emotions (happy, surprise, sad, anger, fear and disgust) by Probo.

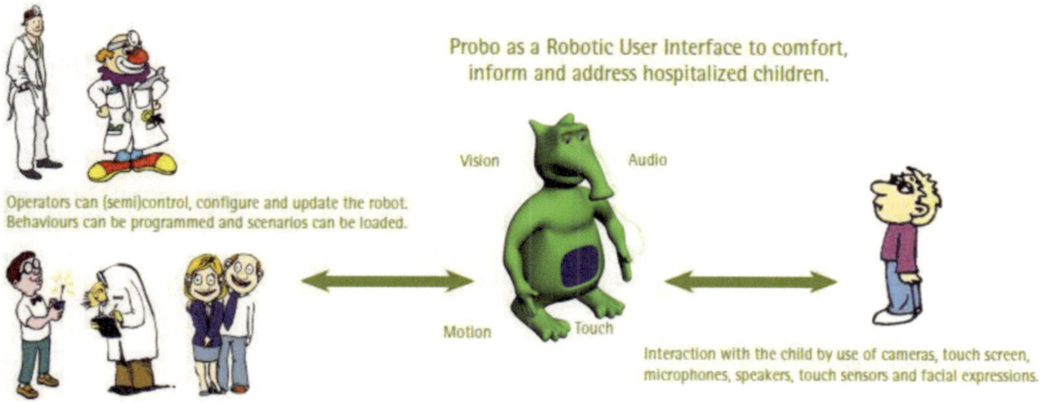

Fig. 3. As a robotic user interface (RUI) between the operators and a child.

4.3 Probo as an interface

At first, Probo is used as a Robotic User Interface (RUI) (Figure 3) interacting with the children and controlled by an operator Goris et al. (2008). The operator can be anyone who wants to communicate with the child, in particularly caregivers and researchers. At the moment there is a shared control between the operator evoking behaviors, emotions and scenarios, and the robot, performing intrinsic (pre-programmed) autonomous reactions. The robot reacts on basic input stimuli and performs pre-programmed scenarios. The input stimuli, that can be referred to as the low-level perceptions, are derived from vision analysis, audio analysis and touch analysis. Those stimuli will influence the attention- and emotion-system, used to set the robot's point of attention, current mood and corresponding facial expression. The vision analysis includes the detection of faces, objects and later also facial features such as facial expressions. Audio analysis includes detecting the direction and intensity of sounds and later the recognition of emotions in speech. Touch analysis gives the location and force of touch. These are classified into painful, annoying or pleasant touch. A larger classification of haptic interactions will be developed. Now, the prototype is being tested as a RUI interacting with children and controlled by an operator. Some basic behaviors are included; e.g. playing animations, controlling the facial expressions and point of attention. The goal for Probo is to gradually increase its autonomy by implementing new cognitive modules in the RUI to finally obtain an intelligent autonomous social robot. In the mean time Probo will serve as a platform to test and implement new modules for input, processing and output. More pictures and videos of the robot in action can be found on http://probo.vub.ac.be.

5. Future work

Gestures play a major role in many aspects of human life. They are a crucial part of everyday conversation. The word gesture is used for many different phenomena involving human movement, especially of the hands and arms. Only some of these are interactive or communicative. The pragmatics of gesture and meaningful interaction are quite complex cf. Kendon (1970), Mey (2001), Kita (2003). Applications of service or "companion" robots that interact with humans, will increasingly require human-robot interaction (HRI) in which the robot can recognize what humans are doing and to a limited extent why they are doing it, so that the robot may act appropriately. Most of the research now being done is focused on the recognition of human gestures from the robot's point of view. In the line of the Probo project we want to focus on the human point of view, how does a human perceive a robot and its gestures. The language of gesture allows humans to express a variety of feelings and thoughts, from contempt and hostility to approval and affection. Social robots will need the ability of gesturing in order to express social-emotional behavior. Therefore more research has to be done to address the following questions. Which gestures are necessary for social interaction and how can they be implemented in robots? The next generation Probo will therefore be equipped with actuated arms and hands.

6. Conclusions

This chapter surveys some of the research trends in social robotics and its applications to human-robot interaction (HRI). The past four years a unique robotic research platform, called Probo, is developed by the Robotics & Multibody Mechanics (R&MM) group to study physical and cognitive human-robot interaction (HRI) with a special focus on children. The robot Probo is designed to act as a social interface, providing a natural interaction while employing human-like social cues and communication modalities. The concept of the huggable robot Probo is a result of the desire to improve the living conditions of children in hospital environment. These children need distraction and lots of information. Probo can be used in hospitals, as a tele-interface for entertainment, communication and medical assistance. Probo has to be seen as an imaginary animal based on the ancient mammoths. By giving the robot a origin, Probo gets an identity in contrast to other similar robots. The huggable and child friendly robot pal Probo is able to communicate naturally with people using nonverbal cues. Therefore Probo uses its actuated head with eyes, eyelids, eyebrows, trunk, mouth and ears. With these parts Probo is able to express its emotions by showing facial expressions and changing its gaze in contrast with other comparable robots such as: Paro, Huggable, AIBO and Necoro. Philip's iCat has facial expression of emotions, but lacks the huggable appearance and warm touch that attracts children. Besides the prototype of the real robot, a virtual model has been developed. With user friendly software this model can be used as an interface between an operator and a child. Probo will emphasize its expression of emotions by the use of a nonsense affective speech. The next generation Probo will be equipped with arms and hands, together with movements of its torso. Probo will then be able to enforce its emotional expressions with gestures and body language. That way Probo becomes even more the ideal robotic user interface of a research platform for experiments concerning cognitive human robot interaction with great opportunities in different disciplines such as robotics, artificial intelligence, design, sociology, psychology, and many more.

7. References

Allen, K., Blascovich, J., Tomaka, J. & Kelsey, R. (1991). Presence of human friends and pet dogs as moderators of autonomic responses to stress in women, *Journal of personality and social psychology* 61(4): 582–589.

Ballarini, G. (2003). Pet therapy. Animals in human therapy., *Acta Biomed Ateneo Parmense* 74(2): 97–100.

Breazeal, C. (2002). *Designing Sociable Robots*, Mit Pr.

Breazeal, C. (2003). Toward sociable robots, *Robotics and Autonomous Systems* 42(3-4): 167–175.

Breazeal, C., Siegel, M., Berlin, M., Gray, J., Grupen, R., Deegan, P., Weber, J., Narendran, K. & McBean, J. (2008). Mobile, dexterous, social robots for mobile manipulation and human-robot interaction, *International Conference on Computer Graphics and Interactive Techniques*, ACM New York, NY, USA.

Burch, M. (1991). Animal-assisted therapy and crack babies: A new frontier, *Pet Partners Program: A Delta Society Newsletter* .

Dautenhahn, K. (1999). Robots as social actors: Aurora and the case of autism, *Proc. CT99, The Third International Cognitive Technology Conference, August, San Francisco*, pp. 359–374.

Fels, D., Shrimpton, B. & Roberston, M. (2003). Kids in hospital, kids in school.

Goris, K., Saldien, J. & Lefeber, D. (2009). Probo: a testbed for human robot interaction, *Proceedings of the 4th ACM/IEEE international conference on Human robot interaction*, ACM New York, NY, USA, pp. 253–254.

Goris, K., Saldien, J., Vanderniepen, I. & Lefeber, D. (2008). The Huggable Robot Probo, a Multi-disciplinary Research Platform, *Proceedings of the EUROBOT Conference*, pp. 22- 24.

Kendon, A. (1970). Movement coordination in social interaction: some examples described., *Acta Psychologica* 32(2): 100.

Kita, S. (2003). *Pointing: Where language, culture, and cognition meet*, Lawrence Erlbaum.

Kozima, H., Michalowski, M. & Nakagawa, C. (2009). A Playful Robot for Research, Therapy, and Entertainment, *Int J Soc Robot* 1: 3–18.

Libin, A. & Cohen-Mansfield, J. (2002). Robotic cat NeCoRo as a therapeutic tool for persons with dementia: A pilot study, *Proceedings of the 8 th International Conference on Virtual Systems and Multimedia, Creative Digital Culture*, pp. 916–919.

Libin, A. & Libin, E. (2004). Person–Robot Interactions From the Robopsychologists' Point of View: The Robotic Psychology and Robotherapy Approach, *Proceedings of the IEEE* 92(11): 1789–1803.

Mehrabian, A. (1968). Communication without words, *Psychology Today* 2(4): 53–56.

Mey, J. (2001). *Pragmatics: an introduction*, Blackwell publishers.

Saldien, J., Goris, K., Yilmazyildiz, S., Verhelst, W. & Lefeber, D. (2008). On the design of the huggable robot Probo, *Journal of Physical Agents* 2(2): 3.

Shibata, T., Mitsui, T., Wada, K., Touda, A., Kumasaka, T., Tagami, K. & Tanie, K. (2001a). Mental commit robot and its application to therapy of children, *Advanced Intelligent Mechatronics, 2001. Proceedings. 2001 IEEE/ASME International Conference on 2.*

Shibata, T., Mitsui, T., Wada, K., Touda, A., Kumasaka, T., Tagami, K. & Tanie, K. (2001b). Mental commit robot and its application to therapy of children, *2001 IEEE/ASME International Conference on Advanced Intelligent Mechatronics, 2001. Proceedings*, Vol. 2.

Shibata, T., Wada, K., Saito, T. & Tanie, K. (2001). Robot Assisted Activity for Senior People at Day Service Center, *Proc. of Int. Conf. on Information Technology in Mechatronics* pp. 71- 76.

Stiehl, W., Lieberman, J., Breazeal, C., Basel, L., Lalla, L. & Wolf, M. (2005). Design of a therapeutic robotic companion for relational, affective touch, *Robot and Human Interactive Communication, 2005. ROMAN 2005. IEEE International Workshop on* pp. 408–415.

Tamura, T., Yonemitsu, S., Itoh, A., Oikawa, D., Kawakami, A., Higashi, Y., Fujimooto, T. & Nakajima, K. (2004). Is an Entertainment Robot Useful in the Care of Elderly People With Severe Dementia?, *Journals of Gerontology Series A: Biological and Medical Sciences* 59(1): 83–85.

Thomas, F. & Johnston, O. (1981). *Disney animation: The illusion of life*, Abbeville Press New York.

van Breemen, A. (2005). iCat: Experimenting with Animabotics, *Proceedings, AISB 2005 Creative Robotics Symposium.*

Wilson, D. & Reeder, D. (2005). *Mammal Species of the World: A Taxonomic and Geographic Reference*, Johns Hopkins University Press.

Wistort, R. & Breazeal, C. (2009). TOFU: a socially expressive robot character for child interaction, *Proceedings of the 8th International Conference on Interaction Design and Children*, ACM New York, NY, USA, pp. 292–293.

4

Scaling Effects for Synchronous vs. Asynchronous Video in Multi-robot Search

Huadong Wang[1], Prasanna Velagapudi[2], Jijun Wang[3], Paul Scerri[2], Michael Lewis[2] and Katia Sycara[2]
[1]*School of Information Sciences, University of Pittsburgh*
[2]*Robotics Institute, Carnegie Mellon*
[3]*Quantum Leap Innovations, Newark, DE*
USA

1. Introduction

Practical applications of robotics can be classified by two distinct modes of operation. Terrestrial robotics in tasks such as surveillance, bomb disposal, or pipe inspection has used synchronous realtime control relying on intensive operator interaction usually through some form of teleoperation. Interplanetary and other long distance robotics subject to lags and intermittency in communications have used asynchronous control relying on labor intensive planning of waypoints and activities that are subsequently executed by the robot. In both cases planning and decision making are performed primarily by humans with robots exercising reactive control through obstacle avoidance and safeguards. The near universal choice of synchronous control for situations with reliable, low latency communication suggests a commonly held belief that experientially direct control is more efficient and less error prone. When this implicit position is rarely discussed it is usually justified in terms of "naturalness" or "presence" afforded by control relying on teleoperation. (Fong & Thorpe, 2001) observe that direct control while watching a video feed from vehicle mounted cameras remains the most common form of interaction. The ability to leverage experience with controls for traditionally piloted vehicles appears to heavily influence the appeal for this interaction style.

1.1 Viewpoint for robot control

Control based on platform mounted cameras, however, is no panacea. (Wickins & Hollands, 1999) identify 5 viewpoints used in control, three of them, immersed, tethered, and "plan view" can be associated with the moving platform while 3rd person (tethered) and plan views require fixed cameras. In the immersed or egocentric view (A) the operator views the scene from a camera mounted on the platform. The field of view provided by the video feed is often much narrower than human vision, leading to the experience of viewing the world through a soda straw from a foot or so above the ground. This perceptual impairment leaves the operator prone to numerous, well-known operational errors, including disorientation, degradation of situation awareness, failure to recognize hazards, and simply overlooking relevant information (Darken et al., 2001; McGovern, 1990). A sloped surface, for example,

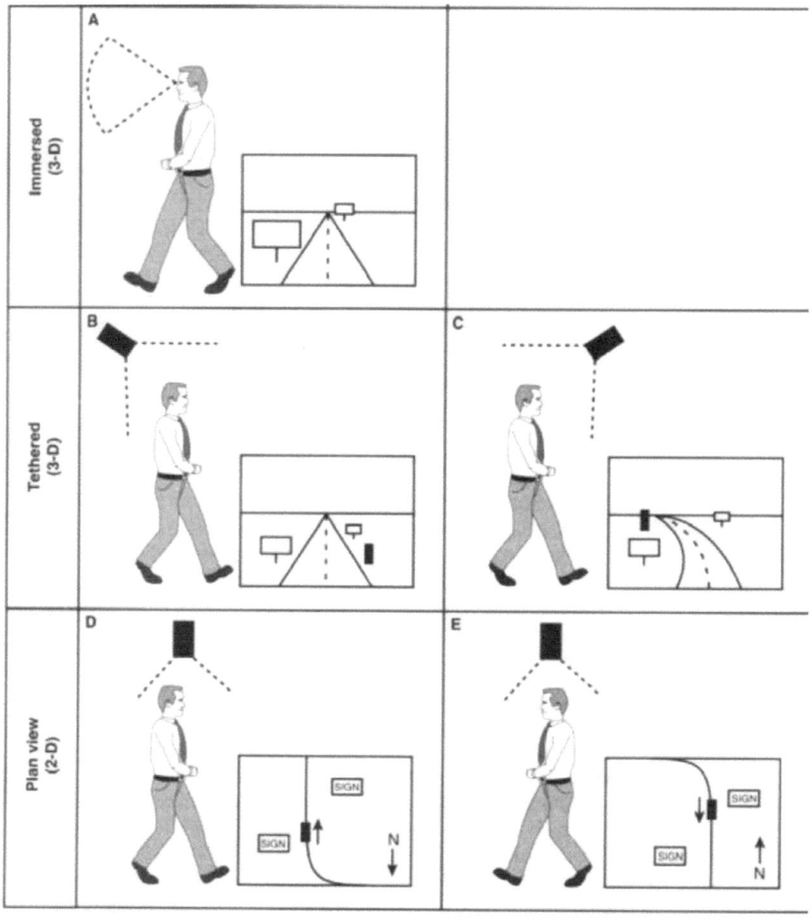

Fig. 1. Viewpoints for control from (Wickens & Hollands, 1999)

gives the illusion of being flat when viewed from a camera mounted on a platform traversing that surface (Lewis et al., 2007). For fixed cameras the operator's ability to survey a scene is limited by the mobility of the robot and his ability to retain viewed regions of the scene in memory as the robot is manoeuvred to obtain views of adjacent regions. A pan-tilt-zoom (PTZ) camera resolves some of these problems but introduces new ones involving discrepancies between the robots heading and the camera view that frequently lead to operational mishaps (Yanco et al., 2004). A tethered "camera" (B, C) provides an oblique view of the scene showing both the platform and its 3D environment. A 3rd person fixed view (C) is akin to an operator's view controlling slot cars and has been shown effective in avoiding roll-overs and other teleoperation accidents (McGovern, 1990) but can't be used anywhere an operator's view might be obstructed such as within buildings or in rugged terrain. The tethered view (B) in which a camera "follows" an avatar (think Mario Brothers©) is widely favored in virtual environments (Milgram, 1997; Tan et al., 2001) for its ability to show the object being controlled in relation to its environment by showing both the platform and an approximation of the scene that might be viewed from a camera mounted on it. This can be simulated for robotic platforms by mounting a camera on a flexible pole giving the operator a partial view of his platform in the environment (Yanco &

Drury, 2004). Because of restriction in field of view and the necessity of pointing the camera downward, however, this strategy is of little use for surveying a scene although it can provide a view of the robot's periphery and nearby obstacles that could not be seen otherwise. The exocentric views show a 2 dimensional version of the scene such as might be provided by an overhead camera and cannot be obtained from an onboard camera. This type of "overhead" view can, however, be approximated by a map. For robots equipped with laser range finders, generating a map and localizing the robot on that map provides a method for approximating an exocentric view of the platform. If this view rotates with the robot (heading up) it is a type D plan view. If it remains fixed (North up) it is of type E. An early comparison at Sandia Laboratory between viewpoints for robot control (McGovern, 1990) investigating accidents focused on the most common of these: (A) egocentric from onboard camera and (C) 3rd person. The finding was that all accidents involving rollover occurred under egocentric control while 3rd person control led to bumping and other events resulting from obstructed or distanced views.

1.2 Multi-robot search

Remotely controlled robots for urban search and rescue (USAR), robots are typically equipped with both a PTZ video camera for viewing the environment and a laser range finder for building a map and localizing the robot on that map. The video feed and map are usually presented in separate windows on the user interface and intended to be used in conjunction. While (Casper & Murphy, 2003) reporting on experiences in searching for victims at the World Trade Center observed that it was very difficult for an operator to handle both navigation and exploration of the environment from video information alone, (Yanco & Drury, 2004) found that first responders using a robot to find victims in a mock environment made little use of the generated map. (Nielsen & Goodrich, 2006) by contrast, have attempted to remedy this through an ecological interface that fuses information by embedding the video display within the map. The resulting interface takes the 2D map and extrudes the identified surfaces to derive a 3D version resembling a world filled with cubicles. The robot is located on this map with the video window placed in front of it at the location being viewed. Result shows that search generated maps to be superior in assisting operators to escape from a maze.

When considering such potential advantages and disadvantages of viewpoints it is important to realize that there are two, not one, important subtasks that are likely to engage operators (Tan et al., 2001). The escape task was limited to *navigation,* the act of explicitly moving the robot to different locations in the environment. In many applications search, the process of acquiring a specific viewpoint—or set of viewpoints—containing a particular object may be of greater concern. Because search relies on moving a viewpoint through the environment to find and better view target objects, it is an inherently egocentric task. This is not necessarily the case for navigation which does not need to identify objects but only to avoid them.

Search, particularly multi-robot search, presents the additional problem of assuring that areas the robot has traversed have been thoroughly searched for targets. This requirement directly conflicts with the navigation task which requires the camera to be pointed in the direction of travel in order to detect and avoid objects and steer toward its goal. These difficulties are accentuated by the need to switch attention among robots which may increase the likelihood that a view containing a target will be missed. In earlier studies (Wang & Lewis, 2007a; Wang & Lewis 2007b) we have demonstrated that success in search is directly related to the frequency with which the operator shifts attention between robots and hypothesized that this might be due to victims missed while servicing other robots. Recent data (Wang et al., 2009), however, suggests that other effects involving situation awareness may be involved.

1.3 Asynchronous Imagery

To combat these problems of attentive sampling among cameras, incomplete coverage of searched areas, and difficulties in associating camera views with map locations we are investigating the potential of asynchronous control techniques previously used out of necessity in NASA applications as a solution to multi-robot search problems. Due to limited bandwidth and communication lags in interplanetary robotics camera views are closely planned and executed. Rather than transmitting live video and moving the camera about the scene, photographs are taken from a single spot with plans to capture as much of the surrounding scene as possible. These photographs taken with either an omnidirectional overhead camera (camera faces upward to a convex mirror reflecting 360°) and dewarped (Murphy, 1995, Shiroma et al., 2004) or stitched together from multiple pictures from a ptz camera (Volpe, 1999) provide a panorama guaranteeing complete coverage of the scene from a particular point. If these points are well chosen, a collection of panoramas can cover an area to be searched with greater certainty than imagery captured with a ptz camera during navigation. For the operator searching within a saved panorama the experience is similar to controlling a ptz camera in the actual scene, a property that has been used to improve teleoperation in a low bandwidth high latency application (Fiala, 2005).

In our USAR application which requires finding victims and locating them on a map we merge map and camera views as in (Ricks, Nielsen, & Goodrich, 2004). The operator directs navigation from the map being generated with panoramas being taken at the last waypoint of a series. The panoramas are stored and accessed through icons showing their locations on the map. The operator can find victims by asynchronously panning through these stored panoramas as time becomes available. When a victim is spotted the operator uses landmarks from the image and corresponding points on the map to record the victim's location. By changing the task from a forced paced one with camera views that must be controlled and searched on multiple robots continuously to a self paced task in which only navigation needs to be controlled in realtime we hoped to provide a control interface that would allow more thorough search with lowered mental workload. The reductions in bandwidth and communications requirements (Bruemmer et al., 2005) are yet another advantage offered by this approach.

1.4 Pilot experiment

In a recent experiment reported in (Velagapudi et al., 2008) we compared performance for operators controlling 4 robot teams at a simulated USAR task using either streaming or asynchronous video displays. Search performance was somewhat better using the conventional interface with operators marking slightly more victims closer to their actual location at each degree of relaxation. This superiority, however, might have occurred simply because streaming video users had the opportunity to move closer to victims thereby improving their estimates of distance in marking the map. A contrasting observation was that frequency of shifting focus between robots, a practice we have previously found related to search performance (Scerri et al., 2004) was correlated with performance for streaming video participants but not for participants using asynchronous panoramas. Because operators using asynchronous video did not need to constantly switch between camera views to avoid missing victims we hypothesized that for larger team sizes where forced pace search might exceed the operator's attentional capacity asynchronous video might offer an advantage. The present experiment tests this hypothesis.

2. Experiment

2.1 USARSim and MrCS

The experiment was conducted in the high fidelity USARSim robotic simulation environment (Lewis et al., 2007) developed as a simulation of urban search and rescue (USAR) robots and environments intended as a research tool for the study of human-robot interaction (HRI) and multi-robot coordination. The MrCS (Multi-robot Control System), a multirobot communications and control infrastructure with accompanying user interface developed for experiments in multi-robot control and RoboCup competition (Wang & Lewis, 2007a) was used with appropriate modifications in both experimental conditions. MrCS provides facilities for starting and controlling robots in the simulation, displaying camera and laser output, and supporting inter-robot communication through Machinetta (Scerri et al., 2004) a distributed mutiagent system. The distributed control enables us to scale robot teams from small to large.

Figures 2 and 3 show the elements of the MrCS involved in this experiment. In the standard MrCS (Fig. 2) the operator selects the robot to be controlled from the colored thumbnails at the top of the screen. Robots are tasked by assigning waypoints on a heading-up map through a teleoperation widget. The current locations and paths of the robots are shown on the Map Data Viewer. In the Panorama interface thumbnails are blanked out and images are acquired at the terminal point of waypoint sequences.

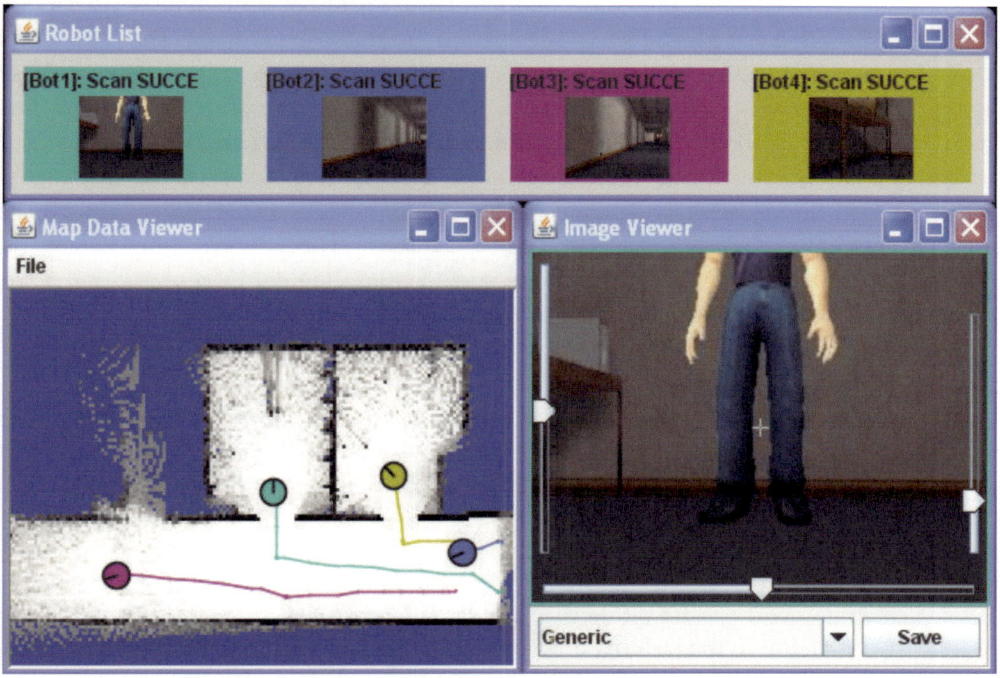

Fig. 2. MrCS components for Streaming Video mode

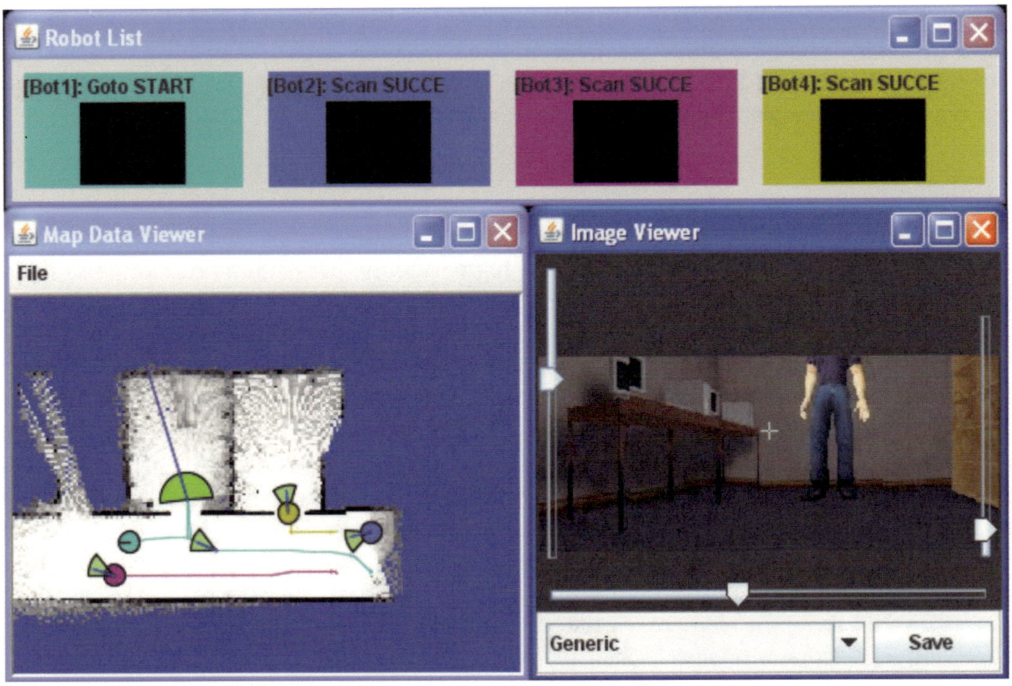

Fig. 3. MrCS Asynchronous Panorama mode

2.2 Method

A large search environments previously used in the 2006 RoboCup Rescue Virtual Robots competition (Balakirsky et al., 2007) was selected for use in the experiment. The environment consisted of maze like halls with many rooms and obstacles, such as chairs, desks, cabinets, and bricks. Victims were evenly distributed throughout the environments. Robots were started at different locations leading to exploration of different but equivalent areas of the environment. A third simpler environment was used for training. The experiment followed a between groups design with participants searching for victims using either panorama or streaming video modes. Participants searched over three trials beginning with 4 robots, then searching with 8, and finally 12. Robots were started from different locations within a large environment making learning from previous trials unlikely.

2.3 Participants and procedure

29 paid participants were recruited from the University of Pittsburgh community. None had prior experience with robot control although most were frequent computer users. Approximately a quarter of the participants reported playing computer games for more than one hour per week.

After collecting demographic data the participant read standard instructions on how to control robots via MrCS. In the following 15~20 minute training session, the participant practiced control operations for either the panorama or streaming video mode and tried to find at least one victim in the training environment under the guidance of the experimenter. Participants then began three testing sessions in which they performed the search task controlling 4, 8, and 12 robots.

3. Results

Data were analyzed using a repeated measures ANOVA comparing streaming video performance with that of asynchronous panoramas. On the performance measures, victims found and area covered, the groups showed nearly identical performance with victim identification peaking sharply at 8 robots accompanied by a slightly less dramatic maximum for search coverage (Fig. 4).

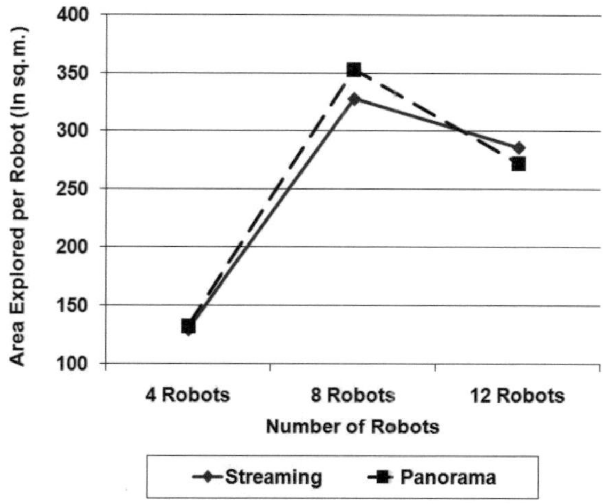

Fig. 4. Area Explored as a function of N robots (2 m)

The differences in precision for marking victims observed in the pilot study were found again. For victims marked within 2m, the average number of victims found in the panorama condition was 5.36 using 4 robots, 5.50 for 8 robots, but dropping back to 4.71 when using 12 robots. Participants in the Streaming condition were significantly more successful at this range, $F_{1,29} = 3.563$, $p < .028$, finding 4.8, 7.07 and 4.73 victims respectively(Fig. 5).

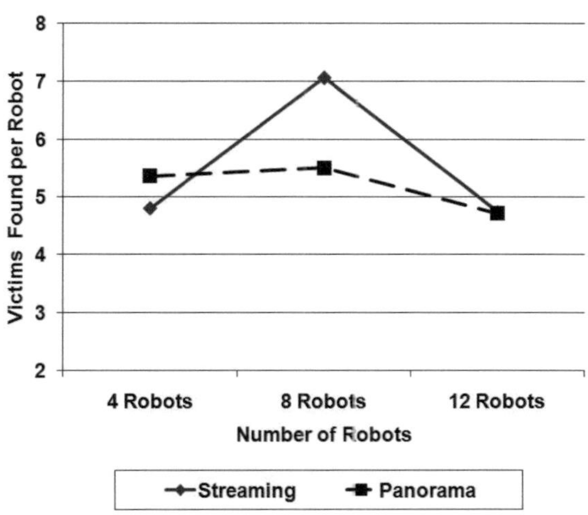

Fig. 5. Victims Found as a function of N robots (within 2 m)

A similar advantage was found for victims marked within 1.5m, with the average number of victims found in the panorama condition dropping to 3.64, 3.27 and 2.93 while participants in the streaming condition were more successful, $F_{1,29} = 6.255$, $p < .0025$, finding 4.067, 5.667 and 4.133 victims respectively (Fig. 6).

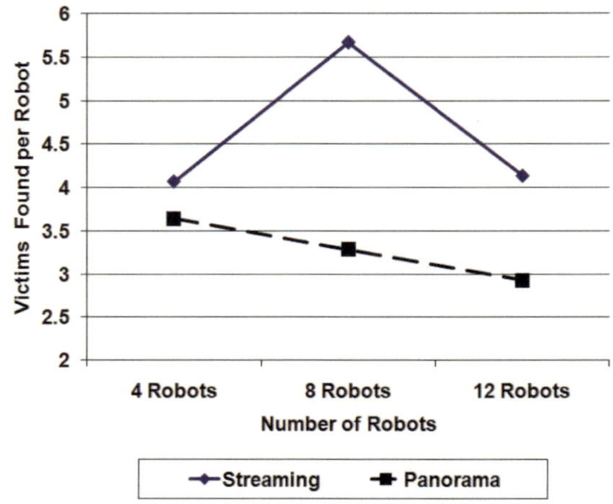

Fig. 6. Victims Found as a function of N robots (within 1.5 m)

Fan-out (Olsen & Wood, 2004) is a model-based estimate of the number of robots an operator can control. While Fan-out was conceived as an invariant measure, operators are noticed to adjust their criteria for adequate performance to accommodate the available robots (Wang et al., 2009; Humphrey et al., 2006).

We interpret Fan-out as a measure of attentional reserves. If Fan-out is greater than the number of robots, there are remaining reserves. If Fan-out is less than the number of robots, capacity has already been exceeded. Fan-out for the panorama conditions increased from 4.1, 7.6 and 11.1 for 4 to 12 robots. Fan-out, however, was uniformly higher in the streaming video condition, $F_{1,29} = 3.355$, $p < .034$, with 4.4, 9.12 and 13.46 victims respectively (Fig.7).

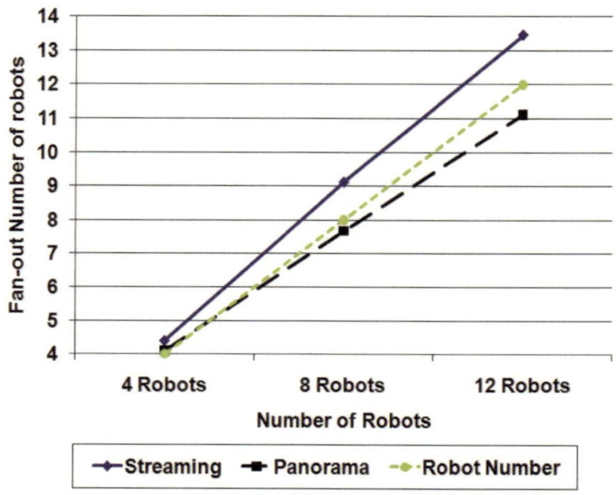

Fig. 7. Fan-out as a function of N robots

Number of robots had a significant effect on every dependent measure collected except waypoints per mission (a *Mission* means all the waypoints which the user issued for a robot with a final destination), which next lowest N switches in focus robot, $F_{2,\ 54} = 16.74$, $p < .0001$. The streaming and panorama conditions were easily distinguished by some process measures. Both streaming and panorama operators followed the same pattern issuing the fewest waypoints per Mission to command 8 robots, however, panorama participants in the 8 robot condition issued observably fewer (2.96 vs. 3.16) waypoints (Fig.8).

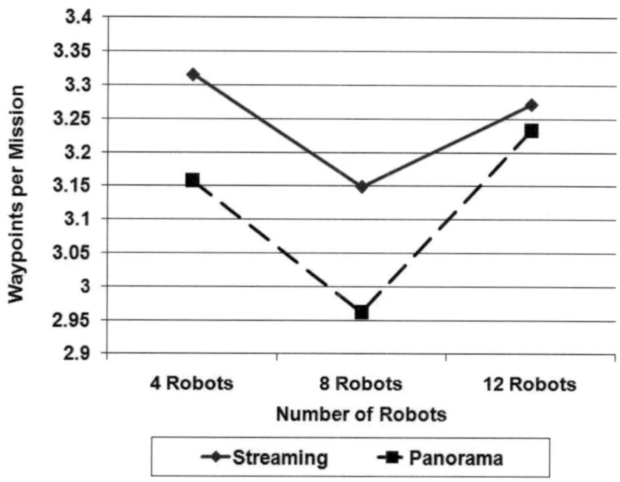

Fig. 8. Waypoints issued per Mission

The closely related pathlength/mission measure follows a similar pattern with no interaction but significantly shorter paths (5.07 m vs. 6.19 m) for panorama participants, $F_{2,54} = 3.695$, $p = .065$ (Fig. 9).

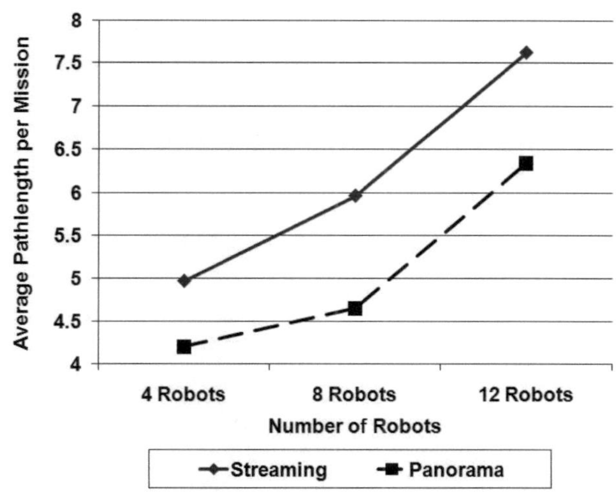

Fig. 9. Waypoints issued per Mission

The other measures like number of missions and switches between robots in focus by contrast were nearly identical for the two groups showing only the recurring significant effect for N robots. A similar closeness is found for NASA-TLX workload ratings which rise together monotonically for N robots (Fig. 10).

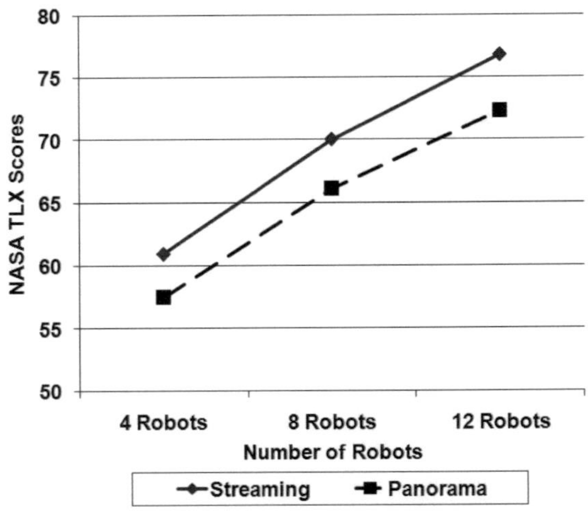

Fig. 10. NASA-TLX Workload

4. Discussion

The most unexpected thing about these data is how similar the performance of streaming and asynchronous panorama participants was. The tasks themselves appear quite dissimilar. In the panorama condition participants direct their robots by adding waypoints to a map without getting to see the robots' environment directly. Typically they tasked robots sequentially and then went back to look at the panoramas that had been taken. Because panorama participants were unable to see the robot's surrounding except at terminal waypoints, paths needed to be shorter and contain fewer waypoints in order to maintain situation awareness and avoid missing potential victims. Despite fewer waypoints and shorter paths, panorama participants managed to cover the same area as streaming video participants within the same number of missions. Ironically, this greater efficiency may have resulted from the absence of distraction from streaming video (Yanco & Drury, 2004) and is consistent with (Nielsen & Goodrich, 2006) in finding maps especially useful for navigating complex environments.

Examination of pauses in the streaming video condition failed to support our hypothesis that these participants would execute additional maneuvers to examine victims. Instead, streaming video participants seemed to follow the same strategy as panorama participants of directing robots to an area just inside the door of each room. This leaves panorama participants' inaccuracy in marking victims unexplained other than through a general loss of situation awareness. This explanation would hold that lacking imagery leading up to the panorama, these participants have less context for judging victim location within the image and must rely on memory and mental transformations.

Panorama participants also showed lower Fan-out perhaps as a result of issuing fewer waypoints for shorter paths leading to more frequent interactions. While differences in switching focus among robots were found in our earlier study (Wang & Lewis, 2007b) the present data (figure 7) show performance to be almost identical.

Our original motivation for developing a panorama mode for MrCS was to address restrictions posed by a communications server added to RoboCup Rescue competition to simulate bandwidth limitations and drop-outs due to attenuation from distance and obstacles. Although the panorama mode was designed to drastically reduce bandwidth and allow operation despite intermittent communications our system was so effective we decided to test it under conditions most favorable to a conventional interface. Our experiment shows that under such conditions allowing uninterrupted, noise free, streaming video a conventional interface leads to somewhat equal or better search performance.

Furthermore, while we undertook this study to determine whether asynchronous video might prove beneficial to larger teams we found performance to be essentially equivalent to the use of streaming video at all team sizes with a small sacrifice of accuracy in marking victims. This surprising finding suggests that in applications that may be too bandwidth limited to support streaming video or involve substantial lags; map-based displays with stored panoramas may provide a useful display alternative without seriously compromising performance.

5. Future work

The reported experiment is one of a series exploring human control over increasingly large robot teams. We are seeking to discover and develop techniques and strategies for allocating tasks among teams of humans and robots in ways that improve overall efficiency. By analogy to computational complexity we have argued that command tasks can also be classified by complexity. Some task-centric rather than platform-centric commands such specifying an area to be searched would have a complexity of O(1) since they are independent of the number of UVs. Others such as authorizing a target or responding to a request for assistance that involve commanding individual UVs would be O(n). Still others that require UVs to be coordinated would have higher levels of complexity and rapidly exceed human capabilities. Framing the problem this way leads to the design conclusion that commanders should be issuing task-centric commands, UV operators should be handling independent UV specific tasks (perhaps for multiple UVs), and coordination among UVs (in accordance with the commander's intent) should be automated to as great an extent as possible.

The reported experiment is one of a series investigating O(n) control of multiple robots. We model robots as being controlled in a round robin fashion (Crandall et al., 2004) with additional robots imposing an additive load on the operator's cognitive resources until they are exceeded. Because O(n) tasks are independent, the number of robots can safely be increased either by adding additional operators or increasing the autonomy of individual robots. In a recent study (Wang et al., 2009a) we showed that if operators are relieved of the need to navigate they could successfully command more than 12 UVs. Conversely, teams of operators might command teams of robots more efficiently if robots' needs for interaction could be scheduled across operators. A recent experiment (Wang et al., 2009b) showed that without additional automation, operators commanding 24 robots were slightly more effective controlling 12 independently. In a planned experiment we will compare these two conditions with navigation automated. In other work we are investigating both O(1) control and interaction with autonomously coordinating robots. We envision multirobot systems requiring human input at all of these levels to provide tools that can effectively follow their commander's intent.

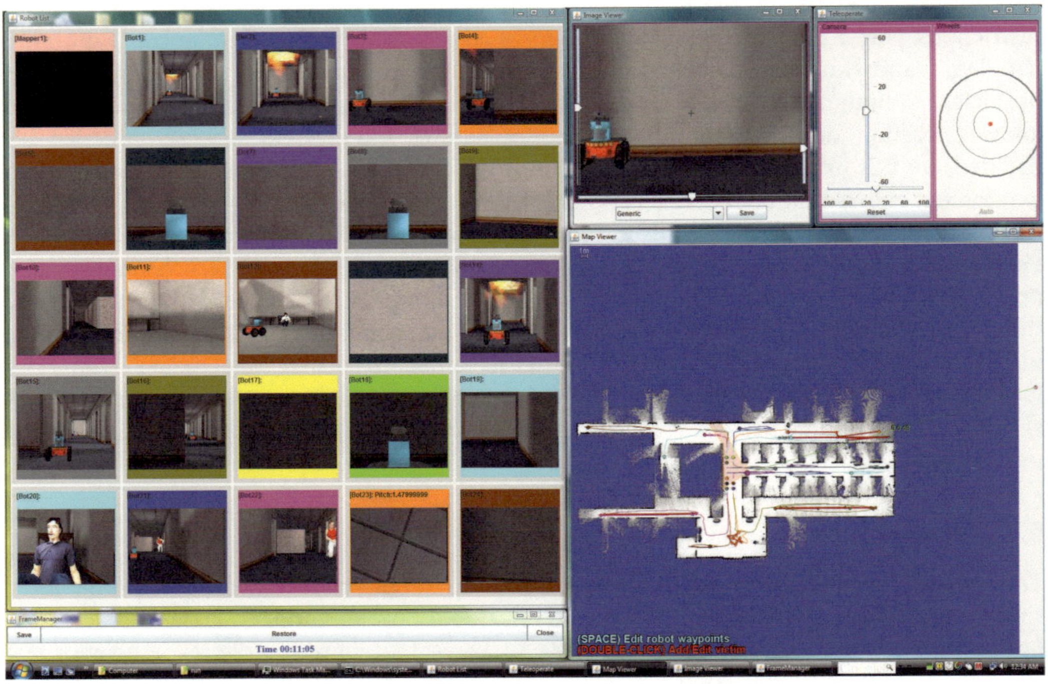

Fig. 11. MrCS interface screen shot of 24 robots for Streaming Video mode

6. Acknowledgements

This work was supported in part by AFOSR grants FA9550-07-1-0039, FA9620-01-0542 and ONR grant N000140910680.

7. References

Balakirsky, S.; Carpin, S.; Kleiner, A.; Lewis, M.; Visser, A., Wang, J., & Zipara, V. (2007). Toward hetereogeneous robot teams for disaster mitigation: Results and performance metrics from RoboCup Rescue, *Journal of Field Robotics*, 24(11-12), 943-967, ISSN: 1556-4959.

Bruemmer, D., Few, A., Walton, M., Boring, R., Marble, L., Nielsen, C., & Garner, J. (2005).. Turn off the television: Real-world robotic exploration experiments with a virtual 3-D display. *Proc. HICSS*, pp. 296a-296a, ISBN: 0-7695-2268-8, Kona, HI, Jan, 2005.

Casper, J. & Murphy, R. (2003). Human-robot interactions during the robot-assisted urban search and rescue response at the world trade center. *IEEE Transactions on Systems, Man, and Cybernetics Part B*, 33(3): 367–385, ISSN: 1083-4419.

Crandall, J., Goodrich, M., Olsen, D. & Nielsen, C. (2005). Validating human-robot interaction schemes in multitasking environments. IEEE Transactions on Systems, Man, and Cybernetics, Part A, 35(4):438–449.

Darken, R.; Kempster, K. & Peterson B. (2001). Effects of streaming video quality of service on spatial comprehension in a reconnaissance task. *Proc. Meeting of The Interservice/Industry Training, Simulation & Education Conference (I/ITSEC)*, Orlando, FL.

Fiala, M. (2005). Pano-presence for teleoperation, *Proc. Intelligent Robots and Systems (IROS 2005)*, 3798-3802, ISBN: 0-7803-8912-3, Alberta, Canada, Aug. 2005.

Fong, T. & Thorpe, C. (1999). Vehicle teleoperation interfaces, *Autonomous. Robots*, no. 11, 9–18, ISSN: 0929-5593.

Humphrey, C.; Henk, C.; Sewell, G.; Williams, B. & Adams, J.(2006). Evaluating a scaleable Multiple Robot Interface based on the USARSim Platform. 2006, *Human-Machine Teaming Laboratory Lab Tech Report.*

Lewis, M. & Wang, J. (2007). Gravity referenced attitude display for mobile robots : Making sense of what we see. *Transactions on Systems, Man and Cybernetics*, Part A, 37(1), ISSN: 1083-4427

Lewis, M., Wang, J., & Hughes, S. (2007). USARsim : Simulation for the Study of Human-Robot Interaction, *Journal of Cognitive Engineering and Decision Making*, 1(1), 98-120, ISSN 1555-3434.

McGovern, D. (1990). Experiences and Results in Teleoperation of Land Vehicles, *Tech. Rep.* SAND 90-0299, Sandia Nat. Labs., Albuquerque, NM.

Milgram, P. & Ballantyne, J. (1997). Real world teleoperation via virtual environment modeling. *Proc. Int. Conf. Artif. Reality Tele-Existence*, Tokyo.

Murphy, J. (1995). Application of Panospheric Imaging to a Teleoperated Lunar Rover, *Proceedings of the 1995 International Conference on Systems, Man, and Cybernetics*, 3117-3121, Vol.4, ISBN: 0-7803-2559-1, Vancouver, BC, Canada

Nielsen, C. & Goodrich, M. (2006). Comparing the usefulness of video and map information in navigation tasks. *Proceedings of the 2006 Human-Robot Interaction Conference*, Salt Lake City, Utah.

Olsen, D. & Wood, S. (2004). Fan-out: measuring human control of multiple robots, *Proceedings of the SIGCHI conference on Human factors in computing systems*, pp. 231-238, ISBN:1-58113-702-8, 2004, Vienna, Austria, ACM, New York, NY, USA

Ricks, B., Nielsen, C., and & Goodrich, M. (2004). Ecological displays for robot interaction: A new perspective. *International Conference on Intelligent Robots and Systems IEEE/RSJ*, ISBN 0-7803-8463-6, 2004, Sendai, Japan, IEEE, Piscataway NJ, ETATS-UNIS.

Scerri, P., Xu, Y., Liao, E., Lai, G., Lewis, M., & Sycara, K. (2004). Coordinating large groups of wide area search munitions, In: *Recent Developments in Cooperative Control and Optimization*, D. Grundel, R. Murphey, and P. Pandalos (Ed.), 451-480, Springer, ISBN: 1402076444, Singapore.

Shiroma, N., Sato, N., Chiu, Y. & Matsuno, F. (2004). Study on effective camera images for mobile robot teleoperation, *In Proceedings of the 2004 IEEE International Workshop on Robot and Human Interactive Communication*, pp. 107-112, ISBN 0-7803-8570-5, Kurashiki, Okayama Japan.

Tan, D., Robertson, G. & Czerwinski, M. (2001). Exploring 3D navigation: Combining speed-coupled flying with orbiting. *CHI 2001 Conf. Human Factors Comput. Syst.*, pp. 418-425, Seattle, WA, USA, March 31 - April 5, 2001, ACM, New York, NY, USA.

Velagapudi, P.,Wang, J., Wang, H., Scerri, P., Lewis, M., & Sycara, K. (2008). Synchronous vs. Asynchronous Video in Multi-Robot Search, *Proceedings of first International Conference on Advances in Computer-Human Interaction (ACHI'08)*, pp. 224-229, ISBN: 978-0-7695-3086-4, Sainte Luce, Martinique, February, 2008.

Volpe, R. (1999). Navigation results from desert field tests of the Rocky 7 Mars rover prototype, *The International Journal of Robotics Research*, 18, pp.669-683, ISSN: 0278-3649.

Wang, H., Lewis, M., Velagapudi, P., Scerri, P., & Sycara, K. (2009). How search and its subtasks scale in N robots, *Proceedings of the ACM/IEEE international conference on Human-robot interaction (HRI'09)*, pp. 141-148, ISBN:978-1-60558-404-1, La Jolla, California, USA, March 2009, ACM, New York, NY, USA.

H. Wang, H., S. Chien, S., M. Lewis, M., P. Velagapudi, P., Scerri, P. & Sycara, K. (2009b) Human teams for large scale multirobot control, Proceedings of the 2009 International Conference on Systems, Man, and Cybernetics (to appear), San Antonio, TX, October 2009.

Wang, J. & Lewis, M. (2007a). Human control of cooperating robot teams, *Proceedings of the ACM/IEEE international conference on Human-robot interaction (HRI'07)*, pp. 9-16, ISBN: 978-1-59593-617-2, Arlington, Virginia, USA, March 2007ACM, New York, NY, USA.

Wang, J. & Lewis, M. (2007b). Assessing coordination overhead in control of robot teams, *Proceedings of the 2007 International Conference on Systems, Man, and Cybernetics*, pp. 2645-2649, ISBN:978-1-60558-017-3, Montréal, Canada, October 2007.

Wickens, C. & Hollands, J. (1999). *Engineering Psychology and Human Performance*, Prentice Hall, ISBN 0321047117, Prentice Hall, Upper Sider River, NJ

Yanco, H. & Drury. J. (2004). "Where am I?" Acquiring situation awareness using a remote robot platform. *Proceedings of the IEEE Conference on Systems, Man, and Cybernetics*, ISBN 0-7803-8566-7, The Hague, Netherlands.

Yanco, H., Drury, L. & Scholtz, J. (2004) Beyond usability evaluation: Analysis of human-robot interaction at a major robotics competition. *Journal of Human-Computer Interaction*, 19(1 and 2):117–149, ISSN: 0737-0024

Yanco, H., Baker, M., Casey, R., Keyes, B., Thoren, P., Drury, J., Few, D., Nielsen, C., & Bruemmer, D. (2006). Analysis of human-robot interaction for urban search and rescue, *Proceedings of PERMIS*, Philadelphia, Pennsylvania USA, September 2006.

Human-Robot Interaction Architectures

5

Handling Manually Programmed Task Procedures in Human–Service Robot Interactions

Yo Chan Kim and Wan Chul Yoon
Korea Advanced Institute of Science and Technology
Republic of Korea

1. Introduction

Although a few robots such as vacuum cleaning robots (Jones, 2006; Zhang et al., 2006), lawn mowing robots (Husqvarna; Friendlyrobotics), and some toy robots (Takara; Hasbro) have single functions or perform simple tasks, almost all other service robots perform diverse and complex tasks. Such robots share their work domains with humans, with whom they must constantly interact. In fact, the complexity of the tasks performed by such robots is a result of their interactions with humans. For example, consider a scenario wherein a robot is required to fetch and carry beverages: the methods of delivery are numerous and vary depending on user requirements such as type of beverage, the needs for a container, etc. For a robot designed to control various household devices such as illuminators, windows, television, and other appliances, several services must be provided in various situations; hence, a question-and-answer interaction or some method to infer the necessity of the services is required.

For service robots to perform these complex behaviors and collaborate with humans, the programming of robot behavior has been proposed as a natural solution (Knoop et al., 2008). Robot behavior can be programmed manually using text-based and graphical systems, or automatically by demonstration or instructive systems (Biggs & MacDonald, 2003). Recently, many researchers have proposed methods for a service robot to learn high-level tasks. The two main methods are (1) learning by observing human behaviors (Argall et al., 2009) and (2) learning by using procedures defined by humans (a support system can be used to define these procedures) (Lego, 2003; Ekvall et al., 2006)..

Manual programming systems are more efficient in creating procedures necessary to cope with various interactive situations than automatic programming systems since the latter require demonstrations and advice for every situation. However, in the process of programming behavior, there exist sub-optimalities (Chen & Zelinsky, 2003), and manual programming systems are more brittle than automatic programming systems.

The sub-optimalities of manual programming systems are as follows: (a) in the writing process, humans can make syntactic errors when describing task procedures. For example, writers often misspell the names of actions or important annotations. However, if the errors do not alter the semantic meaning, the problem can be prevented by writing support systems such as in the case of Lego Mindstorms. (b) Another sub-optimality can occur if humans fail to devise all possible behaviors for situations that a robot will confront. In the example of beverage-delivery errands, a writer may describe a sequence in a scene wherein a robot picks up a cup. However, the writer might possibly omit a sequence in a scene wherein a robot lifts a cup *after* picking up the beverage. It is not easy for humans to infer all possible situations and consequent branching out of behavior procedures; hence, an automated system should be able to support such inference and manage robots by inferring new situations based on the given information. (c) The sequence written by a human may be wrong semantically. Humans can insert wrong actions, omit important actions, and reverse action orders by making mistakes or slips. For example, a procedure for a robot for setting a dinner table might not contain actions for placing a fork; this is an example of omission. Another procedure might consist of actions for placing a saucer after placing a teacup. Some researches have attempted to resolve this problem by synthesizing or evaluating a set of procedures based on pro-conditions and the effects of knowledge of each unit action, for example, such as in the case of conventional planning approaches in artificial intelligence field (Ekvall et al., 2006; Ekvall & Kragic, 2008). Moreover, it is possible to search for wrong sequences in procedures by using rules that deal with sequential relations between actions; such rules can be extracted using a statistical data mining method (Kwon et al., 2008). Despite these efforts, the problem of identifying whether a procedure is natural and acceptable to humans continues to be a difficult problem.

In this chapter, we propose methodologies to mitigate the last two sub-optimalities (b and c) using a programming language that can be used to describe the various task procedures that exist in human–service robot interactions.

2. Scripts, abstract task procedures for service robots

In this section, we explain task procedures that are programmed by humans. These task procedures refer to abstract robot behaviors occurring in service domains (Kim et al., 2007). The script is expressed by using a generic procedural language and can be written by humans, especially non-experts, via graphic-based or text-based interface systems. Each script contains several actions and branch-able states (explained in 2.2).

2.1 Action

Action primitives in scripts are the basic units of scripts. These are black boxes from a user's viewpoint, because the user does not have to possess detailed knowledge of their functioning, even when the units are applied to specific hardware platforms via many different modules. There are two types of action primitives: physical actions such as "move to location of object A" or "open the facing door" and cognitive actions such as "find location of object A" or "decide which beverage to take." Physical actions are performed by physical action executors, and the actions play roles as the goals of the executors (Fig. 1.). When cognitive actions are performed, knowledge inference engines explore or reason the related information. Based on the reasoned information, the Decision Manager asks questions to users. The process of asking question has been explained in our previous report (Kim et al., 2007). Some rewriteable sets of action primitives can be defined as abstract actions and used in the script database.

2.2 Branch-able state

A branch-able state refers to an interaction-related state that determines the characteristic of the script in which it is included. "Does the user want to turn on a television? Yes" or "Is it necessary to use a cup to fill the beverage? No" are examples of the branch-able state. These states are the principal evidences for checking whether a script coincides with the current situations or user's demands when a Script-based Task Planner handles scripts.

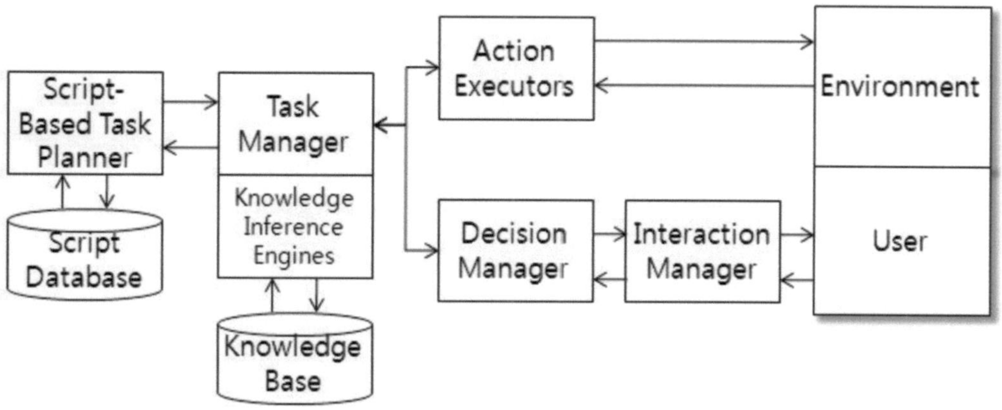

Fig. 1. Configuration diagram of the developed system

2.3 Script

A script is a sequential set of actions and branch-able states. A script database contains the scripts, and it is described in XML. As an example, Fig. 2 shows a script for the delivery of beverages.

```
<script goal="FetchAndCarryBeverage" scriptID="FCB002">
        <action decomposetype="concrete" acttype="cognitive">DecideTargetBeverage</action>
        <action decomposetype="concrete" acttype="cognitive">IdentifyLocationOfBeverage</action>
        <action decomposetype="concrete" acttype="physical">MoveToLocationOfBeverage</action>
        <BranchableProperty>InvisibilityOfBeverage:yes</BranchableProperty>
        <action decomposetype="concrete" acttype="physical">UncoverTargetLoc</action>
        <action decomposetype="concrete" acttype="physical">PickUpTargetBeverage</action>
        <action decomposetype="concrete" acttype="physical">CoverTargetLoc</action>
        <action decomposetype="concrete" acttype="cognitive">DecideNecOfContainer</action>
        <BranchableProperty>NecessityOfContainer:no</BranchableProperty>
        <action decomposetype="concrete" acttype="physical">MoveToDrink</action>
        <action decomposetype="concrete" acttype="physical">DeliverBeverage</action>
</script>
```

Fig. 2. An example of script describing delivery of beverage

3. Related work

To solve the problem of the two sub-optimalities mentioned in the introductory section, it is useful to identify the relationships between several scripts. Some researchers in the field of programming by demonstration analyzed various programmed procedures and derived information that is referable for enhancing the performance of a procedure, for example, the relationships between actions or more abstract procedures (Breazeal et al., 2004; Nicolescu & Matarić, 2003; Ekvall & Kragic, 2008; Pardowitz et al., 2005). Nicolescu and Matarić (2003) represented each demonstration as a directed acyclic graph (DAG) and computed their longest common subsequence in order to generalize over multiple given demonstrations. Ekvall and Kragic (2008) converted sequential relationships between all two states as temporal constraints. Whenever a sequence was added, the constraints that contain contradictions with the constraints of the new sequence were eliminated in order to extract general state constraints. Pardowitz et al. (2005) formed task precedence graphs by computing the similarity of accumulating demonstrations. Each task precedence graph is a DAG that explains the necessity of specific actions or sequential relationships between the actions.

These researches are appropriate for obtaining task knowledge from a small number of demonstrations. However, when a large number of procedures are demonstrated or programmed, these approaches continue to generate one or two constraints. These strict constraints are not sufficient to generate variations in the given demonstrations or to evaluate them.

4. Handling scripts

We propose two algorithms for reducing the sub-optimalities from a large number of scripts. One is an algorithm that generates script variations based on the written scripts. Since the written scripts are composed from a human's imagination, they cannot be systematic or complete. The set of scripts takes either a total-ordered form or a mixture of total-ordered and partial-ordered forms. Our algorithm generates a DAG of all scripts, and hence, it permits the revealing of branches and joints buried among the scripts. The other algorithm is for evaluating the representativeness of a specific script by comparing the given script set. We can generate a sequence that is able to represent entire given scripts. If almost all scripts are semantically correct and natural, the naturalness of a specific script can be estimated by evaluating its similarity with the representative script. Therefore, this algorithm involves an algorithm that generates a representative script and an algorithm that measures similarities with it.

These two algorithms are based on an algorithm for partial order alignment (POA, Lee et al., 2002). Hence, we first explain POA before describing the two algorithms.

4.1 POA algorithm

We utilized an algorithm from multiple sequence alignment (MSA), which is an important subject in the field of Bioinformatics, to identify the relationships among scripts. In MSA, several sequences are arranged to identify regions of similarity. The arrangement can be depicted by placing sequences in a rectangle and inserting some blanks at each column appropriately (Fig. 3.). When we attempt to obtain an optimal solution by dynamic programming, this process becomes an NP-complete problem. Therefore, several heuristics are presented (POA (Lee et al., 2002), ClustalW (Thompson et al., 1994), T-Coffee (Notredame et al., 2000), DIALIGN (Brudno et al., 1998), MUSCLE (Edgar, 2004), and SAGA (Notredame & Higgins, 1996)).

POA is an algorithm that represents multiple sequences as multiple DAGs and arranges them. POA runs in polynomial time and is considered a generally efficient method that produces good results for complex sequence families. Figure 4 shows the strategy of POA. POA redraws each sequence as a linear series of nodes connected by a single incoming edge and a single outgoing edge (Fig. 4b.). By using a score matrix that contains similarity values between letters, POA aligns two sequences by dynamic programming that finds maximum similarity (Fig. 4c.). The aligned and identical letters are then fused as a single node, while the others are represented as separate nodes.

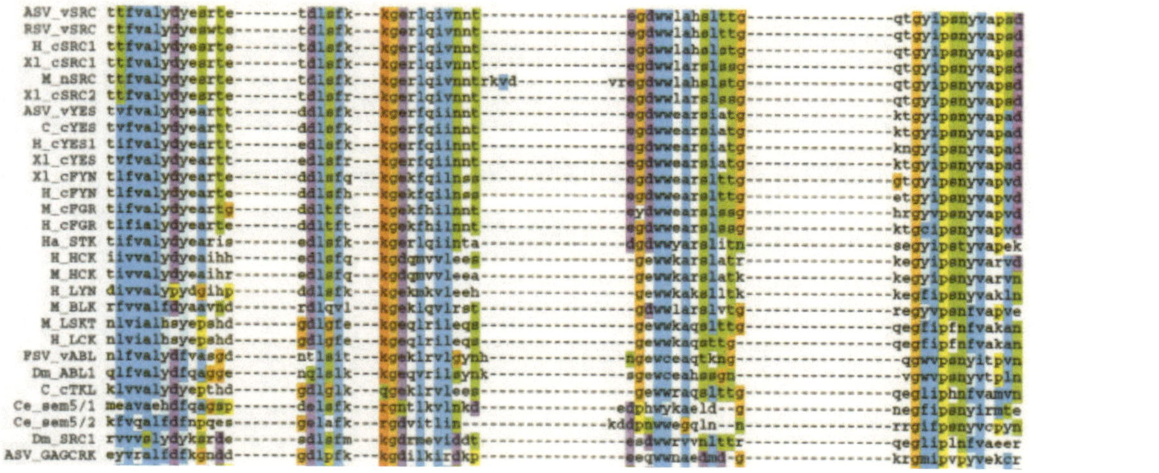

Fig. 3. Example of multiple sequence alignment (MSA) by CLUSTALW (Thompson et al., 1994)

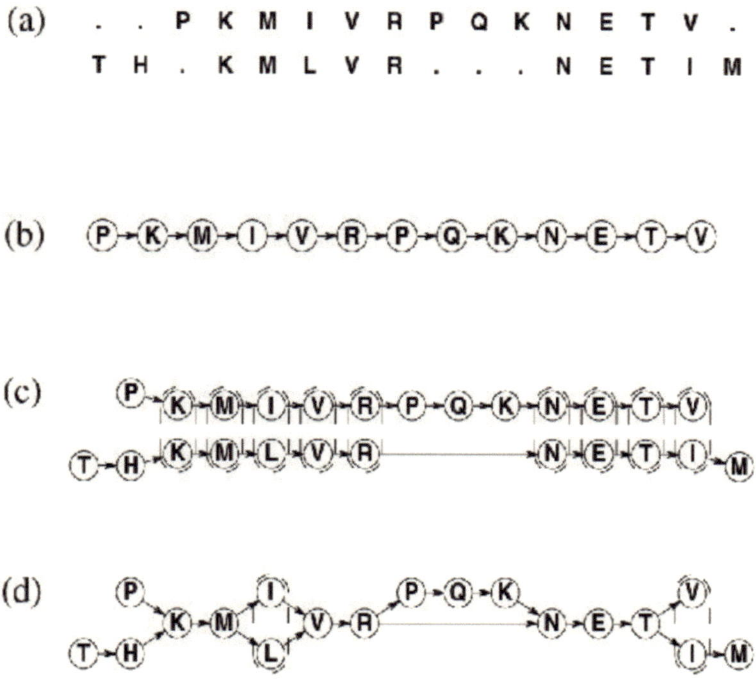

Fig. 4. MSA representation by partial order alignment (POA) algorithm. (a) General representation of MSA, (b) Single representation of POA, (c) Two sequences aligned by POA algorithm, and (d) Aligned result of POA

4.2 Generating script variations

If we regard all scripts as sequences of POA, and all actions and states as nodes or letters of POA, the scripts can be aligned by POA. For example, the parts in two scripts will be aligned as shown in Fig. 5.

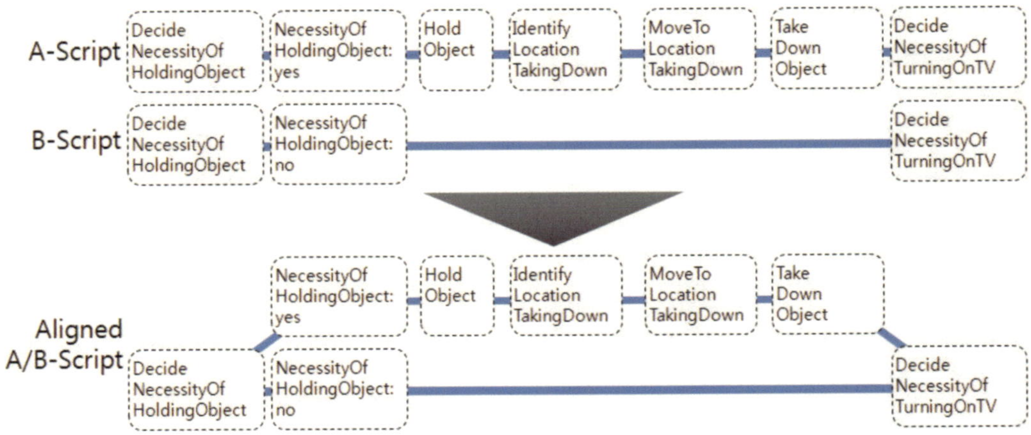

Fig. 5. Two scripts, A and B, are aligned as a form of directed acyclic graph (DAG)

The arranged DAG (ADAG) produced by the POA algorithm allows the generation of script variations and evaluation of script representativeness. Hence, we established a model to generate script variations by using ADAG. This approach attempts to maintain the semantic naturalness of scripts by employing sequential patterns in the given scripts.

The mechanism is as follows: the ADAG of the given scripts is produced by POA. All the paths on the ADAG are searched by employing the "breadth-first" method. New paths that do not have the given sequences still remain, while some deficient scripts are eliminated.

Although script variations are produced by ADAG, there can be deficiencies in some new script variations. Deficiencies in scripts are inspected by using two types of information. One is the basic state relationship of availability. We can predefine each action's preconditions and effects; they are generally domain-independent. For example, a precondition of an action "shifting something down" may be a state wherein the robot is holding the object. By using the information on these states, the model checks whether there is any action that is not satisfied under its preconditions.

The other is user-defined action relationship rules. Any user can pre-describe sequential or associational rules between several actions. Kwon et al. (2008) developed a support system that automatically finds some frequently associative or sequencing actions to aid users to find the action relationship rules. An example of action relationship rules is that a TV channel should not be changed before turning on the TV.

4.3 Evaluating representativeness of scripts

There are paths on which many scripts are overlapped as well as paths on which only one or two scripts are related to the paths on the ADAG. It is possible to link the paths on which many scripts are overlapped; Lee (2003) called the linked paths the consensus sequences of POA (Fig. 6.).

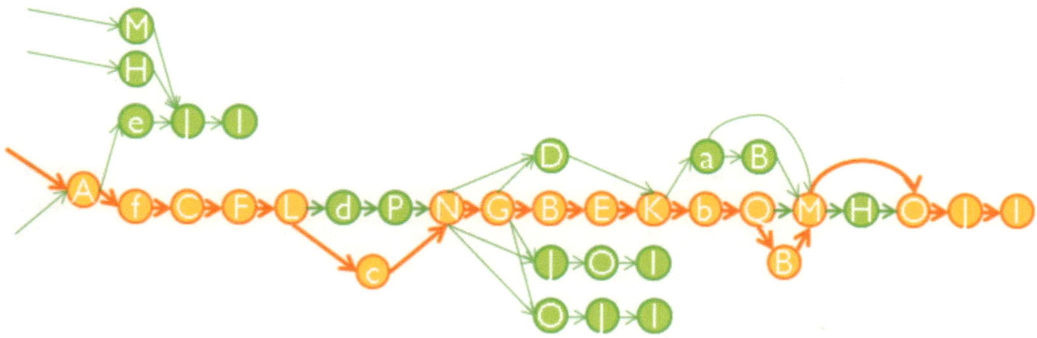

Fig. 6. An example of ADAG and consensus sequence generated from scripts for "greeting user" scenario

The heaviest bundling strategy for discovering consensus sequences is as follows: There are edges between all the actions or nodes on ADAG. The heaviest bundle model attaches 1 *edge_weight* of sequences to every edge on the DAG, and adds each number of aligned edges where two or more sequences are aligned. In such a case, every edge on the DAG has one or more *edge_weight*. While traversing from start nodes to end nodes, the heaviest bundle algorithm finds a path that has the largest sum of *edge_weight* among all paths. The algorithm uses a dynamic traversal algorithm, and an example is shown in Fig. 7. After excluding sequences that contribute to prior consensus generation, the algorithm iterates the consensus generation. Further, the algorithm calculates how many actions are identical to the actions of consensus sequence. We set the exclusion threshold such that scripts coincide over the threshold percentage and the consensus sequences are excluded from each iteration. Iteration continues until no contributed sequence is found.

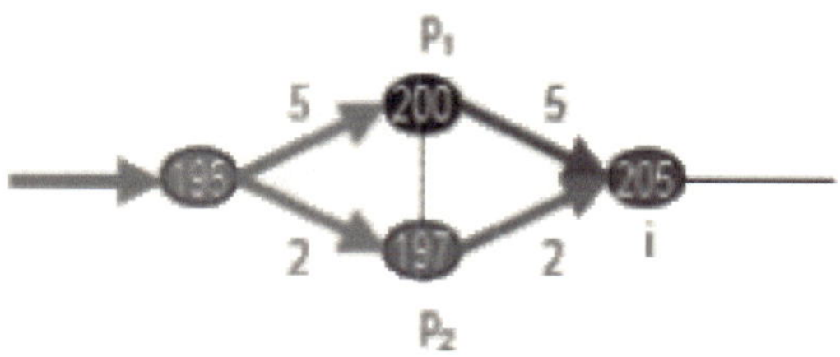

Fig. 7. Dynamic programming for construction of consensus sequences (Lee, 2003)

The representativeness of a script is calculated by computing how the script coincides with the consensus sequences. The equation of representativeness is given as follows:

$$\text{Representativeness} = Coin^i (Thres)^{iteration}, \tag{1}$$

Where i is the index of script; Coin, coincidence variable; and Thres, threshold variable. Coincidence is the number of actions that are identical to those of consensus sequence divided by the total number of actions. Threshold and iteration imply the threshold percentage and the number of iterations in the heaviest bundle algorithm. For example, when the threshold is 80% and a script has ten actions, the script whose nine actions are identical to those of the generated consensus sequence at first iteration has a representativeness value of 0.72(0.9*0.8). If eight actions are the same with the second consensus sequence, the script has a representativeness value of 0.512(0.8*0.8^2).

5. Implementation

To examine the effectiveness of the proposed methodologies, we implemented the algorithms on a set of scripts. We wrote 160 scripts for a "greeting user" task. In the script database, a robot greets a user and suggests many services such as turning on the TV, delivering something, or reporting house status. There are not only very busy scripts but also simple ones. We established a score matrix in which POA scores only identical actions. The system produced 400 new scripts. Two hundred and fifty of them were meaningfully acceptable, for example, the ones human wrote, and the others were eliminated by deficiency inspectors.

We also re-evaluated the representativeness of approximately 160 scripts. Every script was given a value ranging from zero to a positive one. We then added a wrong script in which two actions were inverted from a script having the positive value. The wrong script's representativeness was 0.7, which is lower than that of the original one.

6. Conclusion

The demand for programming systems that do not require complex programming skills to perform tasks is increasing, especially in the case of programming by demonstration. Further, in the manual programming environment, which is more efficient than programming by demonstration, two critical sub-optimalities are present. We applied POA and heaviest bundling to solve the two problems and implemented the applied algorithms. To prevent the problem of writers omitting combinational procedures, an algorithm for script variation generation was proposed. Further, to evaluate how a specific script is semantically acceptable, an automatic evaluation process of representativeness was established. The evaluation of representativeness is a good attempt to estimate the script's naturalness. However, this evaluation only demonstrates that a good script has a high representativeness value; it does not show that a script having a low representativeness value is unnatural. It is still not easy to automatically maintain the semantic naturalness of task plans or evaluate them. We expect that interactive systems that are not only intelligent but also convenient to users will be continuously developed in the future; this is a promising future research direction.

7. Acknowledgement

This work was supported by the Industrial Foundation Technology Development Program of MKE/KEIT. [2008-S-030-02, Development of OPRoS(Open Platform for Robotic Services) Technology].

8. References

Argall, B.D.; Chernova, S.; Veloso, M. & Browning B. (2009). A survey of robot learning from demonstration, *Robotics and Autonomous Systems*, Vol. 57, No. 5, 469-483, 0921-8890

Biggs, G. & MacDonald, B. (2003). A survey of robot programming systems, *Australasian Conference on Robotics and Automation*, Australia, 2003, Brisbane

Breazeal, C.; Brooks, A.; Gray, J.; Hoffman, G.; Kidd, C.; Lieberman, J.; Lockerd, A. & Mulanda, D. (2004). Humanoid robots as cooperative partners for people, *International Journal of Humanoid Robotics*, Vol. 1, No. 2, 1-34, 0219-8436

Brudno, M.; Chapman, M.; Gottgens, B.; Batzoglou, S. & Morgenstern, B. (2003). Fast and sensitive multiple alignment of large genomic sequences, *BMC. Bioinformatics*,Vol. 4, No. 66, 1471-2105, 1-11

Chen, J. & Zelinsky, A. (2003). Programing by demonstration: Coping with suboptimal teaching actions, *The International Journal of Robotics Research*, Vol. 22, No. 5, 299-319, 0278-3649

Edgar, R. C. (2004). MUSCLE: multiple sequence alignment with high accuracy and high throughput, *Nucleic Acids Research*, Vol. 32, No. 5, 1792-1797, 0305-1048

Ekvall, S.; Aarno D. & Kragic D. (2006). Task learning using graphical programming and human demonstrations, *Robot and Human Interactive Communication*, UK, Sept., 2006, Hatfield

Ekvall, S. & Kragic, D. (2008). Robot Learning from Demonstration: A Task-level Planning Approach, *International Journal of Advanced Robotic Systems*, Vol. 5, No. 3, 1729-8806

Friendlyrobotics, Robomow, http://www.friendlyrobotics.com/robomow/

Hasbro, i-dog, http://www.hasbro.com/idog/

Husqvarna, Automower, http://www.automower.com

Jones, J. L. (2006). Robots at the tipping point: the road to iRobot Roomba, *IEEE Robotics & Automation Magazine*, Vol. 13, No. 1, 76-78, 1070-9932

Kim, Y.C.; Yoon, W.C.; Kwon, H.T. & Kwon, G.Y. (2007). Multiple Script-based Task Model and Decision/Interaction Model for Fetch-and-carry Robot, *The 16th IEEE International Symposium on Robot and Human interactive Communication*, Korea, August, 2008, Jeju

Knoop, S.; Pardowitz, M & Dillmann, R. (2008). From Abstract Task Knowledge to Executable Robot Programs, *Journal of Intelligent and Robotic Systems*, Vol. 52, No. 3-4, 343-362, 0921-0296

Kwon, G. Y.; Yoon, W. C., Kim, Y. C. & Kwon, H. T. (2008). Designing a Support System for Action Rule Extraction in Script-Based Robot Action Planning, *Proceedings of the 39nd ISR(International Symposium on Robotics)*, Korea, October, 2008, Seoul

Lee, C. (2003). Generating consensus sequences from partial order multiple sequence alignment graphs, *Bioinformatics*, Vol. 19, No. 8, 999-1008, 1367-4803

Lee, C.; Grasso, C. & Sharlow, M. F. (2002). Multiple sequence alignment using partial order graphs, *Bioinformatics*, Vol. 18, No. 3, 452-464, 1367-4803

Lego (2003). Lego Mindstorms, http://mindstorms.lego.com/Products/default.aspx

Nicolescu, M. N. & Matari´c, M. J. (2003). Natural Methods for Robot Task Learning: Instructive Demonstrations, Generalization and Practice, *In Proceedings of the Second International Joint Conference on Autonomous Agents and Multi-Agent Systems*, Australia, July, 2003, Melbourne

Notredame C.; Higgins D.G. & Heringa J. (2000). T-Coffee: A novel method for fast and accurate multiple sequence alignment, *Journal of Molecular Biology*, Vol. 302, No. 1, 205-217, 0022-2836

Notredame, C. & Higgins, D. G. (1996). SAGA: sequence alignment by genetic algorithm, *Nucleic Acids Research*, Vol. 24, No. 8, 1515-1524, 0305-1048

Pardowitz, M.; Zollner, R. & Dillmann, R. (2005). Learning sequential constraints of tasks from user demonstrations, *IEEE-RAS International Conference on Humanoid Robots*, Japan, December, 2005, Tsukuba

Takara, Tera robot, http://plusd.itmedia.co.jp/lifestyle/articles/0501/20/news030.html

Thompson, J. D.; Higgins, D. G. & Gibson, T. J. (1994). CLUSTAL W: improving the sensitivity of progressive multiple sequence alignment through sequence weighting, position-specific gap penalties, *Nucleic Acids Research*, Vol. 22, No. 22, 4673-80, 0305-1048

Zhang, H.; Zhang, J.; Zong, G.; Wang, W. & Liu R. (2006). SkyCleaner3: a real pneumatic climbing robot for glass-wall cleaning, *IEEE Robotics & Automation Magazine*, Vol. 13, No. 1, 32-41, 1070-9932

6

A Genetic Algorithm-based Approach to Dynamic Architectural Deployment

Dongsun Kim and Sooyong Park
Sogang University
Republic of Korea

1. Introduction

Increasing demands for various and complex tasks on contemporary computing systems require the precise deployment of components that perform the tasks. For example, in service robot systems (Hans et al., 2002; Kim et al., 2008) that have several SBCs (single board computers), users may simultaneously request several tasks such as locomotion, speech recognition, human-following, and TTS (text-to-speech). Each task comprises a set of components that are organized by an architectural configuration. These components execute their own functionality to provide services to the user. To execute components, they must be deployed into computing units that have computing power, such as desktops, laptops, and embedded computing units.

The deployment of components into computing units can influence the performance of tasks. If the system has only one computing unit, every component is deployed in the computing unit and there is no option to vary the deployment to improve the performance. On the other hand, if the system has multiple computing units, performance improvement by varying the deployment can be considered. Different instances of component deployment show different performance results because the resources of the computing units are different. Concentrated deployment into a certain computing unit may lead to resource contention and delayed execution problems. Therefore, the system requires an deployment method to improve performance when the user requests multiple tasks of a system that has multiple computing units.

When determining the deployment of components that comprise the architectural configuration for the tasks, it is important to rapidly and precisely make a decision about deployment. Since there are a large number of candidate deployment instances, even for a small number of computing units and components (i.e., their combinations exponentially increase), the deployment instance selection method must efficiently search for the best deployment instance that provides the most effective performance to the user. The exhaustive search method guarantees to search the best instance; however, it requires a long time for performing search. The greedy search method rapidly finds a solution; however, it does not guarantee to search the best instance.

This study proposes a genetic algorithm-based selection method that searches a set of candidate deployment instances for an optimal instance. This method repeatedly produces generations, and the solution found by the method rapidly converges to the best instance. This method more rapidly and precisely searches an optimal instance than the exhaustive search method and the greedy search method, respectively.

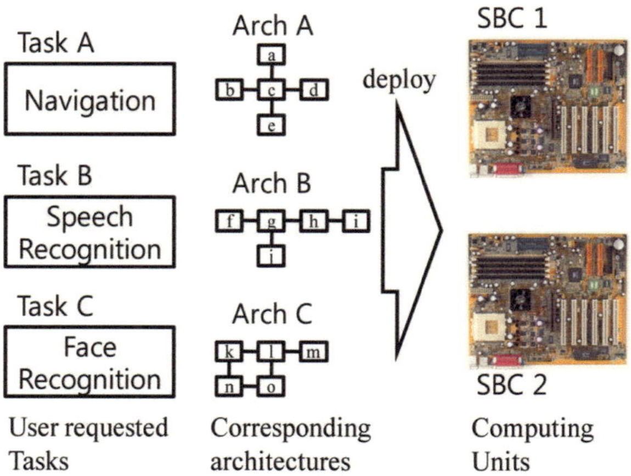

Fig. 1. An example of architectural deployment for required tasks.

This paper is organized as follows: Section 2 illustrates a motivating example that describes the dynamic architectural deployment problem. This problem is formulated as a multiplesack and multidimensional knapsack problem in Section 3. Section 4 describes a genetic algorithm-based approach to the problem. In Section 5, the proposed approach is evaluated in terms of efficiency and accuracy. Section 6 describes a case study conducted to show that our approach can be effectively applied to robot software systems. Section 7 compares our approach to related work. Section 8 provides two issues for discussion. Finally, Section 9 provides conclusion and suggests further studies.

2. Motivating example

Service robot systems such as Care-O-bot (Hans et al., 2002), and home service robot (Kim et al., 2008) have several computing systems termed single board computers (SBCs). An SBC has similar computing power to a desktop or laptop computer. Robot systems (especially service robots) perform their tasks by deploying robot software components into these SBCs, as shown in Figure 1. When a user wants to achieve a specific goal such as navigation, speech recognition, or more complex tasks, the user requests a set of tasks. For each task, the robot system derives its software architecture to handle the requested task. The robot system deploys these architectures to its SBCs to execute the tasks.

Even when the user requests the same tasks, the architectures that perform the tasks can be deployed in different ways. As shown in Figure 2, the consolidated architecture of the architectures shown in Figure 1 can be deployed into two SBCs in different ways. These different instances of deployment can exhibit different results. For example, if components, which consume more CPU time than other components, are deployed into one SBC, then they may lead to resource contention. Resource contention can cause execution delay during task execution. In addition, if two components that require extensive communication between them are deployed into two different SBCs, it may lead to performance degradation.

Performance degradation resulting from inappropriate architectural deployment may lead to more serious problems. For example, the path planning component in the robot navigation architecture and the image analysis component in the face recognition architecture highly consume CPU resources; therefore, if they are deployed into the same SBC, resource contention can occur. This does not allow for the allocation of sufficient CPU

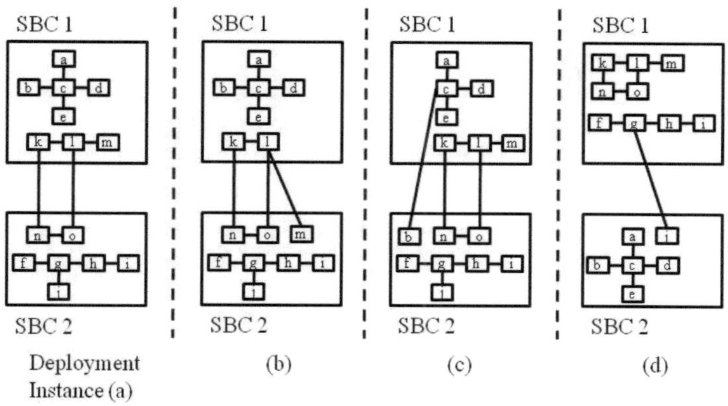

Fig. 2. Examples of architectural deployment instances.

time to the components; the path planning component is not able to respond in the required period of time, and the robot may then collide with the user or a wall.

In the face recognition architecture, the image preprocessing component and the image analysis component frequently interact with each other by passing a large amount of data. It would certainly lead to performance degradation if these two components were deployed into two different SBCs. Suppose that the user simultaneously requests a task that requires the location of the user's face (e.g., human-robot eye contact task). The delay resulting from the inappropriate deployment eventually leads to a malfunction, in which the robot cannot trace the user's face and the required task cannot be performed.

The abovementioned problems can occur when a software system uses multiple computing units (e.g., SBCs) and its elements interact with each other (e.g., software components) with a considerable amount of data. These problems can be serious if the system must deal with real-time tasks that have specific time constraints to achieve the goal of the tasks. This may not only degrade the quality of services but also result in failures of tasks that the user has requested.

To prevent these problems, a software system can search for appropriate deployment instances for every task requested by the user. However, the number of task combinations exponentially increases according to the number of tasks (the set of combinations can be defined by the power set of the set of tasks, i.e., 2^n where n is the number of tasks). Moreover, the number of possible deployment instances is larger than the number of task combinations (when the requested tasks have m components and the system has k computing units, the number of possible deployment instances is k^m where m is strictly larger than the number of tasks). Therefore, it is not efficient to exhaustively search the set of possible deployment instances for an optimal instance at run-time.

It is possible to determine optimal deployment instances for every possible task request prior to system execution, even though the number of possible task requests is large. If a developer has sufficient time to search for optimal deployment instances for every possible task request, then one can find them and record them to exploit them at run-time. However, this method is not applicable if a system has a large number of tasks and a large number of components belonging to the tasks. If the size of the task set is larger than 20 and the average size of the component set in a task request is larger than 30 or 40, it requires a very long time to search an optimal instance, even if it is conducted prior to runtime.

In addition to the size of task and component sets, the set of tasks and their architectural configurations can be dynamically updated at runtime. This implies that a developer should search the sets for optimal deployment again and update the result to the system. This may increase the development time and cost. Moreover, it is impossible to anticipate the configuration of every possible computing unit (e.g., the number of SBCs and their computing power) in operating environments. An operating environment can vary for every user. To deal with this problem, a method is needed to dynamically determine optimal deployment at runtime. This method should decide a deployment instance for every task request in a short time period to prevent the delay in task execution and search for a near-optimal instance.

3. Dynamic architectural deployment

This section formulates the optimal architectural deployment problem. This problem is defined by computing units, their provided resources, architectural configurations, and their required resources. This problem can be modeled by knapsack problems, in which computing units are sacks, components in an architectural configuration are items, and resource consumption efficiency is the value function. The remainder of this section describes the elements of this problem.

3.1 Computing unit

Every software system is executed on a specific computing unit, e.g., desktops, laptops, embedded systems, and other hardware systems that have computing power. When a software system performs its tasks in a computing unit, it uses resources that the computing unit provides, such as CPU time, memory space, and network bandwidth. In particular, software systems that are executed in embedded systems use dedicated resources. For example, in robot systems, the robot software uses sensors (e.g., laser range scanners, ultrasonic sensors, and touch sensors) and actuators (e.g., robot arms, wheels, and speakers). There are two types of resources: sharable and non-sharable.

A sharable resource suggests that a software component consumes a certain amount of the resource. In other words, one component cannot exclusively use the resource and shares the resource with other components. For example, a component consumes a certain amount of main memory space to record its necessary data and another component can simultaneously consume a certain amount of memory to perform its own tasks. However, if components attempt to consume more than the resource can provide, they can experience a resource contention problem, in which the components that request the resource compete for the resource and cannot access it when needed. This implies that the appropriate management of architectural deployment is required.

A non-sharable resource cannot be shared by several components. Only one component exclusively uses a non-sharable resource and other components that require the resource can use the resource only after the currently occupying component releases it. This type of resource is often installed in a specific subset of a computing unit. For example, in general, a robotic system has one wheel actuator and one arm manipulator. They are installed in a specific SBC, respectively (usually, these actuators are installed in separate SBCs). Therefore, components that use these actuators must be deployed in SBCs that have actuators that the components require[1].

---

[1] Remote invocation schema such as RPC (remote procedure call) and RMI (remote method invocation) can facilitate that the component can remotely exploit devices; however, this may lead to performance degradation due to communication burden.

These resources provided by computing units can be defined by the provided resource space that is the Cartesian product of the individual resource dimension r_j. For each computing unit O_i, its provided resource space P_i is formulated as

$$P_i \triangleq \bigotimes_{j}^{n} r_j$$

where n is the number of resources that the system can provide. The provided resource space represents the resource capability that a computing unit can provide.

A resource dimension r_j, which is a sharable resource, is denoted by r_j^s. It can have a certain integer value ($r_j^s \in [v_l, v_u]$ where v_l and v_u are the lower and upper bounds) that represents the maximum amount that the resource can provide. An instance of the j-th sharable resource dimension r_j belongs to the provided resource space of a computing unit and can have a value. For example, the main memory resource dimension of a system is denoted by r^s_{MEM} and its instance has a value of $[Mem_{MIN}, Mem_{MAX}]$ that represents the size of memory installed in the computing unit.

A resource dimension r_j, which is a non-sharable resource, is denoted by r_j^n and can have a Boolean value ($r_j^n \in \{0,1\}$) that represents whether the computing unit provides the resource. For example, the wheel actuator resource dimension of a system is denoted by r^n_{wheel}, and its instance has a value 0 or 1, where 0 represents that the computing unit does not provide the wheel actuator and 1 represents that it does.

The total provided resource space P_{tot} represents the resource capability of all computing units. This is formulated as

$$\begin{aligned} P_{tot} &\triangleq \bigcup_{i=1}^{m} P_i \\ &= \; < \sum_{i}^{m} r_1^i, \sum_{k}^{m} r_2^i, \cdots, \sum_{k}^{m} r_n^i > \end{aligned}$$

where m is the number of computing units, n is the number of resources, and $< r_1^i, r_2^i, \cdots, r_n^i >$ is the provided resource space of the computing unit O_i (i.e., P_i). The total provided resource space is used to determine whether the requested tasks are acceptable in terms of resource capability.

3.2 Architectural configuration

Software architectures represent structural information about the elements of a software system and their relationships. The software architecture of a system comprises software components (elements) that represent system functionality and connectors (relationship) that are responsible for providing a communication medium (Shaw & Garlan, 1996). In software architectures, a component provides an executable code that performs specific functions such as transforming a data set and processing user requests. A connector links two or more components and relays messages between the components. The software architecture organizes components and connectors into a software system.

In this formulation, a software architectural configuration denotes an instance of software architecture in a system. During system execution, components and connectors in an architectural configuration consume resources that the system provides. For example, a component can consume CPU time, memory space, and embedded devices (such as wheel actuators in robotic systems) when it operates in the system. A connector consumes network bandwidth when two components (the connector interconnects) communicate to perform their functionality.

The resource consumption of components and connectors can be defined by the required resource space that is the Cartesian product of the individual required resources. It specifies the amount of resource consumption that a component or connector requires. The required resource space R_i of a component or connector c_i is defined by

$$R_i \triangleq \bigotimes_{j}^{n} r_j$$

where n is the number of resources and r_j represents the j-th resource dimension.

When the j-th resource dimension represents a sharable resource, its instance of the i-th component or connector c_i can have an integer value that represents the required amount of the j-th resource dimension r_j. This formulation assumes that the value represents the average consumption of a component or connector because real-time resource accounting imposes a severe overhead.

The total provided resource space R_{tot} represents the resource requirements of all components and connectors. This is formulated as

$$\begin{aligned} R_{tot} &\triangleq \bigcup_{i=1}^{m} R_i \\ &= <\sum_{i}^{m} r_1^i, \sum_{k}^{m} r_2^i, \cdots, \sum_{k}^{m} r_n^i> \end{aligned}$$

where m is the number of components and connectors, n is the number of resources, and $< r_1^i, r_2^i, \cdots, r_n^i >$ is the required resource space of a component or connector c_i (i.e., R_i). The total required resource space is used to determine whether the requested tasks are acceptable in terms of the resource capability that the system provides.

3.3 Dynamic architectural deployment problem

The dynamic architectural deployment problem can be formulated on the basis of information given in the previous sections. This problem is a combinatorial optimization problem (Cook et al., 1997) in which one searches the problem space for the best combination. Among the various types of combinatorial optimization problems, the dynamic architectural deployment problem can be modeled as a knapsack problem, in which one searches for the best combination of items to be packed in the knapsack. In architectural deployment, components are regarded as the items to be packed and the computing units are regarded as knapsacks that contain the items.

In particular, this problem is a 0−1 knapsack problem, in which items cannot be partially packed into the knapsack because components cannot be decomposed into smaller ones. Further, the dynamic architectural deployment problem has multiple computing units. This implies that the problem should be formulated as a multiple-sack knapsack problem. Additionally, this problem should optimize multidimensional resource constraints. Therefore, this problem is formulated as a multidimensional knapsack problem. Consequently, the dynamic architectural deployment problem is formulated as a 0–1 multiple-sack multidimensional knapsack problem.

A knapsack problem comprises knapsacks, items, and cost/value functions. In dynamic architectural deployment, computing units in the system are knapsacks and components are items, as described in the previous paragraphs. A cost function decides that the selected items meet a certain constraints. In this problem, the cost function determines whether the selected combination in the knapsacks (i.e., the combination is a set of components in computing units) exceeds provided resources. A value function shows how valuable the selected combination in the knapsacks is. The value function of this problem represents the standard deviation of residual resources in all computing units.

The rationale of the value function formulation is based on the component execution pattern. As described in (Keim & Schwetman, 1975), some tasks can consume computing resources in a bursty manner. For example, the face recognition component of a human face-tracing task may temporarily use a large amount of network bandwidth when it requires the next scene of cameras. On the other hand, some other tasks consume computing resources in a steady manner, as when the localizer component of a navigation task continuously identifies the current position of a robot. The former type of resource consumption (bursty consumption) may particularly lead to performance degradation and execution delay problems mentioned in Section 2.

To prevent these problems, the resources of computing units should be managed to tolerate bursty resource consumption. This can be achieved by maximizing the residual resources of computing units. A computing unit can endure bursty consumption if it has sufficient residual resources. This assumption can be valid for sharable resources; however, for non-sharable resources, it is not useful for maximizing residual resources. To deal with this problem, it is assumed that the cost function determines whether the component can use the specified non-sharable resources. In other words, the cost function returns a positive integer if the computing unit O_k does not provide a non-sharable resource r_j (i.e., $r_j^k = 0$) when component c_i that requires a non-sharable resource r_j (i.e., $r_j^i = 1$); otherwise returns 0.

Another issue is the resource consumption of connectors. In this formulation, connectors are not items to be packed into knapsacks; however, they definitely consume resources such as network bandwidth between computing units. It is assumed that connectors consume computing resources, which are used for connecting components, only when the components are deployed into separate computing units. This type of resource consumption is dependently determined by component deployment.

On the basis of the above assumptions, the value function v can be defined by

$$v : A \rightarrow \mathbb{R}$$

$$\begin{aligned} v(a \in A) &= v(< c_1 = O_{c_1}, c_2 = O_{c_2}, \cdots, c_m = O_{c_m} >) \\ &= v(< O_{c_1}, O_{c_2}, O_{c_3}, \cdots, O_{c_m} >) \\ &= v(D_{r_1}(a), D_{r_2}(a), \cdots, D_{r_n}(a)) \\ &= \sum_i^n w_i D_{r_i}(a) \end{aligned}$$

where A is the set of architectural deployment instances, $\mathbb{R}$ is the real number set, c_i is the i-th component, and O_{c_i} is the computing unit, in which c_i is deployed. m is the number of components and n is the number of resource dimensions. w_i is the weight of the i-th resource dimension where $\sum_i^n w_i = 1$. The value function v returns the weighted sum of the residual amount of every resource dimension. Function D_{r_i} represents the residual amount of the resource dimension r_i. This function is defined as

$$D_{r_i} : A \rightarrow \mathbb{R}$$

$$\begin{aligned} D_{r_i}(a) &= D_{r_i}(< O_{c_1}, O_{c_2}, O_{c_3}, \cdots, O_{c_m} >) \\ &= std(d(r_i^1, a), d(r_i^2, a), \cdots, d(r_i^k, a)) \end{aligned}$$

where *std* is the standard deviation function, k is the number of computing units, and d is the residual resource function that returns the residual amount of resource of the i-the resource in computing unit k (i.e., r_i^k when the system selects deployment instance a).
The cost function S, which determines whether the deployment instance violates resource constraints, is defined as

$$S : A \rightarrow \{0,1\}$$

$$\begin{aligned} S(a) &= s(< O_{c_1}, O_{c_2}, O_{c_3}, \cdots, O_{c_m} >) \\ &= \bigcup_j^n s_{r_j}(a) \end{aligned}$$

where s_{r_j} is the function that determines whether architectural deployment instance a exceeds the amount of resource dimension r_j. The cost function returns 0 if no resource dimension exceeds, and 1 if at least one resource dimension exceeds. It determines the excess based on function s_{r_j} that is defined by

$$s_{r_j} : A \rightarrow \{0,1\}$$

$$s_{r_j}(a) = \begin{cases} 0 & \text{if } r_j^k \text{ of } O_k \text{ is equal or less than } r_j \text{ of } a, \forall k, \\ 1 & \text{if } \exists k, r_j^k \text{ of } O_k \text{ is larger than } r_j \text{ of } a \end{cases}$$

On the basis of the value and cost function described above, the goal of this problem is defined by

$$\text{find} \quad a^* = \underset{a \in A}{\arg\min}\, v(a)$$

$$\text{subject to} \quad S(a) \neq 1$$

a^* is the best deployment instance based on the value function and $S(a) \neq 1$ indicates that a deployment instance that violates resource constraints cannot be selected as an actual deployment instance in the system.
As described earlier, this problem is formulated as a 0-1 knapsack problem; however, in general, 0-1 knapsack problems are known as NP-hard problems (Kellerer et al., 2004). In particular, this is a multiple-knapsack, multidimensional, and 0-1 knapsack problem. Moreover, the dynamic architectural deployment problem requires both a better solution and faster search for the solution. To deal with this requirement in the present study, genetic algorithms are applied to the problem. The next section describes our approach in detail.

4. Genetic algorithm based dynamic architectural deployment

A genetic algorithm is applied to the dynamic architectural deployment problem in this study. A genetic algorithm (Holland, 1975) is a metaheuristic search method to approximate an optimal solution in the search space. This method is well known for dealing with combinatorial optimization problems. In a genetic algorithm, the target problem should be represented by a string of genes. This string is called a *chromosome*. By using the chromosome representation, a genetic algorithm generates an initial population of chromosomes. Then, it repeats the following procedure until a specific termination condition is met (usually a finite number of generations): (1) select the parent chromosomes on the basis of a specific crossover probability and perform the crossover; (2) choose and mutate the chromosomes on the basis of a specific mutation probability; (3) evaluate fitness of the offspring; and (4) select the next generation of population from the offspring.
In our approach, a genetic algorithmis used to search for an optimal architectural deployment instance. To apply the above procedure to the problem, architectural deployment instances are encoded into the chromosome representation, mutation and crossover operators for the chromosomes are determined, and the fitness function to evaluate the chromosomes is designed.

4.1 Representing architectural deployment instances in genes

It is important to encode the problem space into a set of chromosomes by a string of genes when applying a genetic algorithm to a certain application. In this approach, architectural deployment instances are encoded into chromosomes because our goal is to find an optimal instance from a set of instances. Components are matched to genes; each gene represents that the corresponding component is deployed into a specific computing unit by specifying the number of the computing unit.
For example, a deployment instance can be represented by a chromosome shown in Figure 3. c_i, e_i, and h_i represent the *i*-th component, deployment instance, and chromosome, respectively. In each instance, the *i*-th digit has a value that indicates the computing unit, in which the *i*-th component is deployed (i.e., the *i*-th digit of the chromosome is 1 if the *i*-th component c_i is deployed in computing unit O_1). When the required number of components in the task that the user requested is *m*, the length of the chromosome is *m*. The string of *m* digits is the *i*-th chromosome h_i of the *i*-th architectural deployment instance e_i. On the basis of the representation described above, our approach configures the search space of the dynamic architectural deployment problem.

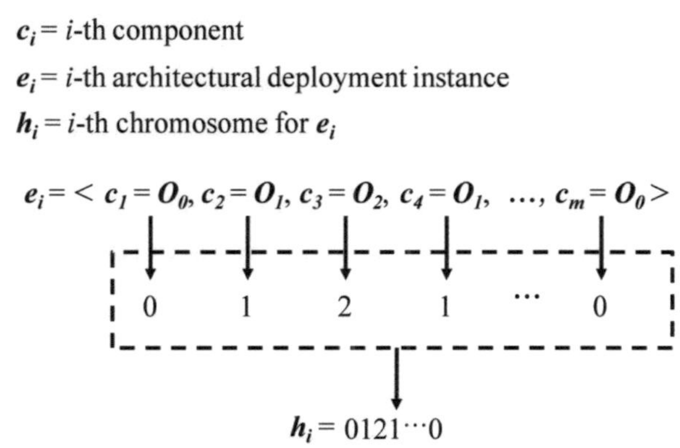

Fig. 3. An example of chromosomes that represent architectural deployment instances.

4.2 Operators

As described earlier in this section, chromosomes that constitute the population (i.e., the search space) are reproduced by crossover and mutation operators. Therefore, designing these operators is an important issue in applying genetic algorithms. In this approach, two-point crossover and digit-wise probabilistic mutations are used.

The two-point crossover operator picks up two chromosomes as parents with crossover probability P_c and chooses two (arbitrary and same) positions of the parents. The operator exchanges digits between the selected positions of the parents. These new chromosomes are offspring. This technique is used because it preserves more characteristics of parent chromosomes than other crossover techniques (other crossover operators may exchange different digits of parents). Further, it is assumed that similar chromosomes may have similar results to the value function.

After producing offspring by the crossover operator, the algorithm should perform mutation. Every digit of offspring produced by the crossover operator is changed to arbitrary values with mutation probability P_m. Note that if the mutation probability is too high, it cannot preserve the characteristics of the parent chromosomes. On the other hand, if the probability is too low, the algorithm may fall into local optima. Offspring produced by crossover and mutation are candidates for the population of the next generation.

4.3 Fitness and selection

After performing crossover and mutation, the next step of our approach is selection. In this step, in general, a genetic algorithm evaluates the fitness values of all offspring, and chromosomes that have better values survive. In this study, the value function described in Section 3.3 is used as a fitness function to evaluate chromosomes, and the tournament selection strategy (Miller et al., 1995) is used as a selection method. The tournament selection strategy selects the best ranking chromosomes from the new population produced by crossover and mutation.

The cost function removes chromosomes that violate resource constraints and that the function specifies. In this approach, this filtering process is performed in the selection step. Even if the best ranking chromosomes have higher values than the value function, chromosomes that the cost function indicates as 1 should be removed from the population. On the basis of the value and cost functions, the approach selects the next population.

The size of the population is an important factor that determines the efficiency of genetic algorithms. If the size is too small, it does not allow exploring of the search space effectively. On the other hand, too large a population may impair the efficiency. Practically, this approach samples at least $k \cdot m$ number of chromosomes, where k is the number of computing units and m is the number of components that the task requires.

By using the described procedure, our approach can find the best (or reasonably good) solution from the search space when the user requests a set of tasks and the task requires a set of components. The next section describes the results of the performance evaluation.

5. Evaluation

This section provides the results of the performance evaluation. First, the performance of exhaustive search and greedy search was measured. Then, the performance of our approach was measured and compared with the results of the earlier experiments.

Every experiment was performed on a set of desktops equipped with Intel Pentium Core 2 2.4 Ghz CPU and 2 GB RAM. Our approach was implemented using Java. A set of components and a set of five computing units were designed. The required resources of the components and the provided resources of the computing units were arbitrarily specified. The total required resources of the components (i.e., R_{tot}) did not exceed the total provided resources of the computing units (i.e., P_{tot}). Under these conditions, the following experiments were conducted.

5.1 Baseline

As a baseline, an experiment was conducted to measure the performance of exhaustive and greedy searches. In this experiment, the elapsed time of the exhaustive and greedy search was measured, and the best and greedy-best chromosomes (i.e., the best combination found by exhaustive search and the greedy combination found by greedy search, respectively) were determined. These results were used as a baseline to compare with the results of our approach. As stated in Section 3.3, the dynamic architectural deployment problem is a combinatorial optimization problem and has time complexity $O(k_m)$, where k is the number of computing units and m is the number of components requested by a task. In this experiment, there are five computing units k and the number of components n is varied from 5 to 13.

As shown in Figure 4, the exhaustive method searches the problem space in 10 seconds when $m < 10$. However, since $m = 10$, the elapsed time to complete searching the problem space exponentially increases. It is not always acceptable for the user to wait for the end of search because it may lead to user intolerance (Schiaffino & Amandi, 2004).

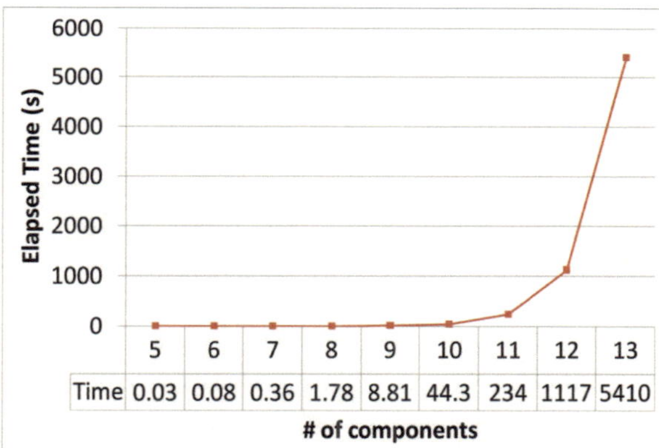

Fig. 4. The performance of the exhaustive search.

Greedy algorithms can be an alternative for reducing the search time to prevent user intolerance. In ubiquitous environments, a greedy algorithm is already adopted to determine better application combinations for a task requested by the user (Sousa et al., 2006). An experiment in which a greedy algorithm searches the same problem space for greedy solutions was also conducted. In every search from $m = 5$ to $m = 13$, the greedy search technique could find greedy solutions in 100*ms*. However, as shown in Figure 5, their quality decreases as the number of components increases. Therefore, an optimal solution cannot be expected using a greedy search.

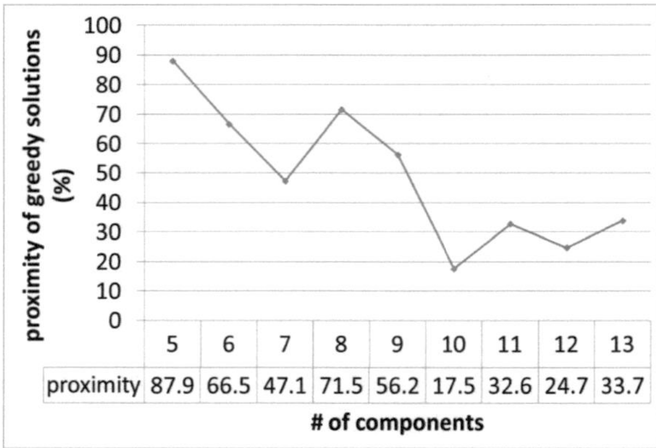

Fig. 5. The proximity of greedy solutions compared to the best solutions found in the exhaustive search.

5.2 Performance of GA-based approach

On the basis of the baseline of the previous experiment, three performance tests were conducted. The first test measured the elapsed time required to search an optimal or near-optimal combination of architectural deployment instances. It is difficult to anticipate the time required to find an optimal solution because genetic algorithms are randomized. However, a near-optimal solution that is close to the best solution (such as the Las Vegas algorithm) can be considered. In the first test, it is assumed that our approach terminates when the difference of the elitist chromosome in the population and the best chromosome (i.e., the best deployment instance found in the exhaustive search) is smaller than 5% of the best chromosomes.

In other words, `if` $Fit(elitist) - Fit(best) < 0.05 \cdot Fit(best)$, `then` terminate the search, where $Fit(a)$ evaluates the fitness (or value) of chromosome a.

As shown in Figure 6, the elapsed time to obtain an approximate chromosome by our approach is very short compared to the time for the exhaustive search. For every number of components, the elapsed time does not exceed 2 seconds and does not proportionally increase. This is because of the randomness of genetic algorithms, as previously stated. Even though this test shows that our approach can find a near-optimal solution in a short time, this type of approximation and termination condition is not feasible in practical systems because it is not possible for a system to anticipate the best solution before executing the search. Therefore, another termination condition is required.

The next test provides another termination condition that is similar to the Monte Carlo simulation (i.e., the termination condition specifies the fixed number of generations). This test was conducted to verify how fast our approach converges to the best architectural deployment instance when the fixed number of generations is used as the termination condition. In this test, the elitist chromosome was recorded for every generation and the results are shown in Figure 7. As shown, the number of generations is fixed as 300 and every search is finished in 500*ms*. Except for the case where the number of components is 12 and 13, our approach gradually converges to over 85% of the best solutions in 300 generations when the number of components is from 5 to 11. This implies that our approach can find near-optimal solutions in short generations. However, when the number of components is 12 and 13, the convergence of the solutions is under 65% of the best solutions. This implies that 300 generations are not sufficient to search the larger problem space of a large number of components.

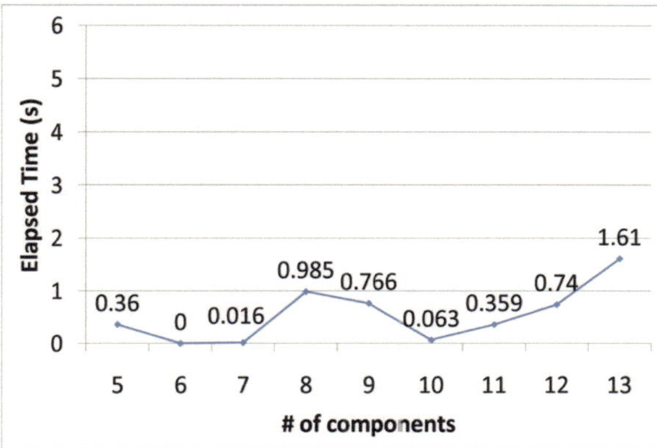

Fig. 6. Elapsed time to obtain an approximate chromosome for each number of components.

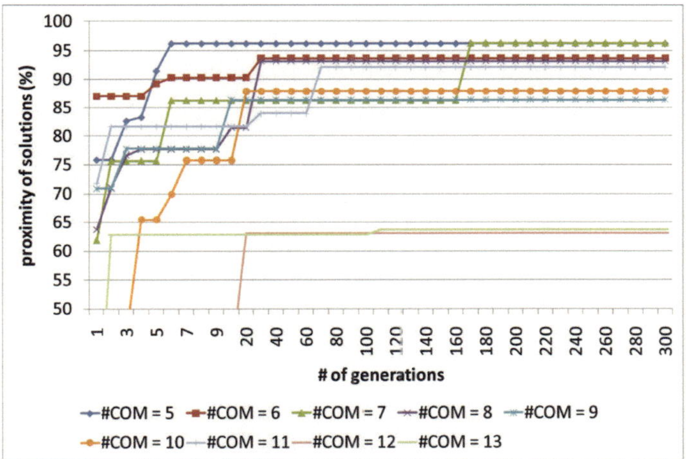

Fig. 7. Ratio of proximity to the best architectural deployment instance (#*COM*= *n* represents the number of components is *n*).

The third test provides an adaptive termination condition. When the number of components is larger than 12, using the fixed number of generation as a termination condition is not feasible; therefore, adaptively increasing the number of generations is applicable. However, simply increasing the number of generations may lead to the problem of the previous test. Hence, it is assumed that the elitist chromosome is the best chromosome with high probability if the elitist has not been changed for a long period. In this test, if the elitist has not been changed in $p \cdot k \cdot m$ generations, the search is terminated, where p is a constant, k is the number of computing units, and m is the number of components. The result of the test is shown in Figure 8 and we set $p = 10$. Each elapsed time in the figure is the average time of 10 test runs. As shown in Figure 8, when the number of components is 20 and 25, the elapsed time to determine that the elitist is the best chromosome (or very close to the best) is smaller than 5 seconds. When the number of components is from 30 to 40, the elapsed time is around 7 seconds. This implies that the required time to determine that the elitist is the best chromosome does not exponentially increase. Therefore, an optimal or near optimal architectural deployment instance can be found in a reasonable time with higher probability than the fixed number of generations, even if the best solution is not known. The constant *p* should be adaptively controlled, as a wait of more than 5 seconds is probably too long for some users.

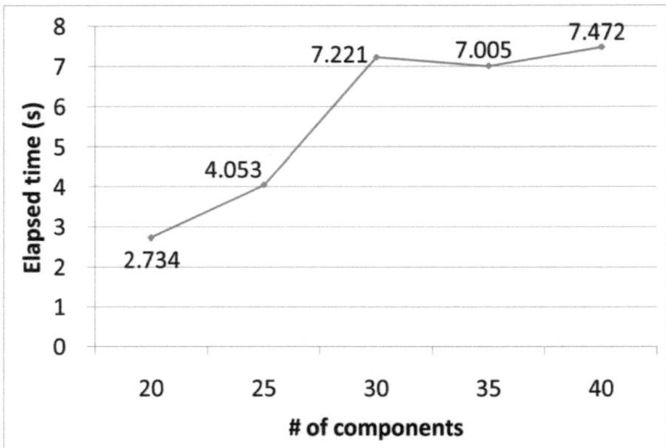

Fig. 8. Required time to perform the search to determine that the elitist is the best.

The evaluation results show that our approach provides optimal, or near-optimal, solutions (which are definitely better than the solutions found by greedy algorithms) in a reasonably short time (which is obviously faster than an exhaustive search). In addition, three different termination conditions are provided and their applicability is discussed. Consequently, the termination condition that adaptively controls the number of generations is practically applicable.

6. Case study

This section describes a case study conducted to show the applicability of our approach. The case study applies our approach to home service robots. In this case study, we assume that the task manager, which manages the robot's operation, requests five tasks to achieve a specific goal requested by the user. *Static deployment*, fixes the location of components that comprise the robot software architecture, obviously has problem. For example, all components in 'Object Recognizer' subsystem architecture must be deployed and executed in the Vision SBC that has cameras. This assumption looked reasonable when the architecture can dominate the SBC's resource. However, this is not appropriate because other tasks can share the resources of the SBC if the tasks *statically* assigned to be deployed into the SBC. Therefore, the system needs dynamic deployment. This case study provides three cases: 1) static deployment to verify the resource contention problem, 2) greedy dynamic deployment, and 3) genetic algorithm-based dynamic deployment.
We conducted the three cases under the following configuration: 1) Two SBCs; one is the Vision SBC that has cameras and pan-tilt gears which controls the robot's head, the other one is called the Main SBC that has wheels, microphones, laser range finders, and speakers. 2) Each SBC is equipped with 1GB of main memory and 2.2 GHz CPU. 3) Two SBCs are connected by 100 Mbps network.

First, we statically deployed autonomous navigator (has five components), TV program recommender (four components), arm manipulator (four components), interaction manager (seven components), and active audition planner (four components) into the Main SBC. At this time the Vision SBC has no resource consumption. Then, we focused on the navigator's behavior. The navigator must receive the robot's current pose (position and direction) and a map around the robot every 200ms to navigate safely. Since all components were deployed in a static manner, the navigator cannot have sufficient computing resource (especially CPU) and cannot retrieve a pose and a map every 200ms. This leads to wrong path planning. Moreover, the robot cannot reach the destination. This situation can be relieved by removing more than one architectures; however, if the goal of the user simultaneously requires all software components, the robot cannot achieve the goal.

To resolve the above problem, we applied dynamic deployment with greedy search. Based on the dynamic architecture reconfiguration framework in our previous research (Kim et al., 2006; Kim & Park, 2006) (Figure 9 shows screen shots of user interfaces in the framework and the robot using the framework), the robot searched for an appropriate deployment instance. The greedy algorithm has $O(nN)$ time complexity where n is the number of components to be deployed and N is the number of SBCs. In this case, we performed deployment 79 times with greedy search. The average time to search a deployment instance was 42ms. This overhead doesn't influence overall performance of the robot software system. However, the found deployment instance was far from the optimal solution. Therefore, we applied our genetic algorithm-based approach described in Section 4.

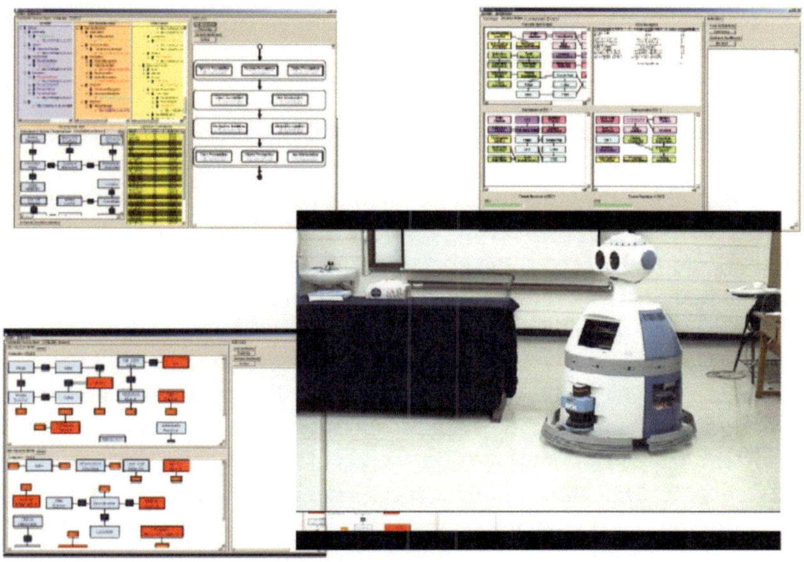

Fig. 9. A capture image of the case study

Using dynamic deployment with our approach, the robot could determine an efficient deployment instance within 3 ~5 seconds. It took more time than the greedy search, but our approach found better (near optimal) deployment instance as shown in Section 5. After deployment, the navigator could have sufficient resources and reliably retrieve the robot's pose and map. Other concurrent tasks were also performed effectively. Consequently, our approach enabled the robot system to perform its services more reliably.

7. Related work

Floch *et al.* (Floch et al., 2006) proposed a utility-based adaptation scheme. This approach assumed that an adaptable application operates on the adaptation middleware that was previously proposed (Hallsteinsen et al., 2004), and the middleware monitors the current user and system context. The system context includes computing resources such as CPU and memory. This approach selects an appropriate component for a specific component type on the basis of the user context (quality attributes) and system context (resource constraints). The differences between this approach and our approach are concerns about resource-efficiency and multiple computing units. This approach evaluates only whether the selected set of components exceeds the provided resources and assumes that the system provides only one computing unit.

Sousa *et al.* (Sousa et al., 2006) described the selection problem in which the system selects an appropriate application for a specific application type. This approach assumes that the user moves around various environments such as home, office, and park. The user needs the same context of his or her task wherever he or she moves. However, the computing power in each environment is different, e.g., a desktop at home or a handheld PC in the street. Therefore, for each environment, the systems must provide a different set of applications with the same context. This approach models the problem as a knapsack problem and solves it by using a greedy algorithm. The difference between Sousa's approach and our approach is the number of computing units and the selection method. This approach assumes that every system has only one computing unit. The greedy algorithm can find a solution in a short time; however, it cannot guarantee that the solution is the best or near-optimal solution.

8. Discussion

The first issue to discuss is the multiobjective (multidimensional) property of the dynamic architectural deployment problem. In general, a multiobjective problem can have a set of solutions that meets the requirements (i.e., Pareto set) (Das & Dennis, 1996). The dynamic architectural deployment problem can also have multiple solutions because it has multidimensional criteria. However, the goal of the problem is to search for only one optimal executable deployment instance rather than a set of possible solutions. In addition, the user or system administrator can specify the priority of dimensions by weight values and searching for an accurate Pareto set is time consuming. Therefore, evaluating the value of instance on the basis of the weighted sum of dimensions is practically applicable to this problem.

In the formulation, it is assumed that the constraint of non-sharable resources (i.e., the required resources of a component) can be acceptable if the computing unit provides the non-sharable resources. This assumption is acceptable only if the computing resource has a scheduling scheme for the non-sharable resource. Fortunately, most computing resources have specific scheduling schema, such as FIFO-like spooling for printer devices and round robin-like access mechanism for disk devices. Therefore, the assumption can be practically applicable to computing systems.

9. Conclusion and future work

The efficient deployment of components into multiple computing units is required to provide effective task execution, as the user simultaneously requests multiple and more complex tasks to the computing system. For example, service robots are requested to simultaneously perform several tasks and have multiple computing units. In such systems, inefficient deployment may lead to the malfunction of the system or the dissatisfaction of the user.

In this paper, a motivating example to deal with this problem was provided that described the problem in detail and formulated it as a multiple-sack multidimensional knapsack problem. To efficiently solve this problem, a genetic algorithm-based approach was proposed and the performance of the approach (efficiency and accuracy) was evaluated. The results of the performance tests demonstrated that our approach produced solutions more rapidly than exhaustive search and more precisely than greedy search methods.

Possible improvements to our approach include the extension of dimensions and parallel execution. Dimension extension implies that the problem formulate can add quality attributes to the problem space. Similar to computing resources, tasks may require a set of quality attributes, and components may provide various levels of quality. Further, some components may provide different quality levels in different computing units.

As stated in the formulation, our approach assumes that the system has multiple computing units. This indicates that our approach can be performed in parallel. The most time-consuming step is the selection process, and an individual chromosome evaluation by the value and cost function can be executed in independent computing units. Future work could focus on how the population of chromosomes can be efficiently divided.

10. References

Cook, W. J., Cunningham, W. H., Pulleyblank, W. R. & Schrijver, A. (1997). *Combinatorial Optimization*, Wiley.

Das, I. & Dennis, J. (1996). Normal-boundary intersection: An alternate method for generating pareto optimal points in multicriteria optimization problems, *Technical report*, Institute for Computer Applications in Science and Engineering.

Floch, J., Hallsteinsen, S. O., Stav, E., Eliassen, F., Lund, K. & Gjørven, E. (2006). Using architecture models for runtime adaptability, *IEEE Software* 23(2): 62–70.

Hallsteinsen, S., Stav, E. & Floch, J. (2004). Self-adaptation for everyday systems, *WOSS '04: Proceedings of the 1st ACM SIGSOFT workshop on Self-managed systems*, ACM Press, New York, NY, USA, pp. 69–74.

Hans, M., Graf, B. & Schraft, R. D. (2002). Robotic home assistant care-o-bot: Past - present - future, *Proceedings of IEEE Int.Workshop on Robot and Human interactive Communication (ROMAN2002)*, pp. 380–385.

Holland, J. H. (1975). *Adaptation in Natural and Artificial Systems*, University ofMichigan Press, Ann Arbor.

Keim, J.W. & Schwetman, H. D. (1975). Describing programbehavior in a multiprogramming computer system, *ANSS '75: Proceedings of the 3rd symposium on Simulation of computer systems*, IEEE Press, Piscataway, NJ, USA, pp. 21–26.

Kellerer, H., Pferschy, U. & Pisinger, D. (2004). *Knapsack Problems*, Springer.

Kim, D. & Park, S. (2006). Designing dynamic software architecture for home service robot software, *in* E. Sha, S.-K. Han, C.-Z. Xu,M. H. Kim, L. T. Yang & B. Xiao (eds), *IFIP In- ternational Conference on Embedded and Ubiquitous Computing(EUC)*, Vol. 4096, pp. 437– 448.

Kim, D., Park, S., Jin, Y., Chang, H., Park, Y.-S., Ko, I.-Y., Lee, K., Lee, J., Park, Y.-C. & Lee, S. (2006). Shage: A framework for self-managed robot software, *Proceedings of Workshop on Software Engineering for Adaptive and Self-Managing Systems(SEAMS)*.

Kim, J., Choi, M.-T., Kim, M., Kim, S., Kim, M., Park, S., Lee, J. & Kim, B. (2008). Intelligent robot software architecture, *Lecture Notes in Control and Information Sciences* 370: 385- 397.

Miller, B. L., Miller, B. L., Goldberg, D. E. & Goldberg, D. E. (1995). Genetic algorithms, tournament selection, and the effects of noise, *Complex Systems* 9: 193–212.

Schiaffino, S. & Amandi, A. (2004). User - interface agent interaction: personalization issues, *International Journal of Human-Computer Studies* 60(1): 129 – 148.

Shaw, M. & Garlan, D. (1996). *Software Architecture: Perspectives on an Emerging Discipline*, Prentice Hall.

Sousa, J., Poladian, V., Garlan, D., Schmerl, B. & Shaw, M. (2006). Task-based adaptation for ubiquitous computing, *Systems,Man, and Cybernetics, Part C: Applications and Reviews, IEEE Transactions on* 36(3): 328–340.

7

Comparison an On-screen Agent with a Robotic Agent in an Everyday Interaction Style: How to Make Users React Toward an On-screen Agent as if They are Reacting Toward a Robotic Agent

Takanori Komatsu
International Young Researcher Empowerment Centre, Shinshu University
Japan

1. Introduction

Communication media terminals, such as PDAs, cell phones, and mobile PCs, are devices that are used globally and are a part of our daily lives. Most people have access to such media terminals. Various interactive agents, such as robotic agents (for example, Gravot et al., 2006; Imai et al., 2003) and embodied conversational agents (ECA) appearing on a computer display (for example, Cassell et al., 2002; Prendinger and Ishizuka, 2004), are now being developed to assist us with our daily tasks. The technologies that these interactive agents can provide will soon be applied to these widespread media terminals.

Some researchers have started considering the effects of these different agents appearing on these media terminals on users' behaviours and impressions, especially for comparisons of on-screen agents appearing in a computer display with robotic agents (Shinozawa et al., 2004; Powers et al., 2007; Wainer et al, 2006). For example, Shinozawa et al. (2004) investigated the effects of a robot's and on-screen agent's recommendations on human decision making. Powers et al. (2007) experimentally compared people's responses in health interview with a computer agent, a collocated robot and a remote robot projected. And Wainer et al. (2006) measured task performance and participants' impression of robot's social abilities in a structured task based on the Towers of Hanoi puzzle.

Actually, most of these studies reported that most users stated that they felt much more comfortable with the robotic agents and that these agents were much more believable interactive partners compared to on-screen agents. In these studies, the participants were asked to directly face the agents during certain experimental tasks. However, this "face-to-face interaction" does not really represent a realistic interaction style with the interactive agents in our daily lives. Imagine that these interactive agents were basically developed to assist us with our daily tasks. It is assumed that the users are engaged in tasks when they need some help from the agents. Thus, it is expected that these users do not look at the agents much but mainly focus on what they are doing. I called this interaction style "an everyday interaction style." I assumed that this interaction style is much more realistic than the "face-to-face interaction" on which most former studies focused.

I then experimentally investigated the effects of an on-screen agent and a robotic agent on users' behaviours in everyday interaction styles. And moreover, I discussed the results of this investigation and focused in particular on how to create comfortable interactions between users and various interactive agents appearing in different media.

2. Experiment 1: Comparison of an on-screen agent with a robotic agent

2.1 Experimental setting

I set up an everyday interaction style between users and agents in this experiment by introducing a dual task setting; that is, one task was obviously assigned to participants as a dummy task, and the experimenter observed other aspects of the participants' behaviours or reactions, of which the participants were completely unaware.

First, the participants were told that the purpose of this experiment was to investigate the computer mouse trajectory while they played the puzzle video game "picross" by Nintendo Co., Ltd. (Fig. 1). The picross game was actually a dummy task for participants. While they were playing the picross game, an on-screen or robotic agent placed in front of and to the right of the participants talked to them and encouraged them to play another game with it. The actual purpose of this experiment was to observe the participants' behavioural reactions when the agent talked to them.

Fig. 1. "picross[1]" puzzle game by Nintendo Co., Ltd.

2.2 An on-screen agent and a robotic agent

The on-screen agent I used was the CG robot software "RoboStudio" by NEC Corporation (the left of Fig. 2), and the robotic agent was the "PaPeRo robot" by NEC Corporation (the right of Fig 2). RoboStudio was developed as simulation software of the PaPeRo, and it was

[1] http://www.nintendo.co.jp/n02/shvc/bpij/try/index.html

equipped with the same controller used with PaPeRo. Therefore, both the on-screen and the robotic agents could express the same behaviours.

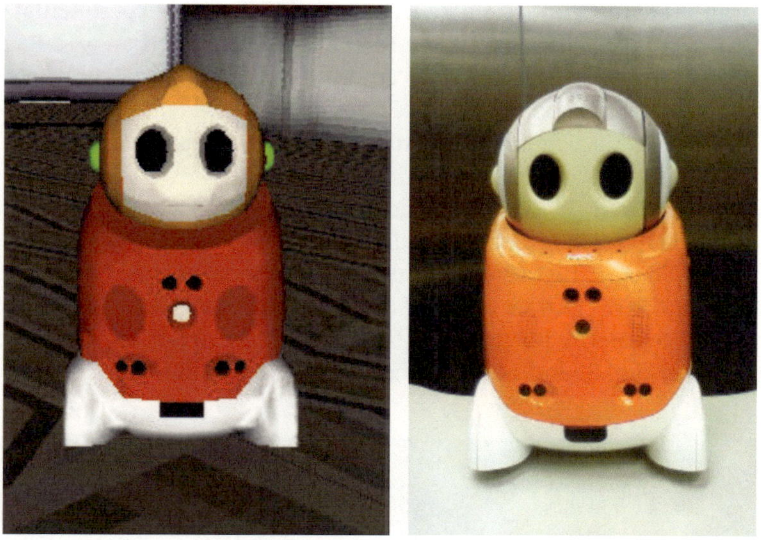

Fig. 2. On-screen agent "RoboStudio[2]" by NEC Corporation (left), robotic agent "PaPeRo[3]" by NEC Corporation (right).

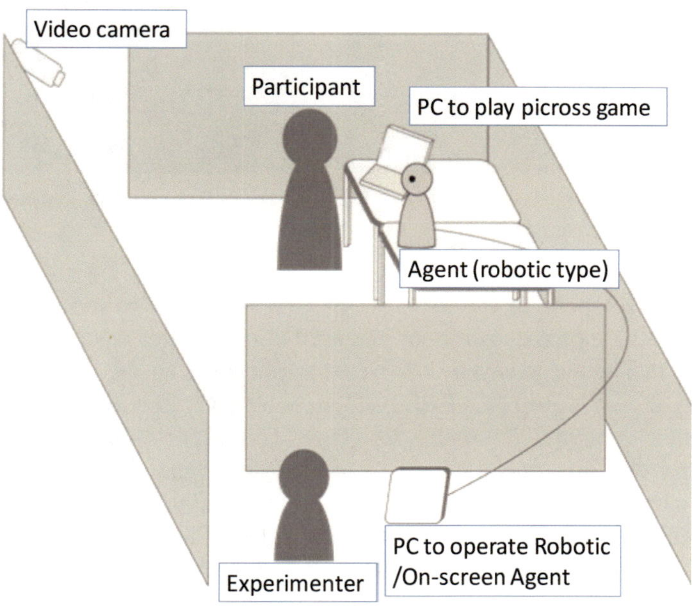

Fig. 3. Experimental Setting.

[2] http://www.necst.co.jp/product/robot/index.html
[3] http://www.nec.co.jp/products/robot/intro/index.html

2.3 Participants

The participants were 20 Japanese university students (14 men and 6 women; 19-23 years old). Before the experiment, I ensured that they did not know about the PaPeRo robot and RoboStudio. They were randomly divided into the following two experimental groups.

- **Screen Group** (10 participants): The on-screen agent appeared on a 17-inch flat display (The agent on the screen was about 15 cm tall) and talked to participants. The agent's voice was played by a loudspeaker placed beside the display.
- **Robotic Group** (10 participants): The robotic agent (It was about 40 cm tall) talked to participants.

Both the robotic agent and the computer display (on-screen agent) were placed in front of and to the right of the participants, and the distance between the participants and the agents was approximately 50 cm. The sound pressure of the on-screen and robotic agent's voice at the participants' head level was set at 50 dB (FAST, A). These agents' voices were generated by the TTS (Text-To-Speech) function of RoboStudio. The overview of the experimental setting is depicted in Fig. 3, and pictures showing both experimental groups are shown in Fig. 4.

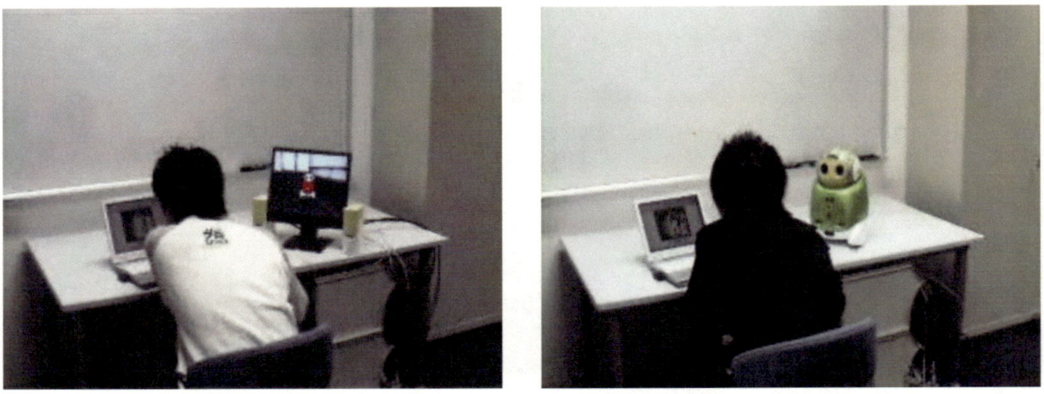

Fig. 4. Experimental Scene: participants in Screen group (left) and in Robotic group (right)

2.4 Procedure

First, the dummy purpose of the experiment was explained to the participants, and they were asked to play the picross game for about 20 minutes after receiving simple instructions on how to play it. The game was a web-based application, so the participants used the web browser to play it on a laptop PC (Toshiba Dynabook CX1/212CE, 12.1 inch display).

The experimenter then gave the instruction that "This experiment will be conducted by this agent. The agent will give you the starting and ending signal." After these instructions, the experimenter exited the room, and the agent started talking to the participants, "Hello, my name is PaPeRo! Now, it is time to start the experiment. Please tell me when you are ready." Then, when the participants replied "ready" or "yes," the agent said, "Please start playing the picross game," and the experimental session started. The agent was located as described earlier so that the participants could not look at the agent and the picross game simultaneously.

One minute after starting the experiment, the agent said to them "Umm...I'm getting bored... Would you play Shiritori (see Fig. 5 about the rules of this game) with me?" Shiritori is a Japanese word game where you have to use the last syllable of the word spoken by your

opponent for the first syllable of the next word you use. Most Japanese have a lot of experience playing this game, especially when they are children. If the participants acknowledged this invitation, i.e., said "OK" or "Yes," then the Shiritori game was started. If not, the agent repeated this invitation every minute until the game was terminated (20 minutes). After 20 minutes, the agent said "20 minutes have passed. Please stop playing the game," and the experiment was finished. Here, Shiritori is an easy word game, so most participants could have played this game while continuing to focus on the picross game.
The agent's behaviours (announcing the starting and ending signals and playing the last and first game) were remotely controlled by the experimenter in the next room by means of the wizard of oz (WOZ) manner.

Japanese Last and First Game (*Shiritori)*

Rule:

- Two or more people take turns to play.
- Only nouns are permitted.
- A player who plays a word ending in the mora "*N*" loses the game, as no word begins with that character.
- Words may not be repeated.

Example:

Sakura (cherry blossom)-> *rajio* (radio)-> *onigiri* (rice ball)-> *risu* (squirrel) -> *sumou* (sumo wrestling) -> *udon* (Japanese noodle)

Note: The player who played the word *udon* lost this game.

Fig. 5. Rules of Shiritori from Wikipedia[4].

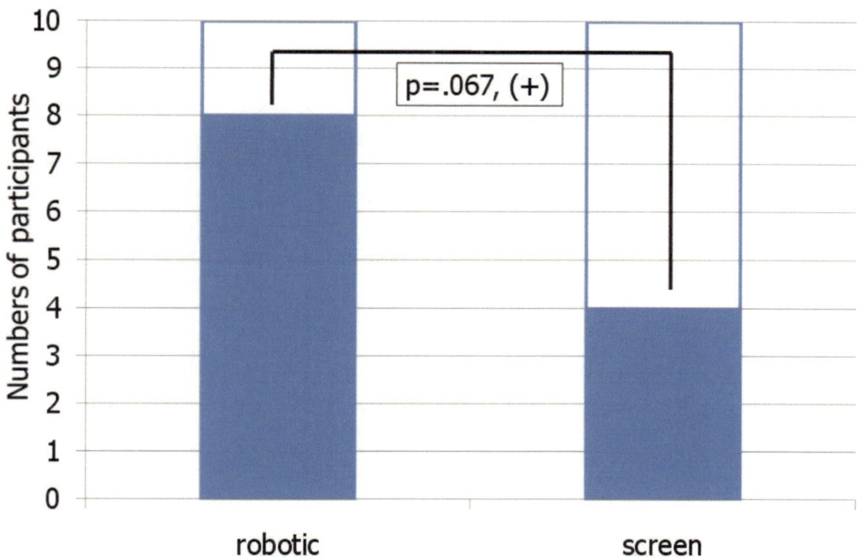

Fig. 6. Rate of participants acknowledging or ignoring the agent's invitation to play Shiritori.

[4] http://en.wikipedia.org/wiki/Shiritori

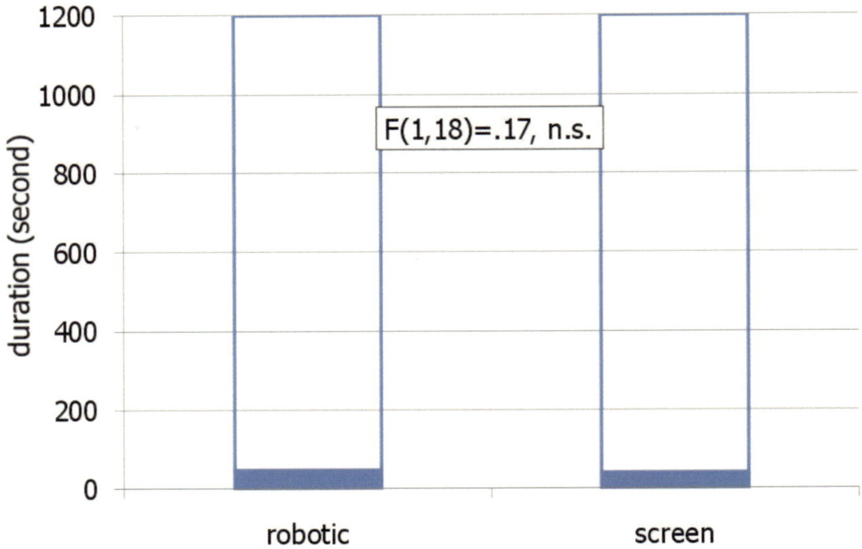

Fig. 7. Duration of participants looking at picross game or agent

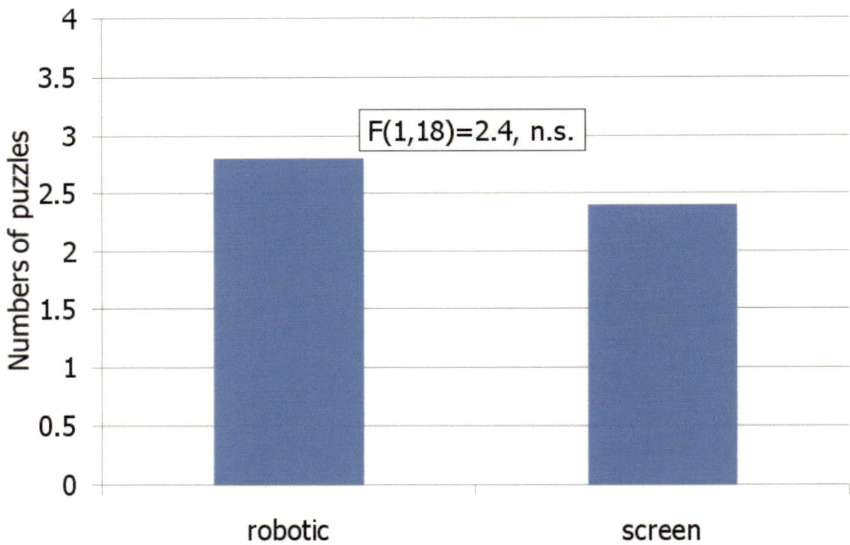

Fig. 8. How many puzzles they succeeded in solving.

2.5 Results

In this experiment, I assumed that the effects of different agents on the participants' impressions would directly reflect on their behaviours. I then focused on the following behaviours: 1) whether the participants acknowledged the agent's invitation and actually played the Shiritori game, 2) whether the participants looked at the agent or the picross game during the game, 3) how many puzzles the participants succeeded in solving.

1. **Whether the participants acknowledged the agent's invitation and actually played the Shiritori game:** In the robotic group, eight out of the 10 participants acknowledged

the agent's invitation and actually played the Shiritori game with the agent. However, in the screen group, only four out of the 10 participants did so (Fig. 6). Fisher's exact test showed a significant tendency between these two experimental groups ($p=0.067$, $p<.1$ (+)).

2. **Where the participants looked (agent or picross game):** In the robotic group, the participants' average duration of looking at the robotic agent was 46.3 seconds. In the screen group, the average duration was 40.5 seconds (Fig. 7). These results revealed that most participants in both groups concentrated on looking at the picross game during this 20-minute (1,200 seconds) game. The results of an ANOVA showed no significant differences between the two groups on this issue ($F(1,18)=0.17$, n.s.).
3. **How many puzzles the participants succeeded in solving**: In the robotic group, the participants' average number of puzzles solved was 2.8, and in the screen group, it was 2.4 (Fig. 8). The results of an ANOVA showed no significant differences between these two groups ($F(1,18)=2.4$, n.s.).

The results of this experiment are summarized in the following:

- **Screen group**: The participants in the screen group showed the same achievement level on the picross game with the robotic group (Fig. 7). They did not look at the on-screen agent much during the experiment (Fig. 8), and they also did not acknowledge the invitation from the on-screen agent (Fig. 6).
- **Robotic group**: The participants in the robotic group showed the same achievement level on the picross game with the screen group (Fig. 7). They also did not look at the robotic agent much (Fig. 8). However, most of them acknowledged the robotic agent's invitation and actually played the Shiritori game (Fig. 6).

3. Summary of experiment 1

The results of this experiment showed that most participants acknowledged the robotic agent's invitation for the Shiritori game, while many neglected the on-screen agent's invitation. The participants in both groups showed nearly the same attitudes toward the picross game; that is, they did not look at the agent much but concentrated on the task, and they achieved nearly the same level on the picross game. The participants in the robotic group interacted with the robotic agent (playing the Shiritori game) without neglecting their given tasks (playing the picross game). Therefore, the robotic agent was also appropriate for interacting with users in an everyday interaction style, which is much more similar to the interaction we encounter in our daily lives compared with the style observed in a typical face-to-face interaction setting. Actually, these results are similar to those of former studies that focused on face-to-face interaction; that is, these studies argued that the robotic agents are much more comfortable and believable interactive partners than on-screen agents.

Let me consider why the participants acknowledged the robotic agent's invitation even though they were not really looking at the robotic agent. Beforehand, I reviewed Kidd and Breazeal's (2004) investigation. They conducted an experiment comparing a physically present robot with a robot appearing on television as live TV output. The results were that the participants did not show different behaviours and impressions of these different robots. They then concluded that "*it is not the presence of the robot that makes a difference, rather it is the fact that the robot is a real, physical thing, as opposed to a fictional animated character on screen, that leads people to respond differently.*" Therefore, the participants' beliefs or mental models about a "robot" based on their expectation or stereotypes (such as "A robot would be nice to talk to") would affect their attitudes toward an interaction with a robotic agent. More specifically, these beliefs and mental models would lead to making the participants assign certain types of personality or characteristics to this robotic agent and would then cause the participants to acknowledge the robotic agent's invitation, even though they did not look at the robotic agent much.

Based on the results of this experiment and the discussion in terms of Kidd & Breazeal's argument mentioned in the above, I would like to investigate the contributing factors that could make users react toward an on-screen agent as if they were reacting toward a robotic agent. Revealing such factors would enable utilizing on-screen agents as interactive partners, especially when robotic agents cannot be used, e.g., in a mobile situation with PDAs or cell phones. If so, I could argue that "on-screen agents are also suitable for interactive agents."

Specifically, I focused on the following two contributing factors: **1) whether users accepted an invitation from a robotic agent** and **2) whether an on-screen agent was assigned an attractive personality or character for the users**. The reason I focused on the first factor was that I assumed that participants who accepted an invitation from a robotic agent would also accept one from an on-screen agent that had a similar appearance. And the reason I focused on the second factor was based on the Kidd & Breazeal's argument (2004) that is about the users' beliefs and mental model issues. I then assumed that assigning an attractive character for an on-screen agent would cause the users to construct beliefs and mental models about the on-screen agent and would also lead to them behaving as if they were behaving toward the robotic agent.

I then conducted a consecutive experiment to investigate the effects of these two factors on the participants' behaviours, especially, on whether the participants accepted or ignored the invitation of the on-screen agent.

4. Experiment 2: How to make users react toward an on-screen agent as if they are reacting toward robotic agent?

4.1 Setting

The setting of this Experiment 2 was nearly same with one of the Experiment 1. However, the picross game was projected on the 46-inch LCD not 12-inch LCD like in Experiment 1 due to the participants' comfort game playing.

4.2 Participants

40 Japanese undergrads participated (20 – 23 years old; 18 men and 22 women). These participants were randomly divided into the following four experimental groups. Actually, this experimental setting was a 2 (whether the users accepted the invitation of the robotic agent; so-called, with/without the robotic agent) x 2 (whether the on-screen agent was assigned an attractive character for the users; so-called, with/without character) factorial design. Note that these participants did not participate in Experiment 1.

- **Group 1** (10 participants): This group was without the robotic agent and without the character setting. An on-screen agent appearing on a 17-inch flat display (the agent on the screen was about 15 cm tall) conducted the experiment for ten minutes.
- **Group 2** (10 participants): This group was also without the robotic agent but had the character setting. The same on-screen agent in Group 1 conducted the experiment.

However, just before the experiment, the participants were passed a memo about the on-screen agent. The memo stated that the agent had a very active character, like a child, and really liked talking with people.

- **Group 3** (10 participants): This group had the robotic agent but was without the character setting. A robotic agent (about 40 cm tall) conducted the experiment. After five minutes passed, the robotic agent made error sounds, and the experimenter immediately replaced the robotic agent with an on-screen agent. Then, the on-screen agent conducted the experiment for the remaining five minutes.
- **Group 4** (10 participants): This group had the robotic agent and the character setting. The experimental procedure was the same as that for Group 3. However, just before the experiment, the participants were passed Group 2's memo about the robotic agent.

4.3 Procedure

First, the dummy purpose of the experiment was explained to the participants, and they were asked to play the picross game for about 10 minutes after receiving simple instructions on how to play it. The game is a web-based application, so the participants used a web browser that was projected on a 46-inch LCD screen.

The experimenter then gave the instruction that "This experiment will be conducted by this agent because the presence of a human experimenter would affect the results. The agent will give you the starting and ending signal." Actually, the on-screen agent appeared on a 17-inch computer display for Group 1 and 2, while the robotic agent was prepared for Group 3 and 4. After these instructions were given, the participants in Group 2 and 4 were passed a memo that described the agents' character, and the experimenter exited the room. Then, the agent started talking to the participants, "Hello, my name is PaPeRo! Now, it is time to start the experiment. Please tell me when you are ready." Then, when the participants replied "ready" or "yes," the agent said, "Please start playing the picross game," and the experimental session started. Actually, the agent was located as described earlier (placed in front of and to the left of the participants) so that the participants could not look at the agent and the picross game simultaneously.

One minute after starting the experiment, the agent said to them "Umm…I'm getting bored…Would you play Shiritori with me?" If the participants accepted this invitation, i.e., the participants said "OK" or "Yes," then the Shiritori game was started. If not, the agent repeated this invitation every two minutes until the game was terminated.

In the cases of Group 3 and 4, after five minutes passed, the robotic agent made error sounds, and it automatically shut down. The experimenter then immediately entered the room and said to the participants "I'm really sorry about this problem… To tell you the truth, this robot has not been working very well in the last few days. I will arrange the experimental setting so that you can continue. Please wait a minute in the next room." While the participants waited in the next room, the experimenter hid the robotic agent where the participants could not see it, and placed the 17-inch computer display for the on-screen agent in the exact same place with the setting of Group 1 and 2. The participants could not see what the experimenter was doing because the participants were waiting in the next room. Afterward (about two minutes later), the participants were asked to go back to the experimental session, and the experimenter said to them "The emergency situation has been taken care of, so please continue the experiment for the remaining five minutes." The experimenter did not mention that the robotic agent was changed to an on-screen agent.

After the experimenter exited the room, the on-screen agent said "Now, it is time to start the experiment. Please tell me when you are ready," and the same experimental procedure was restarted. After 10 minutes in all four groups, the on-screen agent said "The experiment is now finished. Please stop playing the game." Figs 9 and 10 show a picture taken during the actual experimental with the on-screen agent and the robotic agent conducting it. The experimental procedures of the experiment are depicted in Fig. 11. And moreover, the overview of the experimental setting is depicted in Fig. 12.

Fig. 9. Experiment with on-screen agent (Group 1 and 2, and the last five minutes in Group 3 and 4).

Fig. 10. Experiment with robotic agent (the first five minutes in Group 3 and 4).

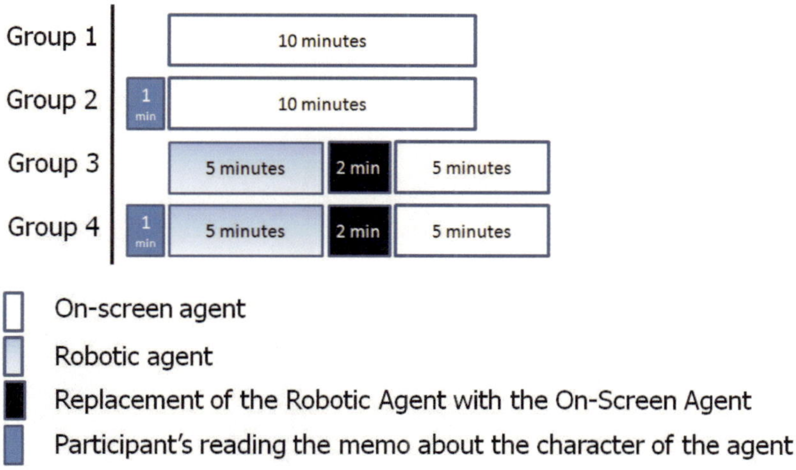

Fig. 11. Experimental procedure in each group. The participants' behaviors during the part depicted by the white rectangle were analyzed (i.e., when the on-screen agent was conducting the experiment.).

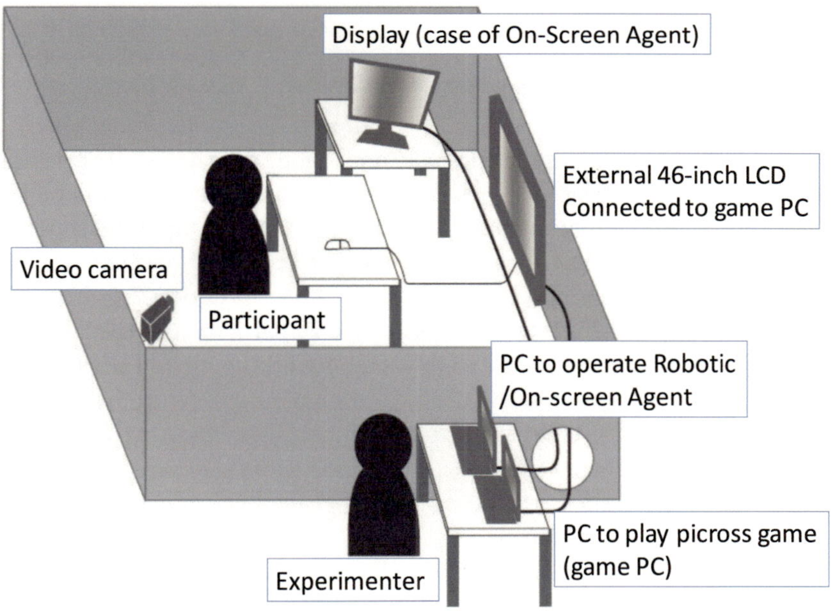

Fig. 12. Experimental Setting.

4.4 Results

I then focused on the following three types of participant behaviours to investigate the effects of the two contributing factors on how these factors contribute to make users react toward an on-screen agent as if they are reacting toward a robotic agent: 1) whether the participants accepted the on-screen agent's invitation and actually played the Shiritori game, 2) how much time the participants spent looking at the agent or at the picross game during the experiment, and 3) how many puzzles the participants succeeded in solving. To investigate these behaviours, I focused on the participants' behaviours when the on-screen agent was conducting the experiment; that is, the full 10 minutes in Group 1 and 2, and the last five minutes in Group 3 and 4.

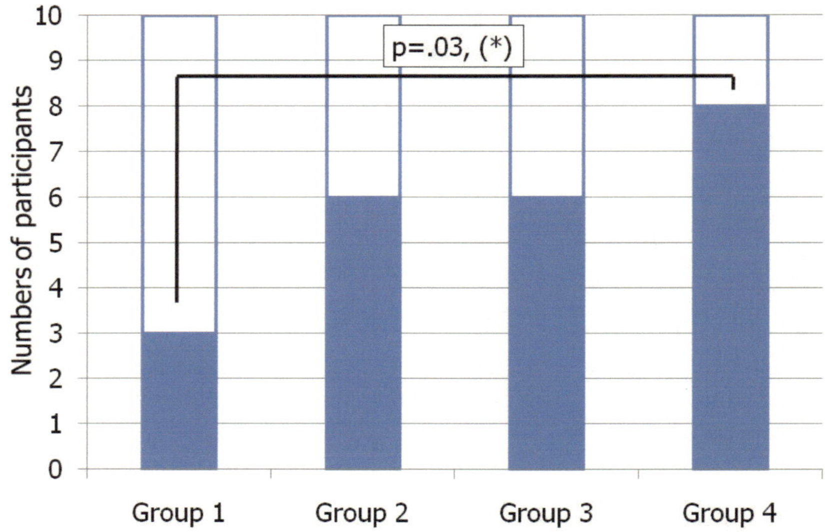

Fig. 13. Numbers of participants who accepted the invitation of on-screen agent in each group.

1) Whether the participants accepted the on-screen agent's invitation and actually played the Shiritori game?: First, I investigated how many participants accepted the on-screen agent's invitation in each experimental group (Fig. 13). In Group 1 (without the robotic agent and without the assigned character), three out of the 10 participants accepted the agent's invitation and actually played the Shiritori game. In Group 2 (without the robotic agent and with the assigned character), six out of the 10 participants accepted the invitation and played the Shiritori game. In Group 3 (with the robotic agent and without the assigned character), six participants accepted and played, and in Group 4 (with the robotic agent and with the assigned character), eight participants did so. Moreover, in Group 3 and 4, all participants who accepted the invitation of the robotic agent also accepted the invitation of the on-screen agent, while no participants who neglected the invitation of robotic agent accepted the invitation of the on-screen one.

A Fisher's exact probability test was used to elucidate the effects of the two contributing factors by comparing Group 1 with the other groups. The results of the comparison of Group 1 with Group 2 showed no significant difference between these two groups (one-sided testing: $p=.18>.1$, n.s.), and the results of the comparison of Group 1 and Group 3 also showed no significant difference between them (one-sided testing: $p=.18>.1$, n.s.). However, the results of the comparison of Group 1 with Group 4 showed a significant difference (one-sided-testing: $p=.03<.05$, (*)). The results of the comparison of Group 2 and 3 with Group 4 showed no significant differences (one-sided testing: $p=.31>.1$, n.s.).

Thus, the results of this analysis clarified that two contributing factors actually had an effect on making the participants react to an on-screen agent as if it were toward a robotic agent (Group 4), while only one of the two factors did not have such effects (Group 2 and 3).

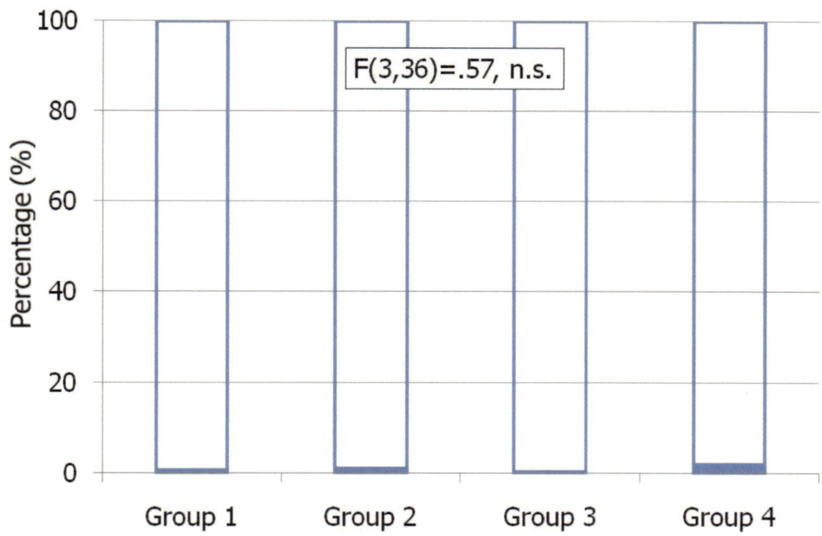

Fig. 14. Duration rates that participants' looked at on-screen agent.

2) Amount of time the participants spent looking at the agent or at the picross game during the experiment: Next, I investigated the amount of time that the participants looked at the on-screen agent. If the participants were not inclined to behave naturally toward the agent, they would look at the agent a lot. In Group 1, the participants looked at the on-screen agent for about 4.15 seconds during the 10 minute experiment, in group 2 they looked at it for about 7.31 seconds during the 10 minute experiment. In group 3, they looked at it for about 1.39 seconds during the five minute experiment, and in group 4, they looked at it for about 6.04 seconds over five minutes.

To elucidate the statistical differences between these four experimental groups, I utilized a duration rate that was calculated by dividing the amount of time that the participants looked at the on-screen agent by the total duration of the experiment (Fig. 14); that is, the duration rate was 0.69% in Group 1, 1.21% in Group 2, 0.47% in Group 3, and 2.01% in Group 4. Then, a two-way ANOVA (between factors: experimental groups) of the duration rates between four experimental groups was conducted. The results of the ANOVA showed no significant differences in the duration rates between these four groups ($F(3,36)=0.57$, n.s.). Therefore, the results of this analysis showed that the two contributing factors did not affect the participants' behaviours regarding how much time they spent looking at the on-screen agent or at the picross game. This means that these participants focused on playing the picross game.

3) How many puzzles the participants succeeded in solving?: Finally, I investigated how many puzzles the participants succeeded in solving in the picross game. If the participants could not behave naturally toward the agent, they would look at the agent a lot, and their performance on the dummy task would suffer. In Group 1, the participants solved on average 2.6 puzzles during the 10 minute experiment, in group 2, they solved 2.6 puzzles during the 10 minute experiment. In group 3, they solved 1.1 puzzles during the five minute experiment, and in group 4, they solved 1.5 puzzles during five minutes.

To elucidate the statistical differences between these four experimental groups, I utilized the solving rate over five minutes for each experimental group (Fig. 15); that is, the solving rate was 1.3 [puzzles/5 min] in Group 1, 1.3 [puzzles/5 min] in Group 2, 1.1 [puzzles/5 min] in Group 3, and 1.5 [puzzles/5 min] in Group 4. Then, a two-way ANOVA (between factors: experimental groups) for the solving rates between the four experimental groups was conducted. The results of the ANOVA showed no significant differences in the solving rates between these four groups ($F(3,36)=0.12$, n.s.).

Therefore, the results of this analysis showed that the two contributing factors did not affect the participants' behaviours regarding the number of puzzles solved. This also means that the participants focused on playing the picross game.

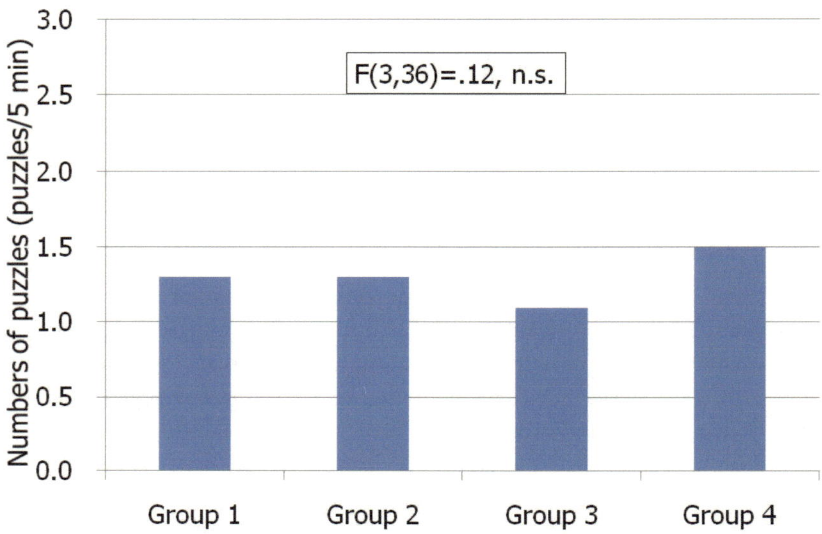

Fig. 15. Numbers of puzzles participants solved.

The results of this experiment can be summarized as follows.

- Eight participants in Group 4, six participants in Group 2 and 3, and three participants in Group 1 accepted the on-screen's invitation and played the Shiritori game. In fact, an exact probability test showed a significant difference in the number of participants accepting the invitation between Group 1 and 4 only.
- No significant difference was evident on the amount of time they spent looking at the on-screen agent or at the picross game between the four groups. The maximum duration rate was 2.01% in Group 4, so most participants in all groups concentrated on looking at the picross game during the experiment regardless of whether they played the Shiritori game or not. No significant differences were evident in the number of solved puzzles between the four groups.

These results indicated that the two contributing factors (whether the users accepted the invitation from the robotic agent and whether the on-screen agent was an attractive character for the users) played a significant role in making the participants react toward the on-screen agent as if they were reacting toward the robotic agent. However, one of these two factors was not enough to have such a role. Moreover, these factors did not have an effect on the participants' performance or behaviours on the dummy task.

6. Discussion and conclusions

I investigated the contributing factors that could make users react toward an on-screen agent as if they were reacting toward a robotic agent. The contributing factors I focused on were whether the users accepted the invitation of the robotic agent and whether the on-screen agent was assigned an attractive character for the users. The results showed that the participants who first accepted the invitation of a robotic agent that was assigned an attractive character reacted toward the on-screen agent as if they were reacting to the robotic one; that is, these two factors played a significant role in making the participants react toward the on-screen agent as if they were reacting toward the robotic agent.
The results of this experiment demonstrated that one of the two factors was not enough to make the participant accept the invitation; both factors were required for them to accept. This means that the robotic agent is still required to create an intimate interaction with on-screen agents and that assigning an attractive character for an on-screen agent is not enough. To overcome this issue, I am planning a follow-up experiment to investigate the effects of the media size on the participants' behaviours toward an on-screen agent, e.g., a comparison of an on-screen agent on a 46-inch LCD with an agent on a 5-inch PDA display. The reason is that, in Experiment 2, the on-screen agent's height was only about 15 cm while the robotic one's was about 38 cm in this experiment. Goldstein et al. (2002) previously reported that "*people are not polite towards small computers,*" so I expected that this follow-up experiment would enable us to clarify how to make users react toward on-screen agents as if they were reacting toward robotic agents without utilizing the robotic agent.
Although the results of this experiment suggested that a robotic agent is required to make users behave toward an on-screen agent as if they were reacting toward a robotic agent, introducing such robotic agents into families or into offices is somewhat difficult due to the higher costs and uncertain safety level. Therefore, users should first have the experience of interacting with a robotic agent at special events, such as robot exhibitions, and they should install an on-screen character into their own media terminals, such as mobile phones or normal personal computers, and continue to interact with this on-screen agent in their daily lives. This methodology could be easily applied for various situations where on-screen agents are required to assist users in which robotic agents cannot be used, e.g., in a mobile situation with PDAs, cell phones, or car navigation systems.

7. Acknowledgement

Experiment 1 was mainly conducted by Ms. Yukari Abe, Future University-Hakodate, Japan, and Experiment 2 was mainly conducted by Ms. Nozomi Kuki, Shinshu University, Japan as their Undergrad theses. The detailed results of each experiment was published in Komatsu & Abe (2008), and Komatsu & Kuki (2009a, 2009b), respectively. This work was supported by KAKENHI 1700118 (Graint-in-Aid for Young Scientist (B)) and Special Coordination Funds for Promoting Science and Technology from the Ministry of Education, Culture, Sports, Science and Technology (MEXT), Japan.

8. References

Cassell, J.; Stocky, T.; Bickmore, T.; Gao, Y.; Nakano, Y.; Ryokai, K.; Tversky, D.; Vaucelle, C. & Vilhjalmsson, H. (2002). MACK: Media lab Autonomous Conversational Kiosk, *Proceedings of Imagina02.*

Goldstein, M.; Alsio, G. & Werdenhoff, J. (2002). The Media Equation Does Not Always Apply: People are not Polite Towards Small Computers. *Personal and Ubiquitous Computing*, Vol. 6, 87-96.

Gravot, F.; Haneda, A.; Okada, K. & Inaba, M. (2006). Cooking for a humanoid robot, a task that needs symbolic and geometric reasoning, *Proceedings of the 2006 IEEE International Conference on Robotics and Automation*, pp. 462 - 467.

Imai, M.; Ono, T. & Ishiguro, H. (2003). Robovie: Communication Technologies for a Social Robot, *International Journal of Artificial Life and Robotics*, Vol. 6, 73 - 77.

Kidd, C. & Breazeal, C. (2004). Effect of a robot on user perceptions, *Proceedings of the 2004 IEEE/RSJ International Conference on Intelligent Robots and Systems*, pp. 3559 - 3564.

Komatsu, T. & Abe, Y. (2008). Comparison an on-screen agent with a robotic agent in non-face-to-face interactions. *Proceedings of the 8th International Conference on Intelligent Virtual Agents (IVA2008)*, pp. 498–504.

Komatsu, T. & Kuki, N. (2009a). Can Users React Toward an On-screen Agent as if They are Reacting Toward a Robotic Agent?, *Proceedings of the 4th ACM/IEEE International Conference on Human-Robot Interaction (HRI2009)*, pp.217-218.

Komatsu, T. & Kuki, N. (2009b). Investigating the Contributing Factors to Make Users React Toward an On-screen Agent as if They are Reacting Toward a Robotic Agent, *Proceedings of the 18th IEEE International Symposium on Robot and Human Interactive Communication (RO-MAN2009)*, to appear.

Powers, A.; Kiesler, S.; Fussel, S. & Torrey, C. (2007). Comparing a computer agent with a humanoid robot, *Proceedings of the 2nd ACM/IEEE International Conference on Human-robot Interactions*, pp. 145 - 152.

Prendinger, H. & Ishizuka, M. (2004). *Life-Like Characters*, Springer.

Shinozawa, K.; Naya, F.; Yamato, J. & Kogure, K. (2004). Differences in effects of robot and screen agent recommendations on human decision-making, *International Journal of Human-Computer Studies*, Vol. 62, 267 - 279.

Wainer, J.; Feil-Seifer, D, J.; Sell, D, A. & Mataric, M, J. (2006). Embodiment and Human-Robot Interaction: A Task-Based Perspective, *Proceedings of the 16th IEEE International Conference on Robot & Human Interactive Communication*, pp. 872 - 877.

Assistive Robotics

8

Development of a Virtual Group Walking Support System

Masashi Okubo
Doshisha University
Japan

1. Introduction

Recently, we often see the people walking around both in the town and countryside with their friends for health keeping in the morning and evening in Japan. Most of them make a group for walking. One of the reasons why they make a group for walking is that group walking helps them to continue the exercise. But, sometimes they can't exercise out of their houses because of weather condition, for example, the rain, snow and so on. Then, they have exercise machine, for example, stepper and walking machine in their home. However, it is hard to keep motivation to exercise alone at home. From this viewpoint, the cycling machine with display, which offers a virtual space and avatar that supports the users exercise, has been developed (IJsselsteijn et al., 2004).
In this research, we have proposed the walking system for health keeping with partners using shared virtual space through the Internet. The system consists of a computer connected to the Internet and a sensor for extracting the user's motion. The proposed system can provide the moving images and footsteps based on the user's step to the users. And in the case of paired use, the voices, motions and footsteps are sent to each other.

2. System configuration

The system configuration of proposed system is shown in Fig. 1. It consists of a stepper for health keeping, a personal computer connected to the Internet and a sensor for measuring the user's motion. The user can walk around the street in the virtual space by using the stepper with sensor. The positional data from sensor attached on the stepper are sent to the PC through the serial interface. When more than one person uses the system, the positional data are sent to each other to actualize the walking with partners

2.1 Hardware configuration

In this research, a stepper for diet tool is used. A sensor is attached on the heel of the tool, and the positional data x, y and z are sent to the PC. This kind of stepper is used in this research shown in Fig.1. However, any kind of walking machine can be used instead of this tool as far as the user's motion can be measured. The magnetic sensor (POLHEMUS, FASTRAK) is used for measuring user's motion. It can measure the positional data x, y, z, axes shown in Fig.2, elevation and roll. The PC (DELL XPS) used in this research has

Pentium D 3.46GHz CPU, 2.0GB memory, GeForce 7800 GTX ×2(SLI) graphic cards and Windows XP OS. It makes the virtual space and moves the user's viewpoint in the virtual space based on the positional data from the magnetic sensor.

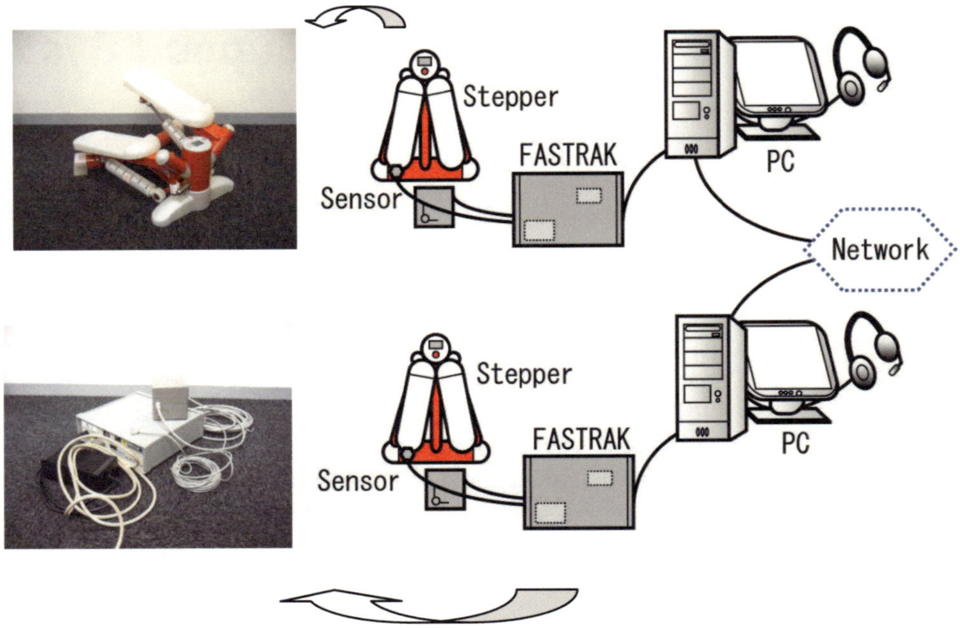

Fig. 1. System Configuration

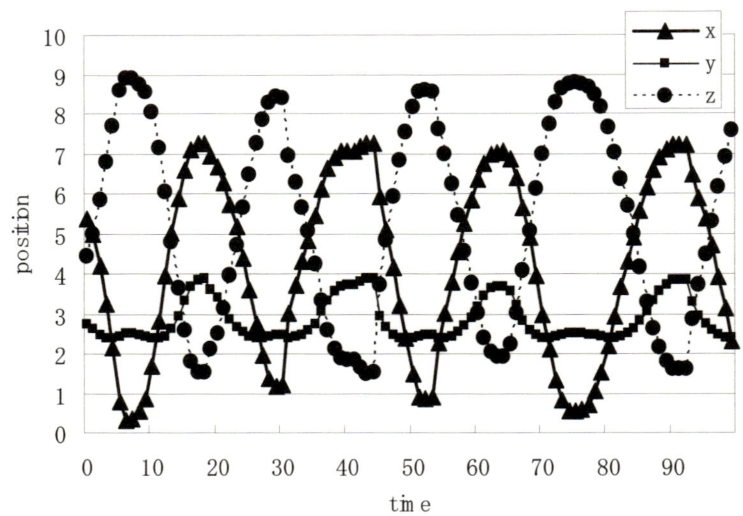

Fig. 2. Example of positional data obtained from magnetic sensor.

2.2 Software configuration

VC++ 2005 has been used to make the program for extracting the user's motion from the positional data, making the virtual space and moving viewpoint in the virtual space based on the user's motion. DirectX loads the x-file such as a walkway, and draws the image. When more than one person use the system, each user's motion data is sent to the partner's PC through socket communication. Fig. 3 shows the examples of image given to user; (a) is for alone use and (b) is for paired use.

(a) alone use (b) paired use

Fig. 3. Examples of user's viewpoint in case of alone use (a) and paired use (b).

In addition, the proposed system can offer the footsteps to user based on the user's walking rhythm (Kobayashi et al., 2006), (Miwa et al., 2001). The footsteps are given to user from speaker when the user switches his step. The footstep sounds are made in the system based on the positional data obtained from magnetic sensors which are put on the steppers. Fig.4 shows the example of user's motion and how to determine the timing to sound the footsteps. The system pays attention to the threshold values which are slight smaller than maximum position and bigger than minimum position. The system will not sound the footstep momentarily after the system does it. Therefore, the system doesn't sound it in case of 9, but in case of 1 to 8.

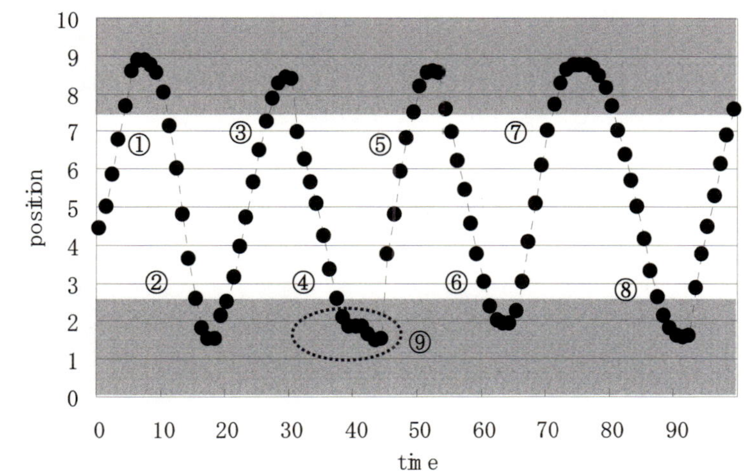

Fig. 4. Timing to sound the footsteps.

More than two people can use this system in the same time through the network. They can share the virtual space and communicate with partners while walking exercise. The proposed system sends the user's footsteps rhythms to each other, and users can feel the partner's presence and motion.

3. System evaluation

We performed the sensory evaluation to evaluate the proposed system's usability both in the case of alone use and paired use. The subjects urged to compare the conditions with moving image and/or footsteps in the virtual space with it without moving image nor footsteps. In the case of paired use, the effectiveness of partner's footsteps on the feeling of partner's presence was estimated.

3.1 Experimental method

The 20 subjects in their 20's performed the exercise under each experimental condition as follows:

Alone walking with

(a-1) TV
(a-2) Only moving image
(a-3) Footsteps and moving image

Paired walking with

(p-1) Only voice chat with remote partner
(p-2) Voice chat and moving image
(p-3) Voice chat, moving image and footsteps

Fig.5 shows the example of experimental scene. After the experiment, the subjects answered some questionnaire.

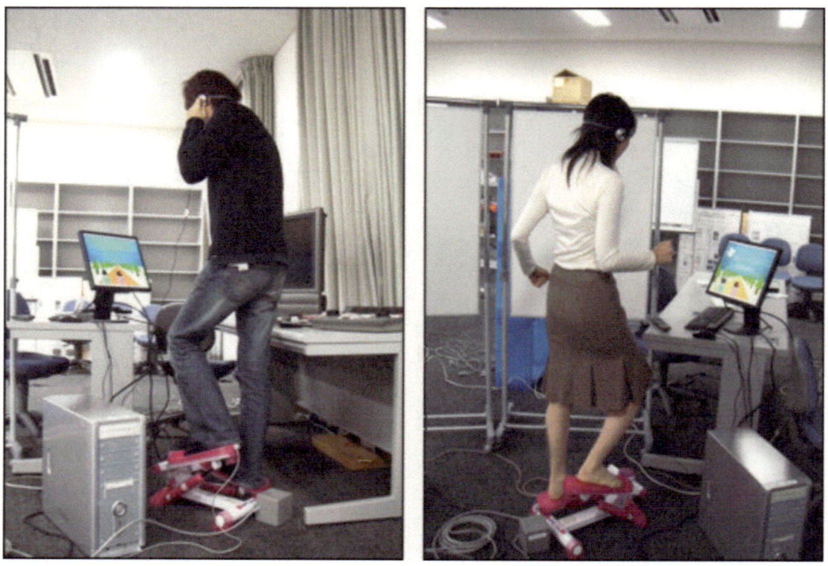

Fig. 5. Example of experimental scene.

In case of alone use, first, to familiarize the subjects with the use of the system, the subjects were walking with the experimental system in 1 min. Secondly the subject urged to perform the exercise by himself under two conditions out of (a-1) to (a-3) and answer which condition he preferred. The experiments were performed with three combinations of three experimental conditions. After the experiment, the subjects answered some questionnaire.
In the case of partner's use, the two subjects in different rooms performed the system with his partner, and the subject urged to perform the exercise with his remote partner undo two conditions out of (p-1) to (p-3) and each subject answered which condition he preferred. The experiments were performed with three combinations of three experimental conditions.

3.2 Experimental results

Table 1 and 2 show the results of paired comparison. The number in the table shows that of subjects who preferred the line condition to the row condition. Most of subjects prefer the exercise with TV (a-1) in the case of alone use.
The Bradley-Terry model was assumed to evaluate the preference of the condition quantitatively, defined as follows (Okubo & Watanabe, 1999);

$$P_{ij} = \frac{\pi_i}{\pi_i + \pi_j} \qquad \sum_i \pi_i = const.(=100) \tag{1}$$

Where π_i: intensity of i, Pij : probability of judgment that i is better than j
Here, π_i shows the intensity of preference of the experimental condition. The model enables to determine the preference based on the paired comparison (see Fig.6 and 7).
The Bradley-Terry model assumed by using the result of paired comparison. And to approve the matching of the model, the goodness-of-fit test and likelihood ratio test were applied to this Bradley-Terry model. As a result, the matching of the model was consistent.

	(a-1)	(a-2)	(a-3)	Total
(a-1)		14	15	29
(a-2)	6		5	11
(a-3)	5	15		20

Table 1. Result of paired comparison in case of alone use.

	(p-1)	(p-2)	(p-3)	Total
(p-1)		7	6	13
(p-2)	13		13	26
(p-3)	14	7		21

Table 2. Result of paired comparison in case of paired use.

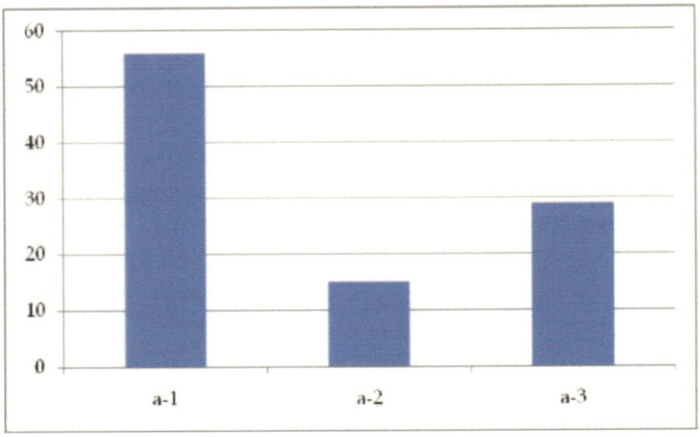

Fig. 6. Bradley-Teery model for paired comparison in case of alone use.

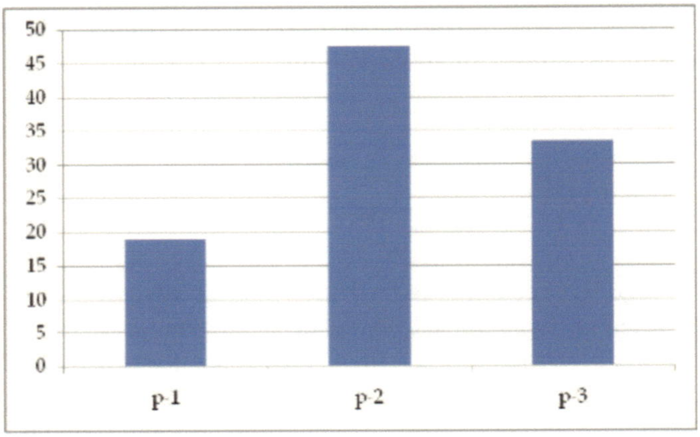

Fig. 7. Bradley-Teery model for paired comparison in case of paired use.

3.3 Answers for questionnaires

After the experiments, some questionnaires about the system usability and the experimental conditions were asked to the subjects. In the questionnaires, the subjects were asked which experimental conditions were preferred between (a-1): the alone walking with TV and (p-1): paired walking with only voice chat. The result is shown in Table 3.

(a-1)		even		(p-1)
4	1	1	4	10

Table 3. Comparison (p-1) with (a-1).

However the experimental condition (a-1) is most preferred one in case of alone walking and (p-1) is worst preferred in case of paired walking, the subjects tend to prefer the paired walking. These results indicate that the paired walking tend to be preferred to the alone walking even in the virtual space.

Moreover, 14 subjects out of 20 answered that they prefer the paired walking to the alone walking. It shows the importance of partners to keep the motivation for exercise.

4. Future works

A diversity of virtual space must be important, especially in case of alone exercise. This is indicated in the result of experiment. Therefore, we have tried to make the virtual space with diversity. Fig.8 shows the example of the virtual space in which the car across the road and unknown people are walking the street randomly. On the other hand, for encouraging communication with the partner, speech driven avatar named InterActor will be applied (Watanabe et al., 2004).

Fig. 8. Example of virtual space with diversity.

Moreover, there is a limitation in a diversity of computer graphics. And we have to think the utilization of video movies in place of computer graphics (Fig.9).

(a) composite video images with CG avatar

(b) simple virtual space

Fig. 9. Utilization of video movies.

5. Conclusions

In this paper, we have proposed the group walking system for health keeping with partners using shared virtual space through the Internet to keep the motivation. And the effectiveness of moving images, footsteps and conversation with partners on the dull exercise by using proposed system is demonstrated. In the case of alone use, the subjects tend to prefer the exercise watching TV, secondly the system with moving image and footsteps based on their steps to that with nothing. On the other hand, in the case of paired use, the subjects tend to prefer the condition with voice chat and virtual images. From the result of questionnaires, most subjects tend to prefer the paired walking to the alone walking with watching something. As a result of sensory evaluation and questionnaires, the effectiveness of proposed system is demonstrated.

6. References

IJsselsteijn W., Kort Y., Westerink J., Jager M. & Bonants R. (2004), Fun and Sports: Enhancing the Home Fitness Experience; Entertainment Computing - ICEC 2004, pp. 46-56.

KOBAYASHI T., MIYAKE Y., WADA Y. & MATSUBARA M. (2006), Kinematic Analysis System of Walking by Acceleration Sensor: An estimation of Walk-Mate in post-operation rehabilitation of hip-joint disease, Journal of the Society of Instrument and Control Engineers, Vol.42, No.5, pp.567-576(in Japanese).

Miwa Y., Wesugi, S., Ishibiki C. and Itai S. (2001), Embodied Interface for Emergence and Co-share of 'Ba'. Usability Evaluation and Interface Design.

Okubo M. & Watanabe T., (1999): "Visual, Tactile and Gazing Line - Action Linkage System for 3D Shape Evaluation in Virtual Space", Proc. of the 8th IEEE International Workshop on Robot and Human Communication (RO-MAN '99), pp.72-75.

Watanabe T., Okubo M., Nakashige M. & Danbara R., (2004), InterActor: Speech-Driven Embodied Interactive Actor, International Journal of Human-Computer Interaction, pp.43-60.

9

A Motion Control of a Robotic Walker for Continuous Assistance during Standing, Walking and Seating Operation

Daisuke Chugo[1] and Kunikatsu Takase[2]
[1]Kwansei Gakuin University, Hyogo,
[2]The University of Electro-Communications, Tokyo,
Japan

1. Introduction

In Japan, the population ratio of senior citizen who is 65 years old or more exceeds 22[%] at February 2009 and rapid aging in Japanese society will advance in the future (Population Estimates, 2009). In aging society, many elderly people cannot perform normal daily household, work related and recreational activities because of decrease in force generating capacity of their body. Today, the 23.5[%] of elderly person who does not stay at the hospital cannot perform daily life without nursing by other people (Annual Reports, 2001). For their independent life, they need a domestic assistance system which enable them to perform daily activities alone easily even if their physical strength reduces.

Usually, their daily activities consist of standing, walking and seating operation continuously. Especially, standing up motion is the most serious and important operation in daily life for elderly person who doesn't have enough physical strength (Alexander et al., 1991) (Hughes & Schenkman, 1996). In typical bad case, elderly person who doesn't have enough physical strength will cannot operate standing up motion and will falls into the wheelchair life or bedridden life. Furthermore, if once elderly person falls into such life, the decrease of physical strength will be promoted because he will not use his own physical strength (Hirvensalo et al., 2000).

In previous works, many researchers developed assistance devices for standing up motion (Nagai et al., 2003) (Funakubo et al., 2001). However, these devices are large scale and they are specialized in only "standing assistance". Therefore, the patient has to use other assistance device for their daily activities, for example when they want to walk after standing operation, and these devices are not suitable for family use. Furthermore, these devices assist all necessary power for standing up and they do not discuss the using the remaining physical strength of patients. Thus, there is a risk of promoting the decrease of their physical strength. On the other hand, devices based on the walking assistance system which can assist the standing and walking operation are developed (Chuv et al., 2006) (Pasqui & Bidaud, 2006). However in these devices, the patient has to maintain his body posture using his physical strength and it is difficult operation for elderly.

Therefore, we are developing a rehabilitation walker system with standing assistance device which uses a part of the remaining strength of the patient in order not to reduce their

muscular strength. Our system is based on a walker which is popular assistance device for aged person in normal daily life and realizes the standing motion using the support pad which is actuated by novel manipulator with three degrees of freedom.
From opinions of nursing specialists, required functions for daily assistance are (1) the standing assistance which uses a remaining physical strength of the patient maximally, (2) the posture assistance for safety and stability condition during standing, walking and seating assistance continuously, (3) the position adjustment assistance especially before seating and (4) the seating assistance to a target chair. In our previous work, we developed a force assistance scheme which realizes function (1) (Chugo et al., 2007). Therefore, in next step, for realizing function (2) and (3), we develop an active walker system in this paper. Please note function (4) will be our future work.
In this paper, our key ideas are the following two topics. First topic is a novel stability control scheme during standing up motion using the active walker function. Our active walker coordinates the assisting position cooperating the standing assistance manipulator according to the posture of the patient. Second topic is a seating position adjustment scheme using interactive assistance. Usually, an adjustment operation of the accurate position is difficult for elderly people and this operation has high risk of falling down (Hatayama & Kumagai, 2004). Therefore, this function is most important for walking assistance systems.
This paper is organized as follows: we introduce the mechanical design and controller of our system in section 2; we propose the body stability control scheme in section 3; we propose the seating position adjustment scheme in section 4; we show the result of experiments using our prototype in section 5; section 6 is conclusion of this paper.

2. System configuration

2.1 Assistance mechanism

Fig.1 shows overview of our proposed assistance system. Our system consists of a support pad with three degrees of freedom and an active walker system. The support pad is actuated by proposed assistance manipulator mechanism with four parallel linkages. The patient leans on this pad during standing assistance. Our active walker is actuated by two brushless motors on each front wheel. (We discuss in next paragraph.) Fig.2 shows our prototype. Our prototype can lift up the patient of 1.8[m] height and 150[kg] weight maximum, and it can assist him during walking using actuated wheels.
Fig.3 shows our developed support pad based on the opinions of nursing specialists at a welfare event (Chugo & Takase, 2007). The support pad consists of the pad with low repulsion cushion and arm holders with handles. In general, a fear of falling forward during standing motion reduces the standing ability of elderly person (Maki et al., 1991). Using this pad, a patient can maintain his posture easily during standing up motion without a fear of falling forward. Furthermore, the pad has two force sensors in its body (We discuss in section 3.). Our assistance system can measure its applied force and can estimate a body balance of the patient during standing up motion using these sensors.

2.2 Controller

Our developed control system is shown in Fig.4. Our assistance walker consists of two parts, a standing assistance system and an active walker system. The standing assistance system has three DC motors and three potentiometers in each joint and two force sensors on the arm holder. Motors are connected each joint using worm gears, thus, our manipulator can maintain its posture even if system power is down.

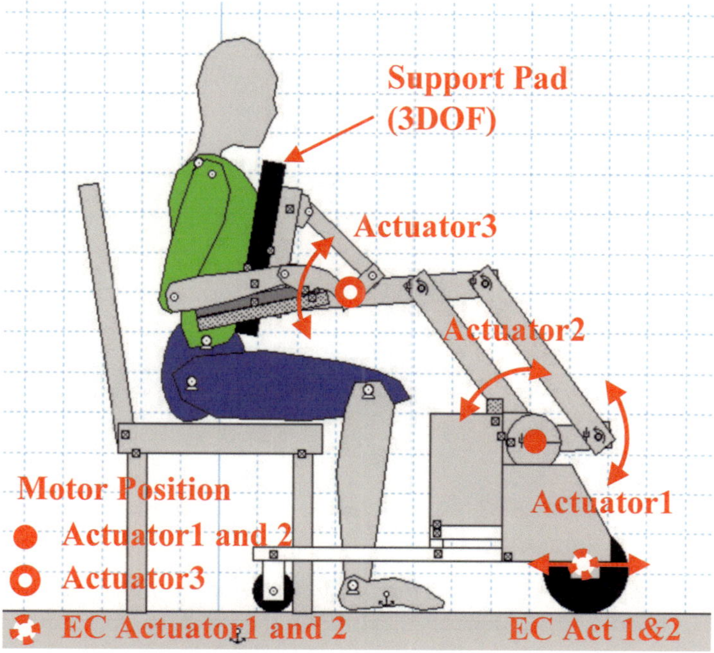

Fig. 1. Overview of our system.

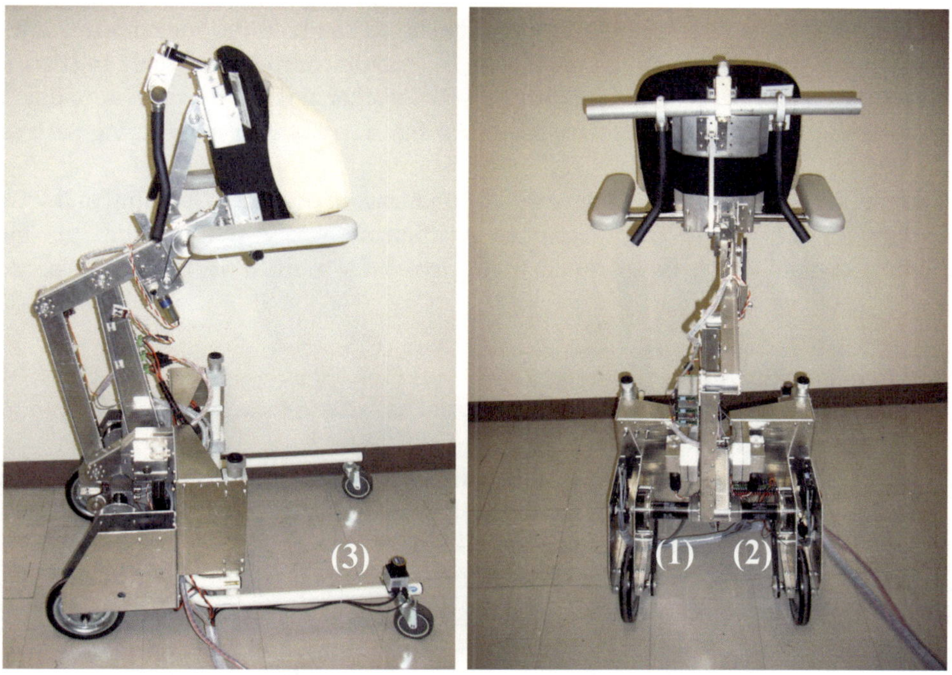

(a) Side View (b) Front View

Fig. 2. Our prototype. Its weight is about 35[kg] without batteries. Our prototype requires an external power supply and control PC. (In future works, we will use batteries and built-in controller.) (1) is EC Actuator 1 and (2) is EC Actuator 2. (3) is LRF.

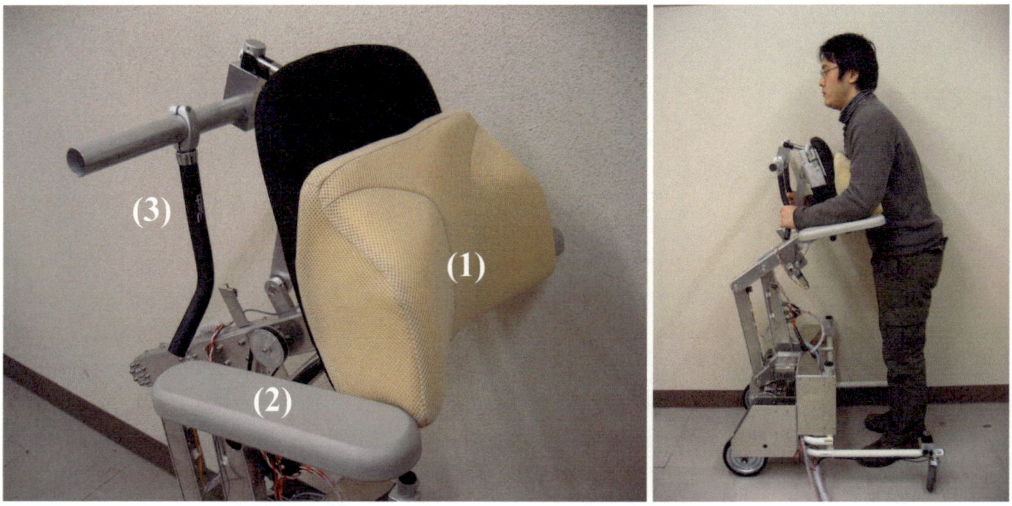

(a) Support Pad (b) Assistance Posture

Fig. 3. Our proposed support pad. (1) is the pad with a low repulsion cushion, (2) is the arm holder and (3) is a handle. Its diameter is 0.24[m] which is easy to grip for the elderly.

The active walker system has two Maxson brushless EC motors and two electromagnetic brakes in each front wheel. ((1) and (2) in Fig.2 (b)) Electromagnetic brakes can stop the walker when the patient seems to fall down. This break can hold the walker when it assists the 150[kg] weight patient maximum. These EC motors can operate with traction force limitation and can follow when the patient push the walker against to the advance direction. These advantageous characteristics are useful for the active walker considering with safety reason.

Our system has two laser range finders which can measure objects within 4[m] and wireless LAN adapter. The system can receive its position data at real time from the indoor positioning system which is equipped in the patient's room. (We discuss in section 4 closely.)

2.3 Problem specification

We questions to nursing specialists about required assistance for aged people in their daily life. Their results are the followings.

- Aged person requires standing, walking and seating assistance continuously by a same device. In typical required case, he stands up from the bed, he walks to the toilet and he sits on it by himself using the assistance system.
- When he stands up, he requires power assistance for reducing the load and he also requires position assistance for maintaining his body balance.
- When he sits on the target seat, he requires the position adjustment assistance. A failure of this motion causes a serious injury, therefore, this assistance is important.

In our previous works, a reducing the load during standing is realized (Chugo & Takase, 2007). Therefore, in this paper, we focus on (1) the assistance for stable posture during standing to walking motion and (2) the position adjustment assistance to target seating position.

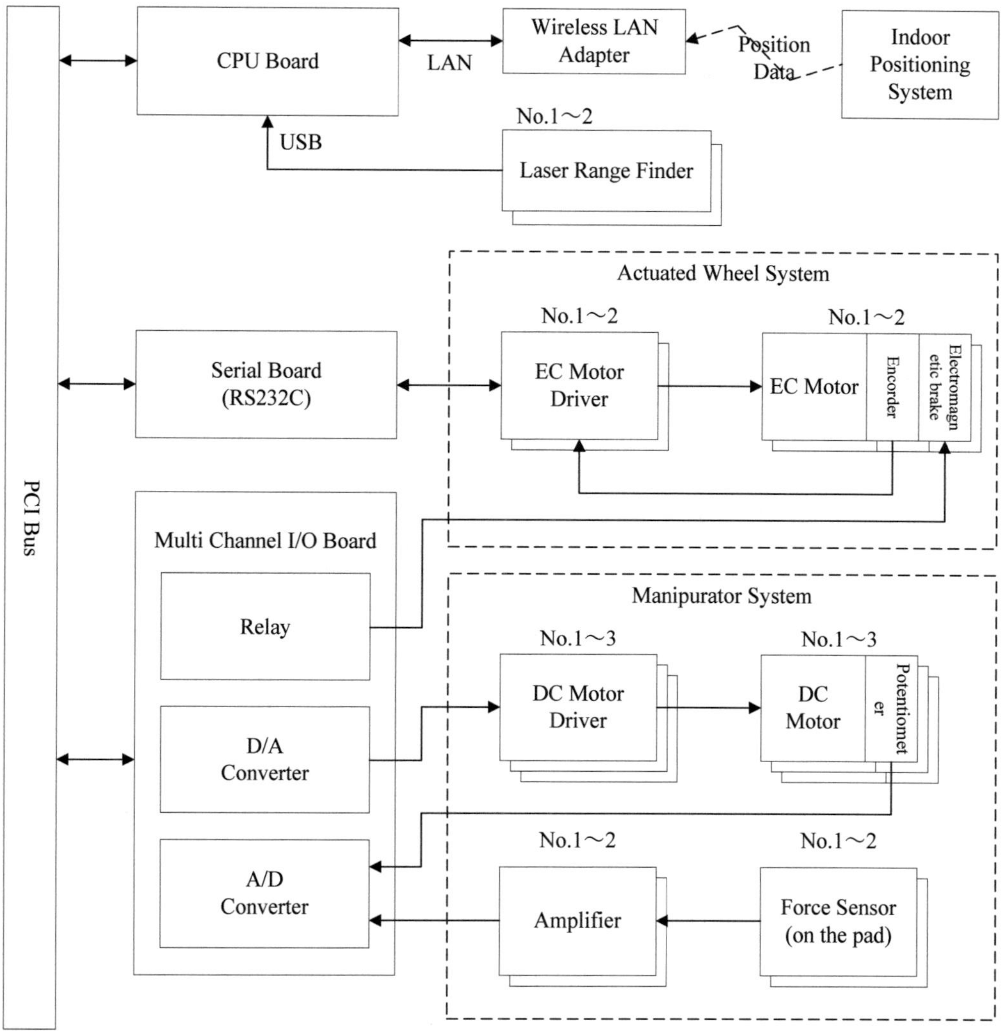

Fig. 4. Overview of our control system.

3. Body stability control

3.1 Motion by nursing specialists

In previous study, many standing up motions for assistance are proposed. Kamiya (Kamiya, 2005) proposed the standing up motion which uses remaining physical strength of the patients based on her experience as nursing specialist. Fig.5(a) shows the standing up motion which Kamiya proposes.

In our previous work, we analyze this standing up motion and find that Kamiya scheme is effective to enable standing up motion with smaller load (Chugo et al., 2006). We assume the standing up motion is symmetrical and we discuss the motion as movement of the linkages model on 2D plane (Nuzik et al., 1986). We measure the angular values among the linkages, which reflect the relationship of body segments. The angular value is derived using the body landmark as shown in Fig.5(b).

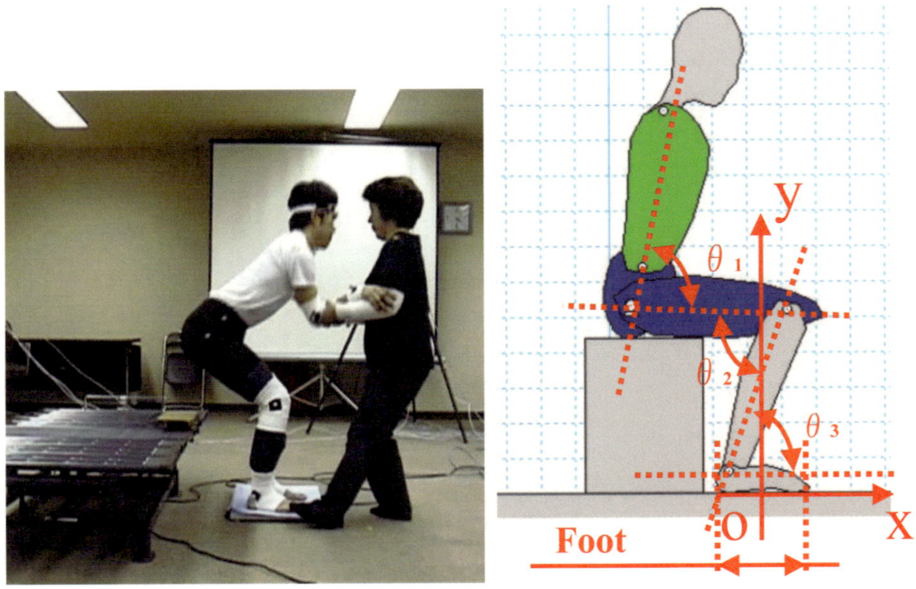

(a) Assistance Motion (b) Coordination

Fig. 5. Standing-up motion with Kamiya scheme. θ_1 shows the angular of the pelvis and the trunk. θ_2 and θ_3 show the angular of the knee and the ankle, respectively.

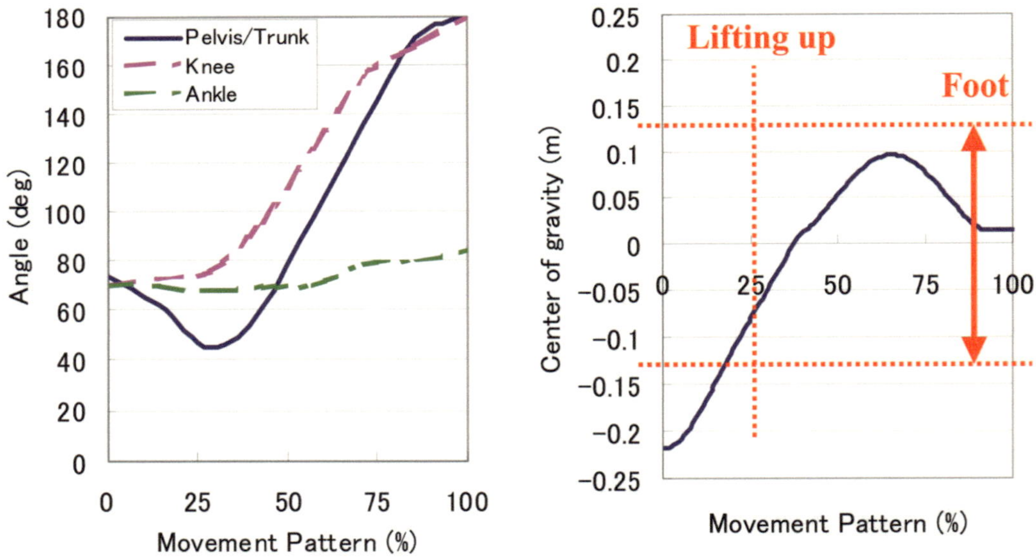

(a) Angular value (b) Position of COG

Fig. 6. Analysis result of standing-up motion with Kamiya scheme. Foot size of human model is 0.26[m] and his foot area is shown by red arrow in (b). Before 25[%] movement pattern, he still sits on a chair.

In order to realize the Kamiya scheme, the trunk needs to incline to forward direction during lifting up from chair as shown in Fig.6(a). Y-axis shows the angular value (Pelvis and trunk, knee, ankle) and X-axis shows the movement pattern (Hughes & Schenkman, 1996) which means the ratio of standing up operation as (1). Fig.6(b) shows the position of the patient's center of gravity (COG) which indicates the body balance of the patient during standing up motion.

$$\hat{s} = \frac{t}{t_s} \tag{1}$$

where t_s is required time to the standing up operation and t is present time.
Generally, inclining the trunk reduces the load of knee during standing up (Schenkman et al., 1990). Furthermore, the position of the patient's center of gravity (COG) is in areas of his foot (± 0.13 [m]) during standing up operation. This means his body balance is maintained. Therefore, this motion is useful for elderly person who doesn't have enough physical strength.

3.2 Required condition

In standing up motion of Kamiya scheme, we can divide the standing up motion into four phases (Nuzik et al., 1986). In first phase, the patient inclines his trunk to forward direction. In second phase, he lifts off from the chair and in third phase (as shown in Fig.5(a)), he lifts the body. In fourth phase, he extends his knee joint completely and ends the standing up motion.
In previous study, for realizing the motion of Kamiya scheme (Kamiya, 2005), the conditions are discussed as follows.

- In third phase, it is required to reduce the knee load.
- In other phases, it is required to maintain the standing up motion with stably posture.

Thus, in our previous work, the assistance manipulator uses a damping control in third phase and in other phases, it uses a position control (Chugo et al., 2007). Using this scheme, the patient's load reduces in third phase and in other phases, the patient is required to use own physical strength. The reference of position control is based on the standing up motion of nursing specialist as Fig.6(a) and the body posture of the patient is maintained during standing up operation. However, in third phase, our assistance system uses a damping control and it cannot guarantee his body balance.
Therefore, it is required to maintain his body balance during standing up motion using the active walker function.

3.3 Force control of active walker

For realizing the required condition, we maintain the COG of the patient as an index of body balance. We use PID controller as (2) and coordinate the COG. The coordination of our system is shown in Fig.7.

$$\tau = k_p e + k_i \int e\,dt + k_d \frac{de}{dt} \tag{2}$$

$$e = x_{COG} - x_{COG}^{ref} \tag{3}$$

$$\mathbf{x}_{COG}^{ref} = \left[x_{COG}^{ref}(0), \cdots, x_{COG}^{ref}(\hat{s}), \cdots, x_{COG}^{ref}(1) \right]^T \tag{4}$$

$$\tau_{out} = \begin{cases} \tau & (|\tau| < \tau_0) \\ \tau_0 & (|\tau| \geq \tau_0) \end{cases} \quad (5)$$

where τ is force reference of the active walker and x_{COG} is the actual position of COG. x_{COG}^{ref} is the position reference of COG (Fig.6(b)) and it is function of the movement pattern $\hat{s}$ as (4). k_p, k_i, k_d are proportional, integral and derivative gain of PID controller, respectively. τ_{out} is actual output force of the active walker as (5). For safety reason, output force is limited to τ_0.

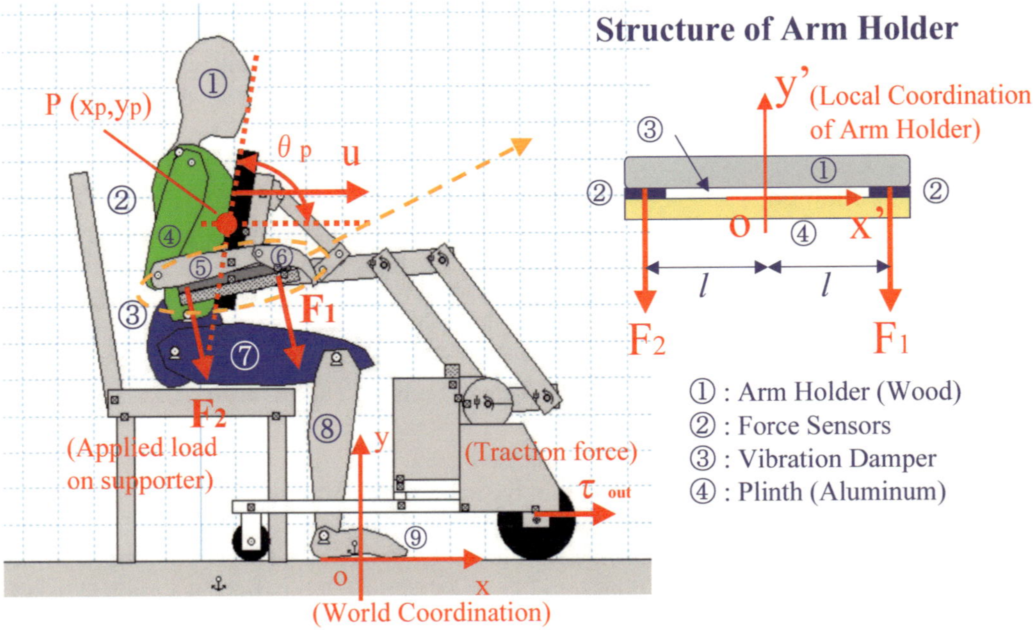

Fig. 7. Coordination and structure of the arm holder.

3.4 Posture estimation

Our proposed force control of the active walker requires the information of the position of COG. Usually, COG is measured using a force plate, however, this device is not suitable in practical use. Thus, it is required to estimate the position of COG with the sensors equipped on this walker. Our proposed walker has two force sensors on its arm holder as Fig.7. F_1 is the applied force of the front part of the arm holder and F_2 is one of its back part. When the COG moves to ahead, F_1 is heavier than F_2 and when it moves to back, F_2 is heavier than F_1. Therefore, using F_1 and F_2, we can estimate the COG.

We discuss the relationship between the COG and applied force to the pad in preliminary experiment using five subjects. Fig.8(a) shows the relationship between the position of COG (which is measured by a force plate) and COG on force sensors as (6) when the angle of support pad is 70[deg].

$$x_{COG}^{f} = \frac{(F_1 - F_2) \cdot l}{F_1 + F_2} \quad (6)$$

where l is distance between two sensors as Fig.7.
Subjects are adult male with the special wearing equipment for the experience of the elderly (Takeda et al., 2001). From these results, both values seem to be proportional as Fig.8(a). In Fig.8(a), we show an approximation line of Subjects A.
Furthermore, Fig.8(b) shows the relationship between the COG on force sensors and the angle of the support pad when the COG of patient is zero. From these results, COG on force sensors moves according to the angle of support pad.
From this experiment, we can estimate the COG comparing with the measuring values of force sensors and these data which are derived in this preliminary experiment.

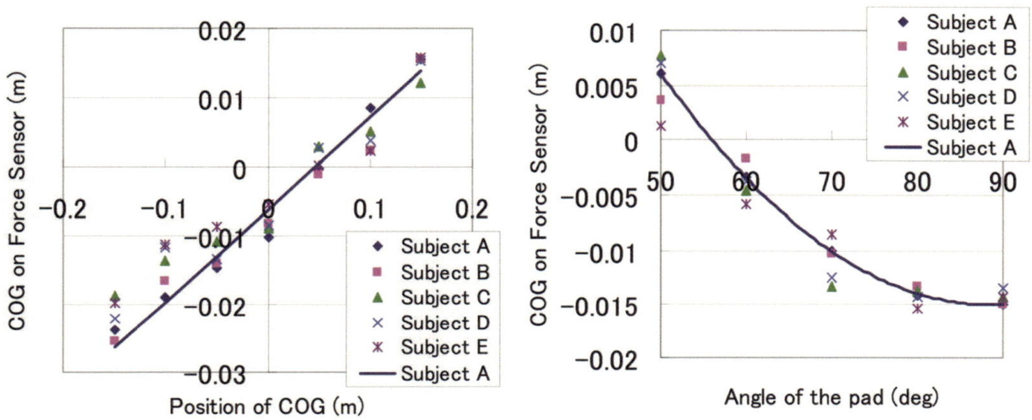

(a) The position of COG (b) The angular value of the pad

Fig. 8. Relationship between the COG and the applied force.

4. Seat position adjustment assistance

4.1 Required condition

In general, a failure of sit down has high risk factors for a falling down (Graafmans et al., 1996). Therefore, we questions to nursing specialists on a required assistance for aged people when they move in their home. Their results are the followings.

- Usually, a patient should walk using a walker without force assistance for his rehabilitation.
- However, it is difficult for him to walk backward with exact position enough for seating. Therefore, system should assist this operation part.
- System should assist him with his operation speed for safety reason. Too strong assistance causes his falling from the walker.

For maximizing a rehabilitation performance, the patient walks himself and the system specializes only a seating position adjustment assistance.

4.2 Assistance algorithm

For realizing the previous condition, we proposed the following algorithm. The flow chat is as shown in Fig.9. The system is inputted the position and shape data of target chair, bed and toilet etc in the room.

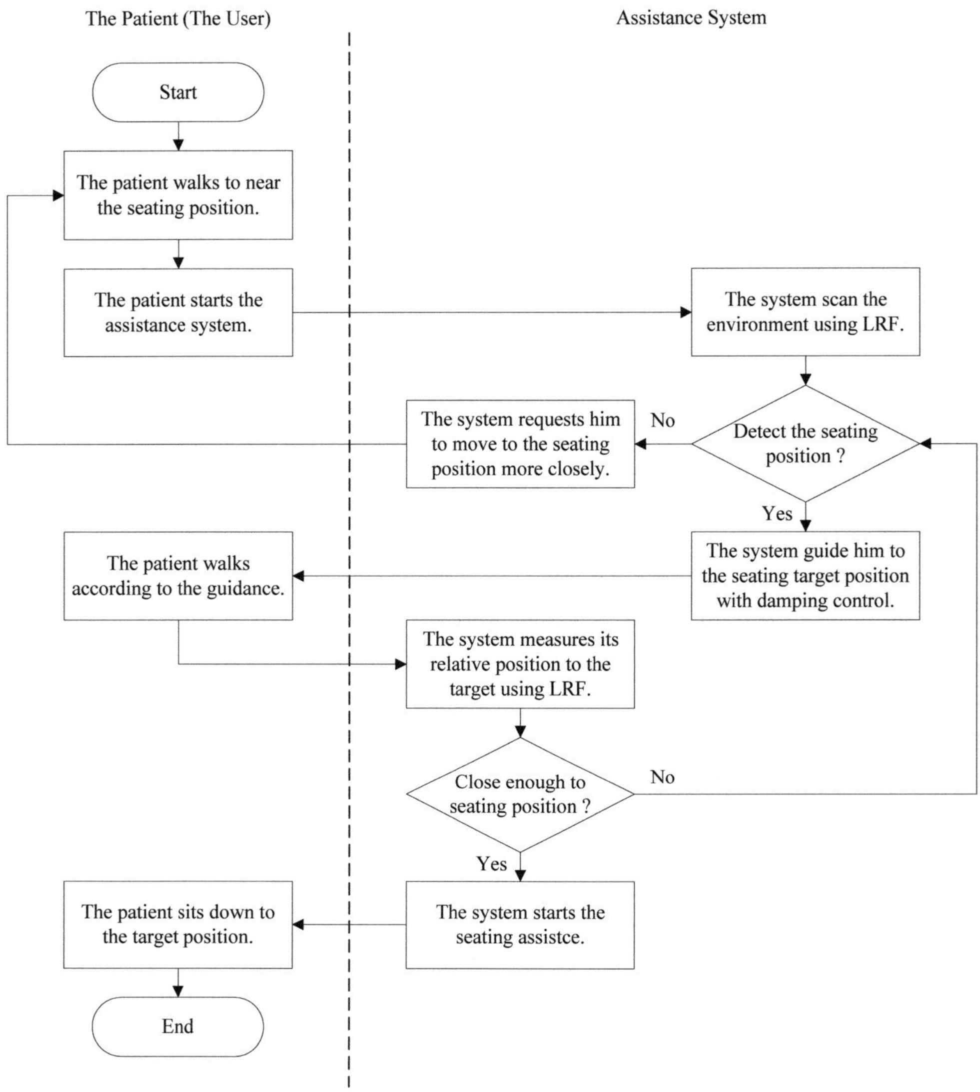

Fig. 9. A flow chart of assistance algorithm.

- System assists only for adjusting the seating position. The user walks by him near the chair, which they want to sit on as Fig.10, and starts the assistance system.
- System receives its position using our developed indoor positioning system (Ohnishi & Takase, 2003) and detects nearest chair (or bed, toilet) as target.
- System scans around the walker and finds the target using suitable LRF. (In case of Fig.10, LRF1 is suitable. The system detects the target using its shape data (Matsushima et al., 2008).)
- If the system finds the target chair, it assists him to the seating position in front of the target chair.
- If the system cannot find the target chair, it announces the user to go around the target chair.

We use our indoor positioning system which uses infrared LED (Ohnishi & Takase, 2003). The system detects infrared LEDs on the walker by the CCD camera with infrared filter on ceiling as Fig.10 and calculates its position using kinematics. The controller of the walker receives its poison data from the server by CORBA network. Using infrared filter, the system is robust and the privacy of the patient is protected because the system only can detect the infrared LEDs. Furthermore, our positioning system can be used in entire room, therefore it will be suitable for room guidance system. (In our future work.)
Our assistance scheme works only at "seating position adjustment", therefore, system requires the patient to move to the target closely. This concept increases rehabilitation performance and realizes the low cost.

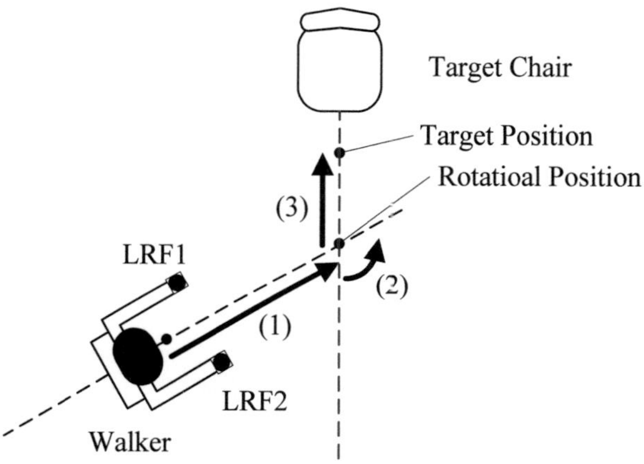

Fig. 10. Path planning to target chair.

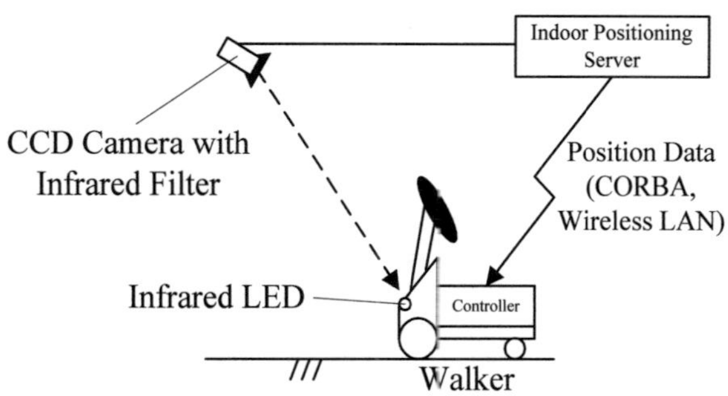

Fig. 11. The indoor positioning system.

4.3 Guiding to seating position

From the opinions of nursing specialists, a patient should go straight path especially when he walks to backward. Therefore, after detecting the target chair, our active walker guides a patient using a following path as Fig.10. The first, the walker moves to backward as allow (1) in Fig.10. The second, it rotates at same spot as allow (2) and finally, it moves to the seating position.

For guiding the patient softly, the system uses a damping control (Chugo et al., 2006) as (7). The coordination is defined in Fig.7.

$$\tau = cv_{ref} - B(u - u_0) \tag{7}$$

where v_{ref} is velocity control reference which is derived using the position information by LRF between the active walker and the target seating position. u is the applied load to the walker and u_0 is the threshold which selects from damping control mode and velocity control mode. c and B are coefficients. ($B = 0\left(if\ u < u_0\right)$)
The output force is limited to τ_0 as (5) for safety reason.

5. Experiment

Here, we verify the performance of our prototype system by the experiment. In this experiment, testers use the special wearing equipment for the experience of the elderly (Takeda et al., 2001). This wear limits the motion of the tester body as elderly.
In this experiment, we test three cases. The first, we test the body stability control during standing up motion. The second, we test the adjustment assistance for the target seating position. Finally, we test the assistance performance for standing up, walking and sitting motion as daily activity.

5.1 Body stability control

In this experiment part, the tester stands using our assistance system which utilize our force control scheme (Chugo et al., 2007). The height of the tester is 1.7[m] and our system lifts him at 30[sec]. For verifying an effectiveness of our proposed scheme, we test two cases. One case uses our proposed body stability control scheme and the other case does not use it for standing assistance. The coordination is defined as Fig.8.
Fig.12 shows the standing assistance posture of the subject and Fig.13 shows the movement of our active walker utilizing proposed control during lifting up as Fig.12. After 25[%] movement pattern, system starts to coordinate the body balance of the patient, because in this period, our assistant manipulator uses force control mode and the posture of the patient is not maintained as references.
Fig.14 shows the COG of the patient and we can verify that our system coordinate the body balance according to the reference of the nursing specialists.
Furthermore, seven aged users test our prototype. They are 67 years old and more, and their care levels of Japanese Long-term Care Insurance System (Population Estimates, 2009) are 1 or 2.
They stand up using our assistance system twice. One case uses our proposed body stability control scheme and the other case does not use it. We question them about the following topics after standing up with our assistance system.

- Question1: Comparing with standing up using only your own strength, do you feel to easy for standing up using our system? We evaluate the inferior feeling is 1 point, same feeling is 3 points and good feeling is 5 points.
- Question2: Do you feel to fear of falling during standing up using our assistance system? We evaluate the inferior feeling is 1 point, same feeling is 3 points and good feeling is 5 points.

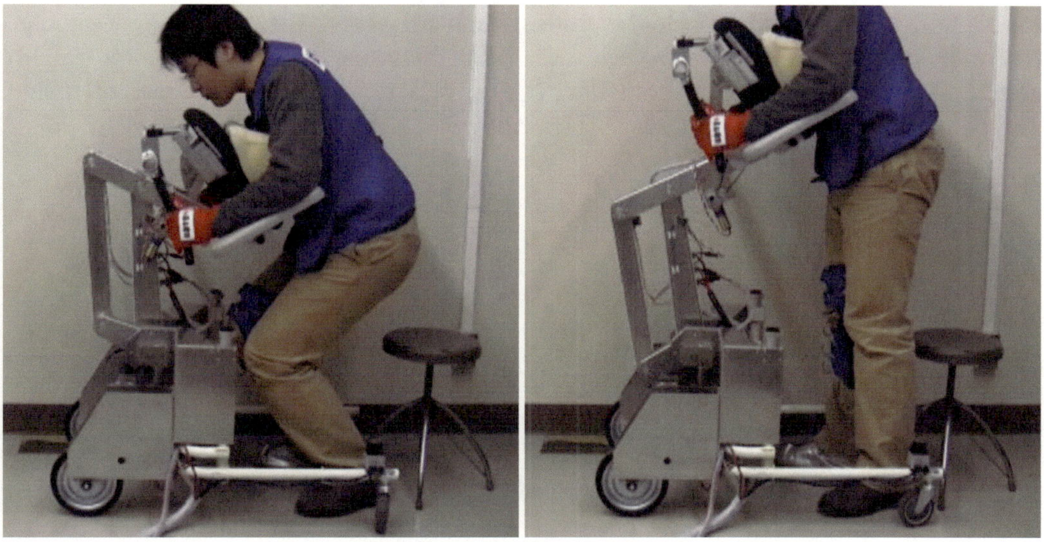

Fig. 12. The posture of the subject during standing assistance.

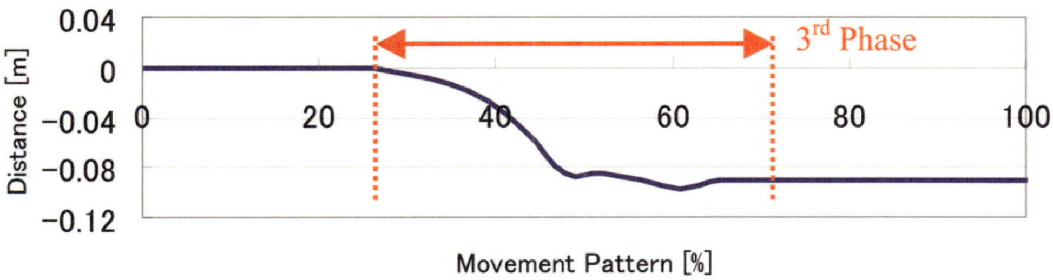

Fig. 13. Movement of the active walker.

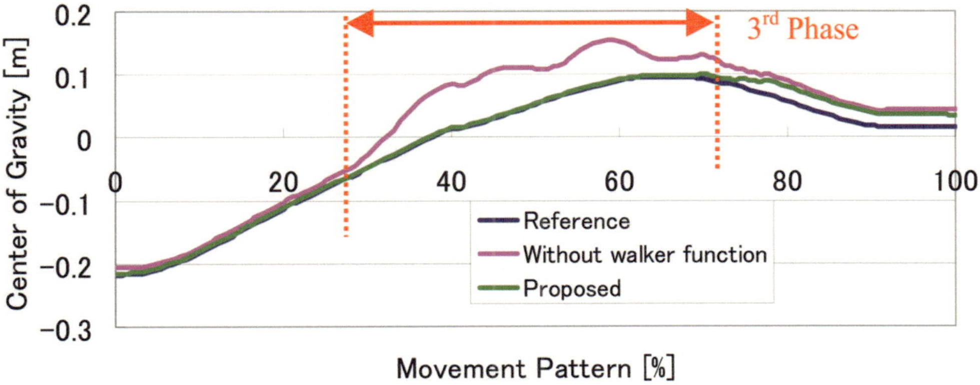

Fig. 14. The position of COG during standing motion.

Fig.15 shows the opinions of elderly testers. Each bar shows the average and standard deviation of evaluation points in each case. From these results, using our body stability control, the patient feels easier for standing up and does not feel the fear of falling. Therefore, we can evaluate our scheme is effective.

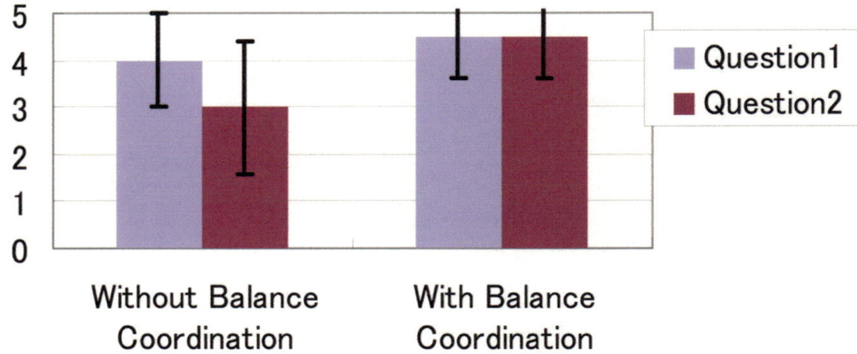

Fig. 15. Organoleptic evaluation of our prototype.

5.2 Seating position adjustment

In this experiment part, the tester with wearing equipment (Takeda et al., 2001) walks to the seating position according to our assistance. In preliminary experiment, we set a suitable seating position (and admissible area) which enables the patient to sit down the target chair easily as Fig.16(a). In this experiment, our proposed system guides him to this target position ten times.

From these results as Fig.16(b), our system can adjust the seating position with enough accuracy for sitting down. Our system uses damping control as (8) for safety reason. Thus, the results have a range. (But they are within admissible area.)

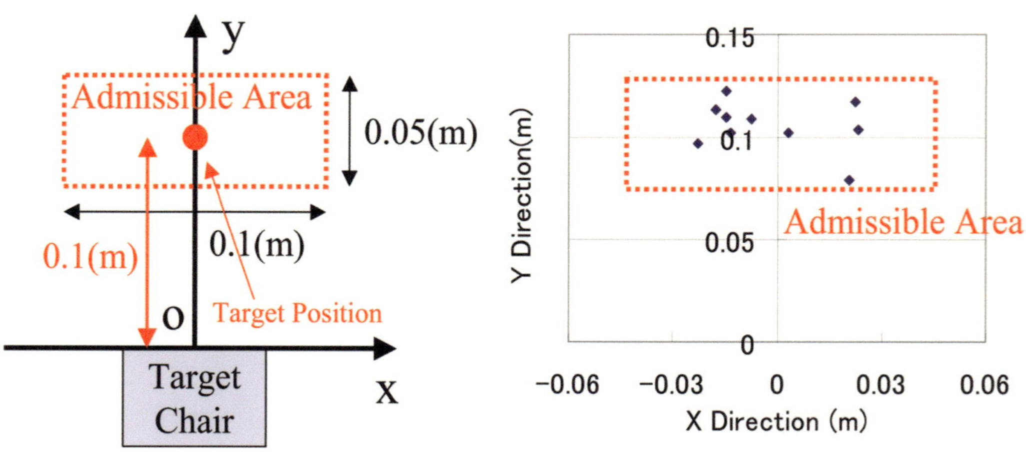

(a) The target seating position (b) The experimental result

Fig. 16. Assistance to the seating position.

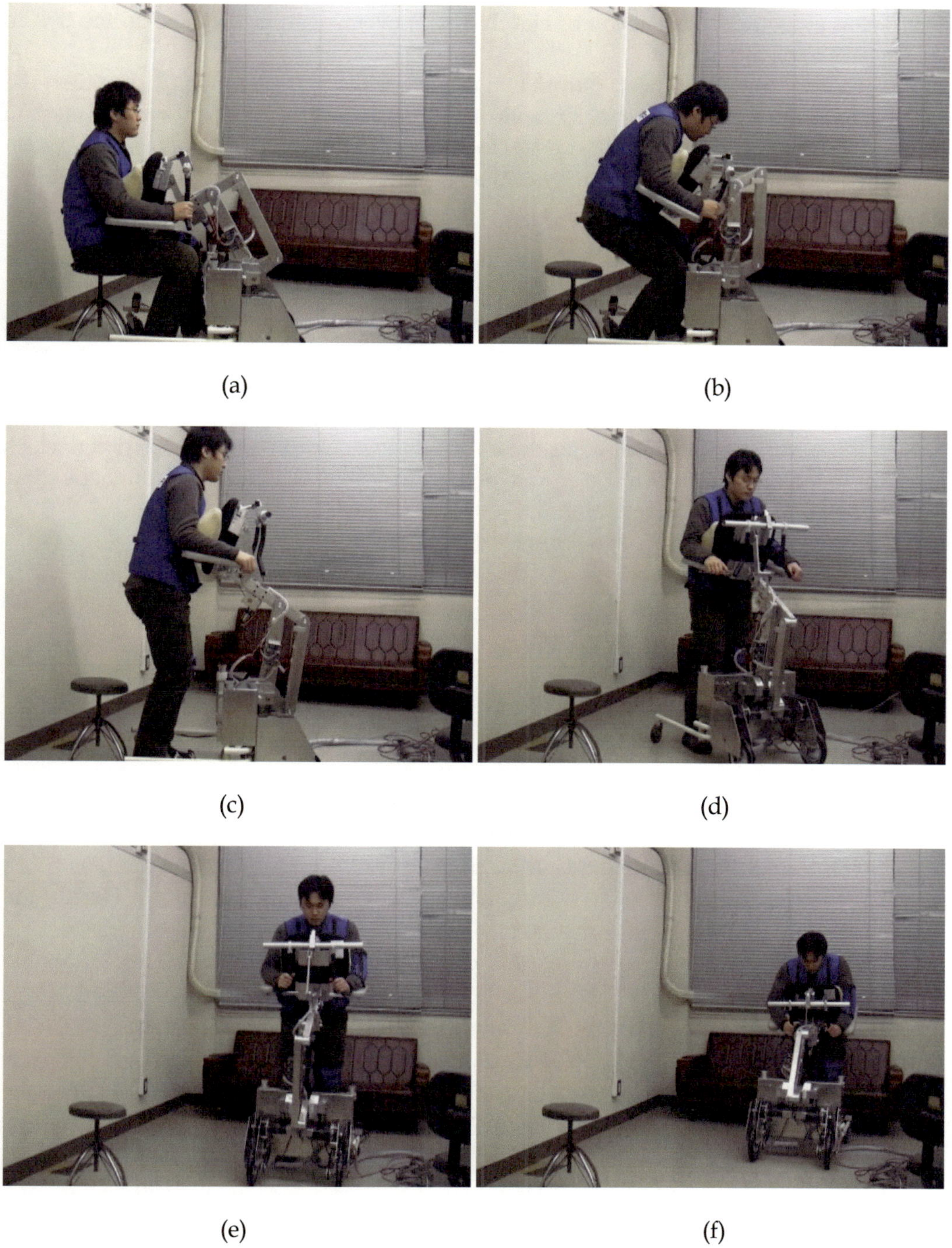

Fig. 17. Assistance demonstration using our prototype.

5.3 General experiment

Here, we verify the general performance of our prototype system by the experiment. This experiment assumes typical action of elderly (the movement from the chair to the bed in same room) in their daily life.

As the result of the experiment, our system can assist the patient as Fig.17. The first, the tester stands up from the left chair with standing assistance of our proposed system (Fig.17(a)-(c)). The second, he walks to near the target bed himself (Fig.17(c)-(d), Our system does not assist him.). The third, our system adjusts the seating position (Fig.17(d)-(e)) and assists the sit down motion (Fig.17(f)).

6. Conclusion

In this paper, we develop an active walker system for standing, walking and seating operation continuously which cooperates the developed standing assistance system with safety and stability. For realizing these conditions, our walker coordinates the assisting position cooperating the standing assistance manipulator according to the posture of the patient. Furthermore, our walker adjusts a seating position when the patient sit down which has high risk for falling down. Using our proposed system, the patient can use standing, walking and seating assistance continuously by a same device.

In our future work, we will discuss the seating assistance operation.

7. Acknowledgement

This work was supported in part by the Sasakawa Scientific Research Grant from The Japan Science Society.

8. References

Statistics Bureau, Ministry of Internal Affairs and Communications, Japan. (2009). Population Estimates by Age (5-Year Group) and Sex of February 1, 2009, http://www.stat.go.jp/data/jinsui/tsuki/index.htm

Ministry of Health, Labour and Welfare, Japan. (2001). Annual Reports on Health and Welfare 2001 Social Security and National Life, http://www.mhlw.go.jp/toukei/saikin/hw/k-tyosa/k-tyosa01/4-3.html

Alexander, N. B.; Schultz, A. B. & Warwick, D. N. (1991). Rising From a Chair: Effects of Age and Functional Ability on Performance Biomechanics. *J. of Geometry: MEDICAL SCIENCES,* Vol.46, No.3, pp.91-98.

Hughes, M. A. & Schenkman, M. L. (1996). Chair rise strategy in the functionally impaired elderly. *J. of Rehabilitation Research and Development,* Vol.33, No.4, pp.409-412.

Hirvensalo, M.; Rantanen, T. & Heikkinen, E. (2000). Mobility difficulties and physical activity as predictors of morality and loss of independence in the community-living older population. *J. of the American Geriatric Society,* Vol.48, pp.493-498.

Nagai, K.; Nakanishi, I. & Hanabusa, H. (2003). Assistance of self-transfer of patients using a power-assisting device. *Proc. of the IEEE Int. Conf. on Robotics and Automation*, Taipei, Taiwan, September 2003, pp.4008-4015.

Funakubo, A.; Tanishiro, H. & Fukui, Y. (2001). Power Assist System for Transfer Aid. *J. of the Society of Instrument and Control Engineers*, Vol.40, No.5, pp.391-395.

Chuv, O.; Hirata, Y.; Wang, Z. & kosuge, K. (2006). Approach in Assisting a Sit-to-Stand Movement Using Robotic Walking Support System. *Proc. of the IEEE Int. Conf. on Intelligent Robots and Systems*, Beijing, China, October 2006, pp.4343-4348.

Pasqui, V. & Bidaud, P. (2006). Bio-mimetic trajectory generation for guided arm movement during assisted sit-to-stand transfer. Proc. of the 9th Int. Conf. on Climbing and Walking Robots, Geneva, Belgium, September 2006, pp.246-251.

Chugo, D.; Matsuoka, W.; Sogmin, J. & Takase, K. (2006). Rehabilitation Walker with Standing-Assistance Device. *J. of Robotics and Mechatronics*, Vol.19, No.6, pp. 604-611.

Hatayama, T. & Kumagai, S. (2004). Falls, physical disability, and mental distress in the elderly, *J. of Health Science*, Vol.26, pp.21-30.

Chugo, D. & Takase, K. (2008). Walker System with Assistance Device for Standing-Up. *Proc. of JSME Conf. on Bio Mechanics*, AIST, Tsukuba, Japan, September 2008, pp.44-47.

Maki, E.; Holliday, P. J. & Topper, A. K. (1991). Fear of falling and postural performance in the elderly, *J. of Gerontology*, No.46, Vol.4, pp.123-131.

Kamiya, K. (2005). Development and evaluation of life support technology in nursing. *Proc. of Proc. of 7th RACE Symp., Research into Intelligent Artifacts for the Generalization of Engineering*, The Univ. of Tokyo, Tokyo, Japan, January 2005, pp.116-121.

Chugo, D.; Okada, E.; Kawabata, K.; Kaetsu, H.; Asama, H.; Miyake, N. & Kosuge, K. (2006). Force Assistance Control for Standing-Up Motion. *Proc. of the IEEE/RAS-EMBS Int. Conf. on Biomedical Robotics and Biomechatoronics*, Pisa, Italy, February 2006, F132.

Nuzik, S.; Lamb, R.; Vansant, A. & Hirt, S. (1986). Sit-to-Stand Movement Pattern, A kinematic Study. *Physical Therapy*, Vol.66, No.11, pp.1708-1713.

Schenkman, M.; Berger, R. A.; Riley, P. O.; Mann, R. W. & Hodge, W. A. (1990). Whole-Body Movements During Rising to Standing from Sitting. *Physical Therapy*, Vol.70, No.10, pp.638-648.

Takeda, K.; Kanemitsu, Y. & Futoyu, Y. (2001). Understanding the Problem of the Elderly through a Simulation Experience – Difference in the Effect between Before and After Clinical Practice –. *Kawasaki Medical Welfare J.*, Vol.11, No.1, pp. 64-73.

Graafmans, W. C.; Ooms, M. E.; Hofstee, H. M. A.; Bezemer, P. D.; Bouter, L. M. & Lips, P. (1996). Falls in the Elderly: A Prospective Study of Risk Factors and Risk Profiles. *American J. of Epidemiology*, Vol.143, No.11, 1129-1136.

Ohnishi, T. & Takase, K. (2003). Study on a Holonomic Omnidirectional Power Wheelchair - Integration of Manual and Automatic Control-. *IEEJ Trans. on Electronics, Information and Systems*, Vol.123, No.6 pp.1109-1116.

Matsushima, S.; Chugo, D. & Takase, K. (2008). AGV Navigation using iGPS, *Proc. of 8th Annual Conf. on System Integration, SICE*, Gifu, Japan, December 2008, pp.365-366.

Sensors and Perception Designed for Human-Robot Interaction

10

Development and Performance Evaluation of a Neural Signal Based Computer Interface

Changmok Choi and Jung Kim
Korea Advanced Institute of Science and Technology (KAIST)
South Korea

1. Introduction

The use of personal computers has drastically increased since the 1990s, and they have been responsible for tremendous achievements in information searching (Internet browsing) and communication (e-mail) around the world. People commonly use standard computer interfaces such as the keyboard and mouse, which are operated through physical contact and movement. These physical interactions inherently involve delicate and coordinated movement of the upper limb, wrist, palm, and fingers. However, there are some people who are not capable of using these interfaces because they have physical disabilities such as spinal cord injuries (SCIs), paralysis, and amputated limbs. In 2005, the Ministry of Health and Welfare in South Korea estimated that there were approximately one million people suffering from motor disabilities in South Korea, and the number has been steadily increasing since 1995. It has also been reported that more than 500,000 individuals are living with SCIs in North America and Europe (Guertin, 2005). If people with disabilities could access computers for tasks such as reading and writing documents, communicating with others, and browsing the Internet, they could become capable of a wider range of activities independently.

Alternative methods for providing individuals with disabilities access to computing environments include direct contact with physical keyboards, such as that shown in Fig. 1 (a); i.e., through the use of mouth sticks and head sticks. However, these devices have the disadvantage of being inaccurate and inconvenient to use. Another notable computer interface is the eye-movement tracking system, shown in Fig. 1 (b). This interface can perform as fast as, or even faster than, a mouse (Sibert & Jacob, 2000). This is because eye-gaze supports hand movement planning (Johansson et al., 2001); therefore, signals due to eye movement are quicker than those due to hand movement. Eye movements, however, as with other passive and non-command inputs (e.g., gestures and conversational speech), are often neither intentional nor conscious. Therefore, whenever a user looks at a point on the computer monitor, a command is activated (Jacob, 1993); consequently, a user cannot look at any point on the monitor without issuing a command. The eye-movement tracking system thus brings about unintended results.

Currently, biomedical scientists are making new advances in computer interface technology with the development of a neural-signal-based computer interface that is capable of directly bridging the gap between the human nervous system and the computer. This neural

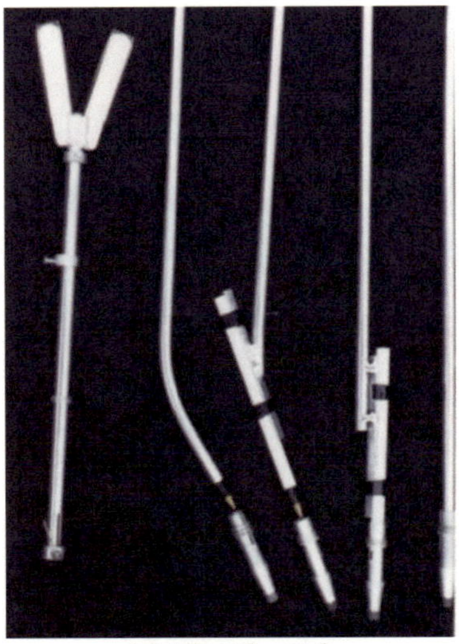

(a) Mouth stick
(http://www.mouthstick.net/)

(b) Eye-gaze interface
(http://www.brl.ntt.co.jp/people/takehiko/freegaze/index.html)

Fig. 1. Alternative computer interfaces for people with physical disabilities.

produce neural signals, because the signals naturally accompany body movements. Second, in this interface, neural signals are produced prior to actual body movements, and thus, the interface is even faster than kinematic and dynamic devices such as force sensors and motion trackers (Cavanagh & Komi, 1979). Such neural interfaces are classified into two categories on the basis of the signal source arriving from the central nervous system (CNS) or the peripheral nervous system (PNS).

Interfaces based on CNS signals, specifically signals from brain activity, have the potential to reveal human thought and are called brain-computer interfaces (BCIs). The major advantage of a BCI is that people with extremely severe motor disabilities such as quadriplegics can access a computer. An electroencephalogram (EEG), which measures brain activity recorded by electrodes placed on the scalp, is a good example of the use of CNS signals (Cheng et al., 2002; Citi et al., 2008; Kennedy et al., 2000; McFarland et al., 2008; Millan Jdel et al., 2004). For end users, the EEG's primary advantage is that it is noninvasive; however, this often results in a low signal-to-noise ratio (SNR), which in turn results in difficulties in accurately representing the users' intentions. In addition to the EEG signals, invasive CNS signals have been studied in recent years, and they capture the activity of individual cortical neurons obtained by microwire arrays that have been surgically implanted within one or more cortical motor areas (Hochberg et al., 2006; Taylor et al., 2002; Wessberg et al., 2000). This method provides better SNRs and spatial resolutions than noninvasive methods; in addition, this approach has been used recently with interesting results for selected quadriplegics (Hochberg et al., 2006). However, these invasive methods cause discomfort to the human and bear the risk of infection. Many issues of BCIs need to be addressed regarding brain map reorganization and chronic usability before making this method functional in extensive clinical experiments (Sanes & Donoghue, 2000).

PNS signals, which extend outside the CNS to serve the limbs and organs, can be used to extract user movement intent. A representative PNS signal is that detected by surface electromyography (sEMG), which is the electrical representation of activity produced by a number of muscle fibers in a contracting muscle and summation of motor unit action potentials. To observe the activities, a sEMG electrode is attached to the skin surface over the muscle; this method avoids any skin incision or percutaneous invasion unlike methods for cortical signal extraction. SEMG has been widely used as an interpretation tool for neural muscular control in neurophysiology studies (d'Avella et al., 2003; Merletti et al., 1999) and rehabilitation (Dipietro et al., 2005; Veneman et al., 2007), and also as an interface tool to detect movement intention of the end user in conjunction with artificial prostheses (Chu et al., 2007; Cipriani et al., 2008) and teleoperation (Fukuda et al., 2003).

In this chapter, we discuss a sEMG-based computer interface that allows people with amputations or SCIs to access a computer without using standard interfacing devices (e.g., a mouse and keyboard). Using the developed interface, a user can move a cursor, click a button, and type text on the computer using only their wrist movement. Furthermore, the efficiency of the interface was quantitatively measured using the Fitts' law paradigm, and the performance of this interface was compared with performances of currently used interfaces using the same test setup and conditions.

2. Materials and methods

2.1 Computer interface overview

The interface was designed to concurrently measure the sEMG signals and control a mouse cursor on a computer screen, as shown in Fig. 2. The activities of four muscles were recorded and amplified 1000 times by bipolar noninvasive surface electrodes (DE-2.1, Delsys, USA) with built-in amplifiers. The electrodes were connected to a data acquisition board (PCI 6034e, National Instruments™, USA), which transmitted the signals to a computer at 1000 Hz. Features were extracted from the measured signals by reducing the randomness of sEMG signals (referred to as feature extraction) and were fed into a pattern recognition program to classify the body movements. The classified movements were translated into predetermined commands and consisted of two-dimensional movements and clicking of a cursor to use the computer. Finally, the cursor was moved or a button was clicked on a computer screen using the classification results, and these processes were repeated by the volitional motor activities of the user with visual feedback.

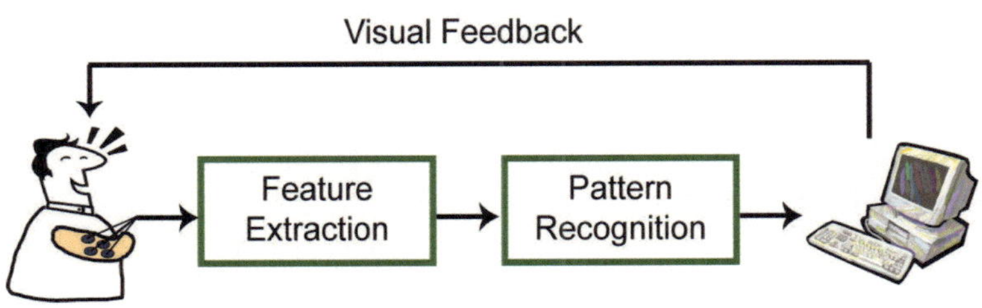

Fig. 2. Computer interface overview.

2.2 Motion and muscle selection

The selection of the target muscles to be used to obtain the signals must satisfy both ease of mapping the signals to computer operating commands and clear observability of the signals on the skin surface. To map the signals to the commands, four different wrist movements (wrist flexion, wrist extension, radial deviation, and ulnar deviation) were chosen, and these movements were mapped to the cursor movement commands (LEFT, RIGHT, UP, and DOWN). The user can intuitively control the cursor through these movements because the direction of the wrist movement corresponds to the direction of the cursor movement. In addition, to CLICK a mouse button and then STOP this movement, the movements of the hand such that it is open (coactivation of the muscles) and at rest were selected, respectively. When the user flexes his/her wrist (wrist flexion), the cursor moves to the left. To maintain the movement, the user must maintain the wrist flexion. The STOP condition occurs when the cursor does not move. Therefore, if the user wants to stop the cursor's movement, he/she should return and maintain the neutral position of the wrist. To observe the cursor movements, four muscles that produce the chosen wrist movements were selected: the flexor carpi ulnaris (FCU), the extensor carpi radialis (ECR), the extensor carpi ulnaris (ECU), and the abductor pollicis longus (APL), as shown in Fig. 3. Their activities were easily observable on the skin surface.

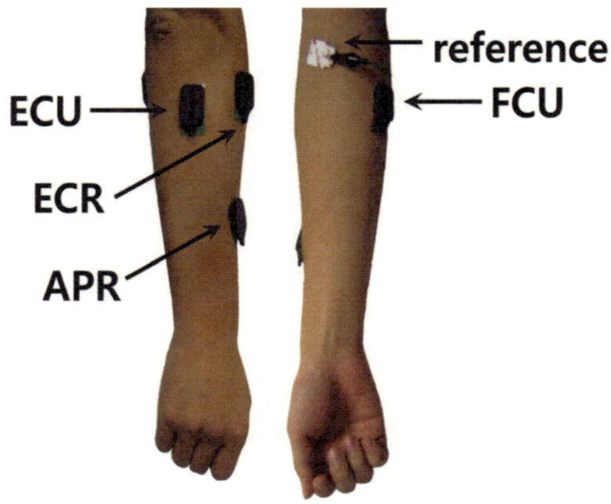

Fig. 3. Myoelectric sites for the sEMG signal extraction. Four muscles were selected to extract volitional motor activities: the flexor carpi ulnaris (FCU), the extensor carpi radialis (ECR), the extensor carpi ulnaris (ECU), and the abductor pollicis longus (APL). Wires were removed from the image for clear expression of the electrode placements.

2.3 Feature extraction

The electrophysiological phenomena at the cell membrane reflect the active state of living cells (Rau et al., 2004). In this sense, sEMG is related to the complex activation of skeletal muscles that results in static and dynamic active force exertion and movement control. The information obtained from sEMG should quantitatively represent the activation of skeletal muscles and highly correlate to the muscle force. Feature extraction (Zecca et al., 2002) converts a raw sEMG signal (which is obtained immediately after the amplification of the signal from the sensor) to a smoothed signal (called also an envelope) related to muscle force or voluntary driving of a muscle.

SEMG signals have been commonly regarded as Gaussian random process, and Hogan and Mann (1980) theoretically showed that the root mean square (RMS) processing shown in equation (1) is a maximum likelihood estimator of the sEMG signal when the magnitude of the raw sEMG signal has a Gaussian distribution.

$$\mathrm{RMS} = \sqrt{\frac{\sum_{i=1}^{N}(M_i - \bar{M})^2}{N-1}} \tag{1}$$

where M_i, N, and $\bar{M}$ are the magnitude of the ith data element, length of the analysis window, and mean of the magnitudes of N data respectively. The function of variance is analogous to a moving average filter excluding the root square term and denominator.
As a moving average filter, the cut-off frequency, f_c, of the low-pass filter was defined in relation to a moving average filter as follows (Smith, 1999):

$$f_c = \frac{f_s}{2N} \tag{2}$$

where f_s and N are the sampling frequency and the analysis window length, respectively. This equation describes how the effectiveness of the low-pass filter increases with a larger window because the cutoff frequency decreases. Since high-frequency components in the signals are effectively reduced, a large window increases the accuracy of the pattern recognition (Englehart & Hudgins, 2003). In contrast, a large window introduces a significant time delay and this delay could become an obstacle for a natural real-time computer interface. Hence, there is a tradeoff between real-time signal processing and the accuracy of the pattern recognition. Recently, Farrell et al. suggested an "optimal controller delay" for the collection and analysis of sEMG signals to maximize the classification accuracy without affecting performance; the maximum calculation time was between 100 and 125 ms (Todd & Richard, 2007). Taking into account this experimental result, the length of the analysis window was set to 100 ms. Thus, the signal processing not only provides effective low-pass filter effects (f_c = 5 Hz) but also prevents significant delays.

2.4 Pattern recognition

Artificial neural networks (ANNs), inspired by biological neural networks, have emerged as an important tool for pattern recognition in much human–computer interface (HCI) research (Barniv et al., 2005; Hiraiwa et al., 1990). An ANN is composed of a number of highly interconnected artificial neurons that are activated by external stimuli and is capable of learning key information patterns in multidimensional domains. There are two primary advantages of using an ANN. First, it is possible to classify data without any knowledge of prior probabilities of patterns belonging to one class or another. Second, because an ANN acts as a black box model, it does not require detailed information such as that of the human muscular-skeleton system. To design the classification network, a set of signals are allowed to flow through the network. The network then adjusts its internal structure until it is stable, at which time the outputs are considered satisfactory. After successful training, the network is preserved and receives new input signals, and then the network processes the data to produce appropriate outputs.
Figure 4 illustrates the structure of an ANN with two hidden layers and 10 hidden neurons for each layer for pattern classification of the six different wrist movements. During the training stage, all subjects were instructed to make the movements in turn, and the signals were recorded. Next, the network was trained using the six groups of features with the desired network responses shown in Table 1. Network tuning was performed using a backpropagation algorithm with a momentum approach (Haykin, 1999).

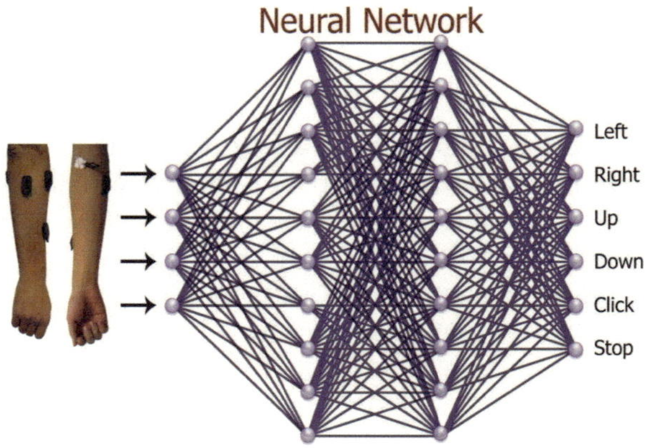

Fig. 4. Structure of the artificial neural network with two hidden layers and ten hidden neurons for each layer. Six neurons are located at the network's output, and each neuron corresponds to a volitional command to control a cursor movement or clicking.

Class of volitional command	Desired network response					
STOP	1	0	0	0	0	0
LEFT	0	1	0	0	0	0
RIGHT	0	0	1	0	0	0
UP	0	0	0	1	0	0
DOWN	0	0	0	0	1	0
CLICK	0	0	0	0	0	1

Table 1. Target vectors for classifying user intentions.

3. Performance evaluation

Fitts' law is a model of human psychomotor behavior derived from Shannon's theorem 17, a fundamental theorem of communication systems (Fitts, 1992), and is the most robust and widely adopted model to emerge from experimental psychology (MacKenzie, 1992). The law reveals an intuitive tradeoff in human movement—the faster we move, the less precise our movements are, or alternatively, the more severe the constraints are, the slower we move. Fitts formulated the tradeoff for three experimental tasks (bar strip tapping, disk transfer, and nail insertion) that are essentially of one paradigm—the hitting of a target over a certain distance. When considering an HCI, this paradigm corresponds to a frequent elemental task—pointing/target selection—and the paradigm can be applied as a predictive model to estimate the time for a user to move a cursor to a button and click it on a graphical interface. It is, therefore, useful to be aware of the effectiveness of a new computer-pointing device and to compare it with others. Since first presented in 1954, Fitts' law has been successfully used in many HCI areas with refinements of its mathematical formulation. It is now a cornerstone of the performance evaluation of pointing devices.

According to Fitts' law, the movement time (MT) required to move a cursor onto a target and the task difficulty (ID, index of difficulty) have the following linear relation.

$$MT = a + b \cdot ID \tag{3}$$

In this form, the reciprocal of b is called the index of performance (IP) and is measured in bits per second (bps). The IP represents how quickly the pointing and clicking can be performed using the computer-pointing device. That is, an interface with a higher IP is better than that with a lower IP, because a high IP indicates that the performance is less affected by a high ID. The ID depends on the width W of the target and the distance D between the cursor and target. To mathematically express this difficulty, the Shannon formulation is used, and the ID is expressed with a unit of bits as follows.

$$ID = \log_2\left(\frac{D}{W}+1\right) \tag{4}$$

Thus, it is obvious that the task becomes more difficult as D increases or W decreases. In this experiment, three different widths (W = 30, 70, and 110 pixels) and three different distances (D = 150, 300, and 450 pixels) were selected in line with previous research (Pino et al., 2003) that evaluated the performance of the commercial assistive pointing device Brainfingers™ (Brain Actuated Technologies, USA), which is based on Fitts' law.

Tests were conducted in two sessions, one each for the developed interface and a mouse (a standard computer interface tool). Five subjects (S1–S5) with intact limbs (five males with an average age of 26.4 years) volunteered and sat comfortably in front of a computer screen that continuously displayed the testbed shown in Fig. 5. The subjects were instructed to point to and click on a rectangular target (a dark rectangle) by moving a cursor, and MT was measured for the task. The targets in this experiment were randomly assigned in each session so that a user could not predict their locations, and the cursor was positioned on the right or left side of the target in accordance with the ID of each session. At the beginning of the experiment, all subjects were instructed to click a dummy target and then click nine targets with different IDs. The duration of the pointing and clicking for each session was measured, and this process was repeated 20 times for each subject.

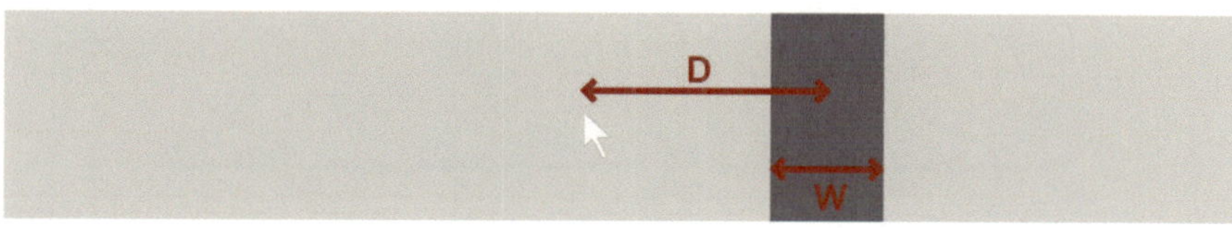

Fig. 5. Snapshot of the testbed for performance evaluation of the developed computer interface.

4. Results

In Fig. 6, the top graphs show the raw recorded sEMG signals from the four muscles (FCU, ECR, ECU, and APL) on the lower arm of the subjects and the bottom graphs represent the features extracted using RMS processing. It is evident that each feature set of the five different computer commands is characterized well for the classification. The features were entered into the ANN as input data, and Fig. 7 shows the ANN output values between 0 and 1. At the end of the ANN, a maximum selector chose the neuron with the largest value, and the neuron was directly matched to the computer commands, as shown in Fig. 8. Recognition accuracy was tested using the same subjects who took part in the Fitts' law test, and Table 2 summarizes the results. All movements were successfully classified, at a rate of over 96% accuracy, and misclassification usually occurred during the period of transition from one gesture to another. In Fig. 6, when a subject had a wrist flexion (LEFT), the FCU muscle was mainly activated. At the end of the movement, the other muscles were suddenly activated and Fig. 8 shows the misclassification at that time. This phenomenon can be explained by the transition from a wrist flexion movement to achieving the neutral position requiring a wrist extension movement.

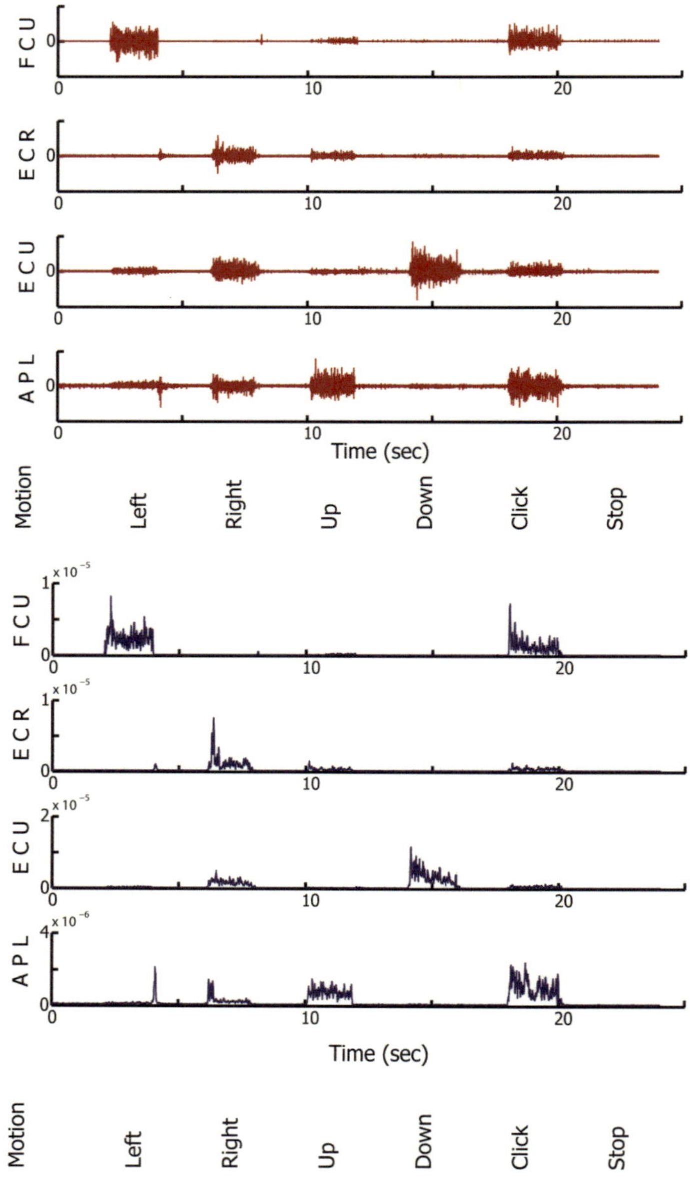

Fig. 6. Recorded raw sEMG signals and their features corresponding to a user's intentions from four muscles: the flexor carpi ulnaris (FCU), the extensor carpi radialis (ECR), the extensor carpi ulnaris (ECU), and the abductor pollicis longus (APL).

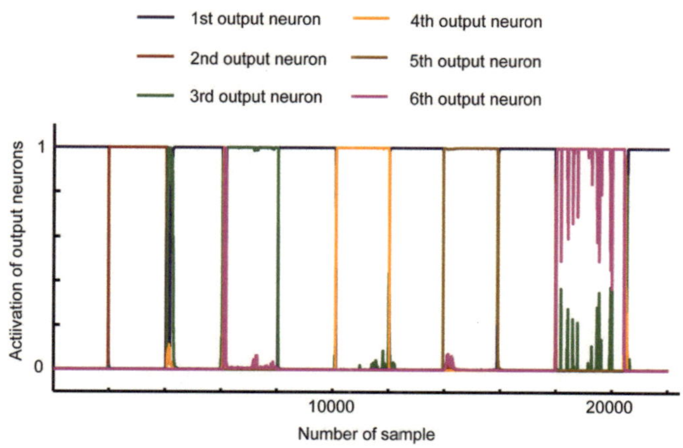

Fig. 7. Output neuron activation at the end of the network.

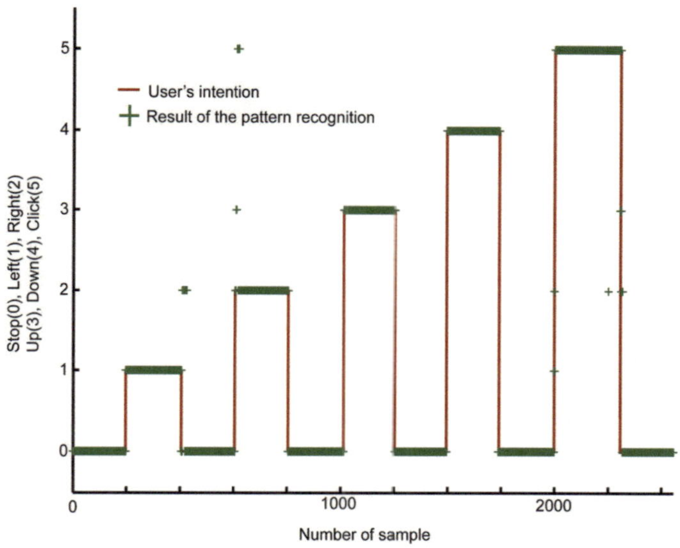

Fig. 8. Pattern recognition results. The red solid line denotes the intended movement of a subject and the green crosses show the recognized results as numerical values 0–5 (0: STOP, 1: LEFT, 2: RIGHT, 3: UP, 4: DOWN, and 5: CLICK).

	S1	S2	S3	S4	S5	Overall
Stop (%)	97.45	97.68	99.47	98.08	97.18	97.97
Left (%)	97.28	90.47	99.47	96.76	97.76	96.95
Right (%)	98.47	98.55	96.27	94.73	94.49	96.50
Up (%)	99.48	94.22	99.94	96.12	91.28	96.21
Down (%)	99.69	95.94	99.53	99.83	93.35	97.67
Click (%)	99.47	92.33	96.75	99.74	97.18	97.10

Table 2. Success rates of the proposed classification method in the discrimination of subject intentions.

Table 3 summarizes the results of the performance evaluation of both the developed interface and the mouse for the five subjects. All movement times were averaged from the 20 data for each subject. Figure 9 shows that the experiment data of the MT and ID for the subject S2 have a linear relationship in accordance with Fitts' law, and the upper and lower lines represent the results for the developed interface and mouse, respectively. From these results, the IP was calculated and is presented in Table 4; the overall IP of the developed

D (pixels)	W (pixels)	ID (bits)	Movement time (MT)									
			Subject 1		Subject 2		Subject 3		Subject 4		Subject 5	
			sEMG	mouse	sEMG	mouse	sEMG	mouse	sEMG	mouse	sEMG	mouse
150	110	1.2410	1920.5	715.9	1520.8	762.7	1925.6	695.8	1729.0	724.9	1664.5	728.2
300	110	1.8981	2770.3	773.2	2443.3	857.4	2693.5	737.7	2466.2	859.6	2480.5	787.9
450	110	2.3479	3437.1	830.0	3175.2	910.5	3251.5	818.4	3479.9	915.8	3578.3	898.2
150	70	1.6521	2073.0	750.4	1680.1	831.4	1881.8	774.1	1717.0	870.5	1762.6	920.1
300	70	2.4021	2825.9	848.3	2536.2	914.2	2557.5	827.6	2614.4	960.0	2656.6	939.6
450	70	2.8931	3763.1	864.1	3186.5	954.2	3353.9	908.2	3338.1	971.0	3630.9	932.9
150	30	2.5850	2694.6	837.7	2366.0	982.2	2007.4	920.8	2721.4	1096.5	2407.1	923.5
300	30	3.4594	3626.5	925.5	3421.5	1081.4	2624.7	957.3	3087.1	1078.0	3255.8	1012.6
450	30	4.0000	3928.0	1015.8	4008.5	1151.0	4221.9	1063.9	3813.3	1215.0	4276.2	1157.0

Table 3. Experiment results of the Fitts' law test for the efficiency of the sEMG interface.

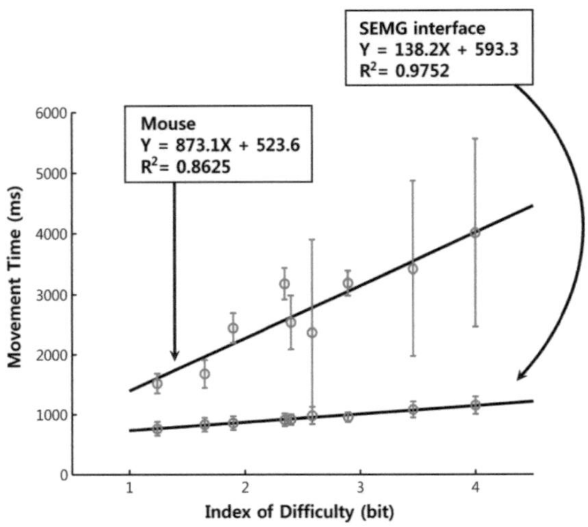

Fig. 9. Relation between movement time (MT) and index of difficulty (ID) from the experiment for subject S2 using a mouse and the developed sEMG computer interface. The gentle slope of the line illustrates the high IP (index of performance) value.

		S1	S2	S3	S4	S5	Overall
sEMG	IP (bps)	1.340	1.145	1.503	1.376	1.129	1.299
	R^2	0.795	0.863	0.554	0.736	0.752	0.740
mouse	IP (bps)	9.600	7.238	7.674	6.343	7.811	7.733
	R^2	0.979	0.975	0.940	0.868	0.824	0.917

Table 4. Experimental results of the efficiency of pointing devices from Fitts' law test using the developed interface and the mouse for five subjects (S1–S5). R^2 is the correlation coefficient.

interface was 1.299 bps, whereas the overall IP of the mouse was 7.733 bps. Pino et al. (2003) and Zhai et al. (2003) reported the IP value of a mouse as 7.048 and 8.445 bps respectively, and thus, the IP value of the mouse obtained in this study is comparable to values in the literature.

5. Discussion and conclusion

The developed interface has the potential to enable people with motor disabilities to interact with a graphic user interface in a natural and intuitive way. The interface was not tested for such individuals; however, it was tested for individuals with bilateral hand amputations and SCIs at the C6–C7 functional levels, who had control of the muscles (FCU, ECR, ECU, and APL). The interface efficiency was quantitatively measured using the Fitts' law test setup, and the performance of the proposed interface was compared with performances of other available interfaces. The ANN of the developed interface provided a high recognition ratio of over 96%, which shows that the computer commands are extracted well from the user. In addition, the IP of the interface was 1.299 bps, compared with the reported IP of 0.386 bps (Pino et al., 2003) for the commercial assistive pointing device Brainfingers. Thus, the performance of the developed interface was approximately three times better than that of the commercial device. The target performance of the alternative interface should be equivalent to the IP of a mouse so that individuals with motor disabilities can use a computer in a manner comparable to people without disabilities. The sEMG interface was, however, not able to perform as well as a mouse; its IP was 7.733 bps. Its performance was between that of a mouse and the commercial interface. Its low efficiency could be attributed to the constant speed of the cursor, which is potentially problematic when the cursor is located far from the target. To solve this problem, a more intelligent technique of producing cursor movements with an adjustable speed is required. A possible feasible way to achieve this is to map the magnitude of the muscular force to speed. However, it is not easy to estimate muscular force. In order to estimate muscular force, the ANN setting should be changed to concurrently estimate both the body movement and muscular force. In addition, cursor movement is restricted to only two directions (horizontal and vertical movements), and it cannot move in a diagonal direction, which has already been mentioned in recent literature (Citi et al., 2008). This issue could be solved if all wrist movements are predicted using 360° of motion, but it is a great challenge to estimate the infinite degrees of freedom in the human muscular-skeletal system.

This study has important implications for future work on the development of assistive computer interfaces, particularly with regard to improving the efficiency of controlling cursor movement. For this purpose, the performance results presented in this paper will be

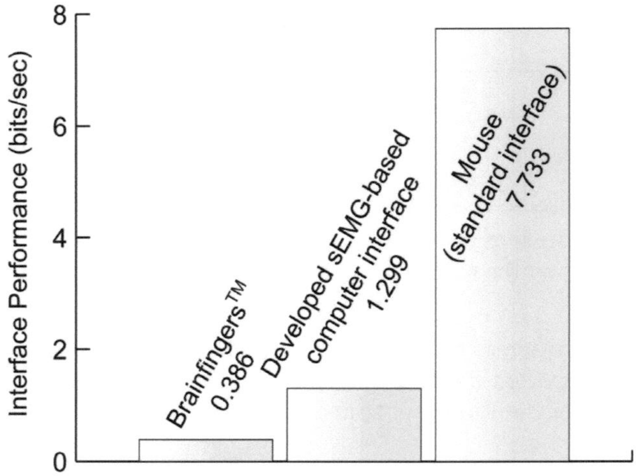

Fig. 10. Comparison of the performances of different interfaces.

analyzed using a design tool to discern the factors that allow a mouse to perform better than other interfaces. The result of such an analysis will enable the development of new assistive computer interfaces with high efficiency. The developed interface methodology can be extended to control various platforms such as bionic robot systems for people with limb disabilities (i.e., disabilities involving exoskeletons and limb prostheses) and teleoperated robotic systems that can perform human tasks in hazardous environments.

6. References

Barniv, Y.; Aguilar, M. & Hasanbelliu, E. (2005). Using EMG to anticipate head motion for virtual-environment applications. *IEEE Trans. Biomed. Eng.*, Vol. 52, No. 6, 1078-93

Cavanagh, P. R. & Komi, P. V. (1979). Electromechanical delay in human skeletal muscle under concentric and eccentric contractions. *Eur. J. Appl. Physiol. Occup. Physiol.*, Vol. 42, No. 3, 159-63, 0301-5548 (print)

Cheng, M.; Gao, X.; Gao, S. & Xu, D. (2002). Design and implementation of a brain-computer interface with high transfer rates. *IEEE Trans. Biomed. Eng.*, Vol. 49, No. 10, 1181-6, 0018-9294 (print)

Chu, J. U.; Moon, I.; Lee, Y. J.; Kim, S. K. & Mun, M. S. (2007). A supervised feature-projection-based real-time EMG pattern recognition for multifunction myoelectric hand control. *IEEE-ASME Transactions on Mechatronics*, Vol. 12, No. 3, 282-290, 1083-4435

Cipriani, C.; Zaccone, F.; Micera, S. & Carrozza, M. C. (2008). On the shared control of an EMG-controlled prosthetic hand: Analysis of user-prosthesis interaction. *IEEE Transactions on Robotics*, Vol. 24, No. 1, 170-184, 1552-3098

Citi, L., Poli, R., Cinel, C. & Sepulveda, F. (2008). P300-based BCI mouse with genetically optimized analogue control. *IEEE Trans. Neural Syst. Rehabil. Eng.*, Vol. 16, No. 1, 51-61, 1534-4320 (print)

d'Avella, A.; Saltiel, P. & Bizzi, E. (2003). Combinations of muscle synergies in the construction of a natural motor behavior. *Nat. Neurosci.*, Vol. 6, No. 3, 300-8, 1097-6256 (print)

Dipietro, L.; Ferraro, M.; Palazzolo, J. J.; Krebs, H. I.; Volpe, B. T. & Hogan, N. (2005). Customized interactive robotic treatment for stroke: EMG-triggered therapy. *IEEE Trans. Neural Syst. Rehabil. Eng.*, Vol. 13, No. 3, 325-34, 1534-4320 (print)

Englehart, K. & Hudgins, B. (2003). A robust, real-time control scheme for multifunction myoelectric control. *IEEE Trans. Biomed. Eng.*, Vol. 50, No. 7, pp. 848-854, 0018-9294

Fitts, P. M. (1992). The information capacity of the human motor system in controlling the amplitude of movement. *Journal of Experimental Psychology: General*, Vol. 121, No. 3, 262-269, 0096-3445

Fukuda, O.; Tsuji, T.; Kaneko, M. & Otsuka, A. (2003). A human-assisting manipulator teleoperated by EMG signals and arm motions. *IEEE Transactions on Robotics and Automation*, Vol. 19, No. 2, 210-222, 1042-296X

Guertin, P. A. (2005). Paraplegic mice are leading to new advances in spinal cord injury research. *Spinal Cord*, Vol. 43, No. 8, 459-61, 1362-4393 (print)

Haykin, S. S. (1999). *Neural Networks: A Comprehensive Foundation*, Prentice Hall, 0-7803349-4-9, Upper Saddle River, NJ

Hiraiwa, A.; Shimohara, K. & Tokunaga, Y. (1990). EEG Topography Recognition by Neural Networks. *IEEE Engineering in Medicine and Biology Magazine*, Vol. 9, No. 3, 39-42, 0739-5175

Hochberg, L. R.; Serruya, M. D.; Friehs, G. M.; Mukand, J. A.; Saleh, M.; Caplan, A. H.; Branner, A.; Chen, D.; Penn, R. D. & Donoghue, J. P. (2006). Neuronal ensemble control of prosthetic devices by a human with tetraplegia. *Nature*, Vol. 442, No. 7099, 164-71, 1476-4687 (electronic)

Hogan, N. & Mann, R. W. (1980). Myoelectric signal processing: optimal estimation applied to electromyography—Part I: derivation of the optimal myoprocessor. *IEEE Trans. Biomed. Eng.*, Vol. 27, No. 7, 382-95, 0018-9294 (print)

Jacob, R. J. K. (1993). Hot topics-eye-gaze computer interfaces: what you look at is what you get. *Computer*, Vol. 26, No. 7, 65-66, 0018-9162

Johansson, R. S.; Westling, G.; Backstrom, A. & Flanagan, J. R. (2001). Eye-hand coordination in object manipulation. *J. Neurosci.*, Vol. 21, No. 17, 6917-32, 1529-2401 (electronic)

Kennedy, P. R.; Bakay, R. A.; Moore, M. M.; Adams, K. & Goldwaithe, J. (2000). Direct control of a computer from the human central nervous system. *IEEE Trans. Rehabil. Eng.*, Vol. 8, No. 2, 198-202, 1063-6528 (print)

MacKenzie, I. S. (1992). Fitts' Law as a Research and Design Tool in Human-Computer Interaction. *Human-Computer Interaction*. Vol. 7, No. 1, 91-139, 0737-0024 (print)

McFarland, D. J.; Krusienski, D. J.; Sarnacki, W. A. & Wolpaw, J. R. (2008). Emulation of computer mouse control with a noninvasive brain-computer interface. *J. Neural Eng.*, Vol. 5, No. 2, 101-10, 1741-2560 (print)

Merletti, R.; Roy, S. H.; Kupa, E.; Roatta, S. & Granata, A. (1999). Modeling of surface myoelectric signals—Part II: Model-based signal interpretation. *IEEE Trans. Biomed. Eng.*, Vol. 46, No. 7, 821-9, 0018-9294 (print)

Millan Jdel, R.; Renkens, F.; Mourino, J. & Gerstner, W. (2004). Noninvasive brain-actuated control of a mobile robot by human EEG. *IEEE Trans. Biomed. Eng.*, Vol. 51, No. 6, 1026-33, 0018-9294 (print)

Pino, A.; Kalogeros, E.; Salemis, E. & Kouroupetroglou, G. (2003). Brain computer interface cursor measures for motion-impaired and able-bodied users, *Proc. Int. Conf.*

Human–Computer Interaction, pp. 1462-1466, 0-8058-4930-0, Crete, Greece, June 2003, Lawrence Erlbaum Associates, Inc, Philadelphia, PA

Rau, G.; Schulte, E. & Disselhorst-Klug, C. (2004). From cell to movement: to what answers does EMG really contribute? *J. Electromyogr. Kinesiol.*, Vol. 14, No. 5, 611-7, 1050-6411 (print)

Sanes, J. N. & Donoghue, J. P. (2000). Plasticity and primary motor cortex. *Annu. Rev. Neurosci.*, Vol. 23, No., 393-415, 0147-006X (print)

Sibert, L. E. & Jacob, R. J. K. (2000). Evaluation of eye gaze interaction, *Proceedings of the CHI 2000 Conference on Human Factors in Computing Systems*, pp. 281-288, The Hague, The Netherlands, April 2000, ACM Press, New York, NY

Smith, S. W. (1999). *The Scientist and Engineer's Guide to Digital Signal Processing*, California Technical Pub., 0-9660176-3-3, San Diego, CA

Taylor, D. M.; Tillery, S. I. & Schwartz, A. B. (2002). Direct cortical control of 3D neuroprosthetic devices. *Science*, Vol. 296, No. 5574, 1829-32, 1095-9203 (electronic)

Todd, R. F. & Richard, F. W. (2007). The optimal controller delay for myoelectric prostheses. *IEEE Trans. Neural. Syst. Rehabil. Eng.*, Vol. 15, No. 1, 111-118

Veneman, J. F.; Kruidhof, R.; Hekman, E. E.; Ekkelenkamp, R.; Van Asseldonk, E. H. & van der Kooij, H. (2007). Design and evaluation of the LOPES exoskeleton robot for interactive gait rehabilitation. *IEEE Trans. Neural. Syst. Rehabil Eng.*, Vol. 15, No. 3, 379-86, 1534-4320 (print)

Wessberg, J.; Stambaugh, C. R.; Kralik, J. D.; Beck, P. D.; Laubach, M.; Chapin, J. K.; Kim, J.; Biggs, S. J.; Srinivasan, M. A. & Nicolelis, M. A. (2000). Real-time prediction of hand trajectory by ensembles of cortical neurons in primates. *Nature*, Vol. 408, No. 6810, 361-5, 0028-0836 (print)

Zecca, M.; Micera, S.; Carrozza, M. C. & Dario, P. (2002). Control of multifunctional prosthetic hands by processing the electromyographic signal. *Crit. Rev. Biomed. Eng.*, Vol. 30, No. 4-6, 459-85, 0278-940X (print)

Zhai, S.; Conversy, S.; Beaudouin-Lafon, M. & Guiard, Y. (2003). Human on-line response to target expansion, *Proc. of the SIGCHI conference on Human Factors in Computing systems*, pp. 177-184

11

Integration of Electrotactile and Force Displays for Telexistence

Katsunari Sato[1], Naoki Kawakami[1], and Susumu Tachi[2]
[1]*The University of Tokyo*
[2]*Keio University*
Japan

1. Introduction

Telexistence (or telepresence) enable us to interact with another human or object in a remote or a virtual place through a robotic system (Tachi & Yasuda, 1994). This technology spreads across the world because of a desire to extend a person's sensing and interacting capability to remote places. In telexistence technologies, a robotic system called haptic display that provides haptic feedback to our hand is essential to touch the remote human or object (Shimoga, 1993a; 1993b). When we communicate or perform a task, a lack of haptic sensation reduces the realism and interactivity. Therefore, there is increasing requirement for haptic display presently.

The haptic feedback can be divided into two types based on the receptor that acquires the sensory information. One type is tactile (cutaneous) feedback, which is acquired by mechanoreceptors that exist at a depth of several millimetres from the skin surface. The other type is force (or kinesthetic) feedback, which is acquired by the proprioceptors that exist in the muscle, tendon, and joint. Based on the characteristics of human perception, it would be appropriate to provide both types of haptic feedback. In particular, a spatially distributed tactile feedback is necessary for dexterous manipulation. The spatially distributed tactile feedback and force feedback help us to perceive the position of the object and improve the stability of hand movements, respectively. For example, while holding a pen, we can pinch it with our fingertips and feel the reactive force; the position of the pen can be determined by tactile sensations.

Thus far, several haptic interfaces have been developed. However, these are not suitable for dexterous manipulation because of inadequate tactile feedback. The tactile display on conventional interfaces provides only a symbolic "contact" sensation of an object. Therefore, we cannot feel the object on our fingertips. It is believed that handling small objects such as pens is difficult without position information. Recently, some systems that can provide spatially distributed tactile sensation of an object have been proposed (Kim, et al., 2006; Methil, et al., 2006; Wagner, et al., 2005). Unfortunately, the systems proposed in these studies are too large for use in dexterous manipulation. A large system limits the workspace, i.e., the movement range of our finger required to manipulate an object. This limitation of workspace complicates manipulations such as pinching.

On the basis of results of conventional studies, we aimed to develop a haptic display for dexterous manipulation. First, we will summarize the requirements for the tactile feedback display intended for dexterous manipulations; the requirements are as follows:

1. The display should provide a highly realistic and intuitive touch sensation, i.e., it should provide not only the contact sensation but also the spatially distributed tactile sensation that humans perceive.
2. The display should be a compact body that does not invade the workspace of our fingers. Compact displays have several advantages over bulky ones during display implementation. Compact body size will help simplify the integration of this display with the force feedback display.

To fulfill these requirements, we used an electrotactile display as the tactile feedback display. We mounted the display on a force display with a wide workspace. This integration provided a haptic display that was suitable for dexterous manipulations.

In this chapter, we introduce a haptic display that integrates a spatially distributed electrotactile feedback and force feedback for telexistence. By integrating the electrotactile and force displays, we can use robotic system to dexterously manipulate an object (Fig. 1). Human interaction with a remote object through the robotic system can be dramatically improved by applying this concept. In section 2, we describe the concept of electrotactile and force integration. In section 3, we show the efficiency of the electrotactile feedback by a shape recognition experiment. In section 4, we describe the construction of a one-fingered haptic display and evaluate the effectiveness of the electrotactile integration. Finally, in section 5, we introduce a multi-fingered robotic hand system that involves the integration of electrotactile and force display for telexistence.

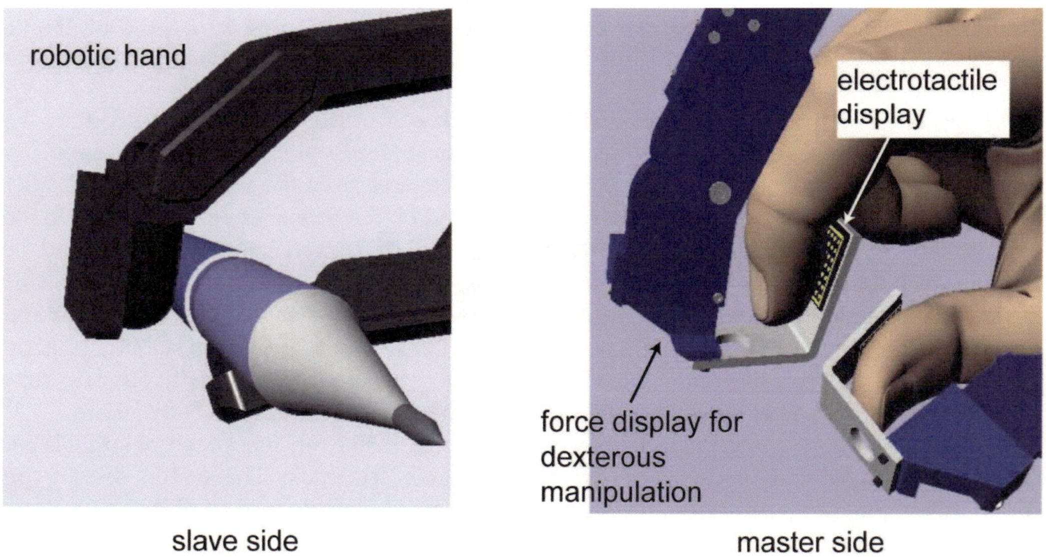

Fig. 1. Conceptual diagram of integration of electrotactile and force displays for dexterous manipulation.

2. Integration of electrotactile and force displays

2.1 Electrotactile display

The electrotactile display that we have developed (Kajimoto, et. al, 2004) can present spatially distributed tactile sensations. It comprises a pin electrode matrix. It directly activates nerve fibers under the skin by passing an electrical current from the surface electrodes (Fig. 2). The electrical currents flow from an electrode to adjacent electrodes through the skin. This display can selectively stimulate each type of receptor and produce vibratory and pressure sensations at an arbitrary frequency. By periodically changing the pin used for stimulation, we can produce the electrotactile stimulus at any points. Therefore, the electrotactile display allows us to perceive touch sensation which help determine position and exact shape of the object. In addition, the electrode plate of this display is small and lightweight. Therefore, it does not affect the workspace. Further, we can easily mount this display on all types of force displays.

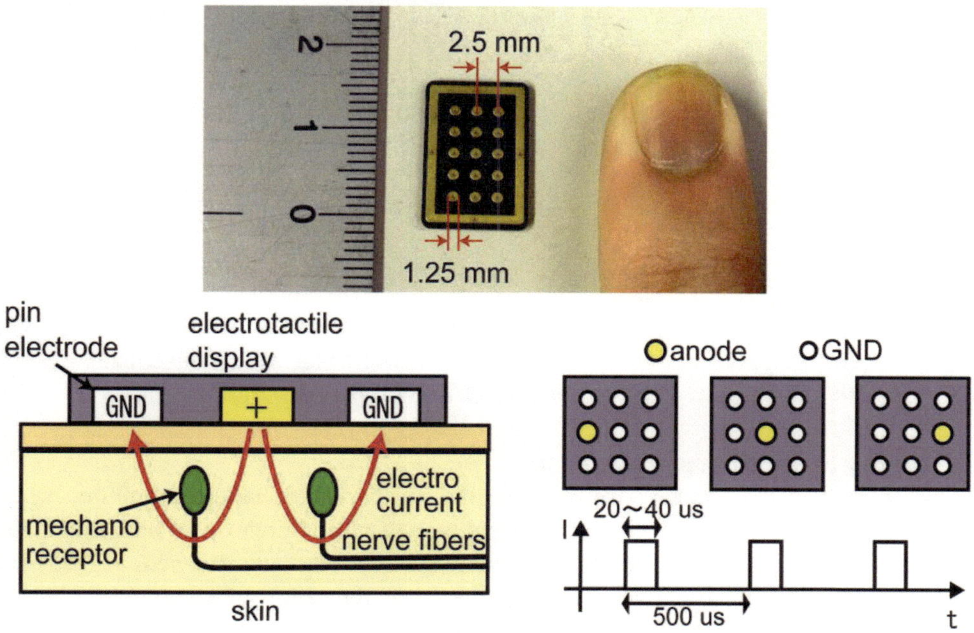

Fig. 2. Electrodes of electrotactile display and method of electrical stimulus.

2.2 Force display

The force display presents the reactive and friction force on object surfaces. It can improve the stability of our hand movements when we manipulate an object. Currently, several types of force displays are used (Bar-Cohen, et al., 2000). In this study, we consider a small-sized display that has multiple degrees of freedom (DOFs) such as PHANToM (SensAble Tec.) and CyberGrasp (Immersion Tec.). Some of these force displays provide a wide workspace and sufficient force feedback to our hand.

2.3 Integration of the displays

When a user touches objects in a remote or virtual environment using our integrated system, he/she can perceive the spatially distributed tactile sensation and reactive force of objects. From these sensations, the user can easily identify the position of the object, its posture, and shape, i.e., he/she can easily recognize the object that he/she touches. For example, from the force sensation of a rounded surface and the tactile sensation of concave-convex surfaces, we can recognize that we are touching a gear (Fig. 3). We believe that this haptic information will also help the user to manipulate objects dexterously.

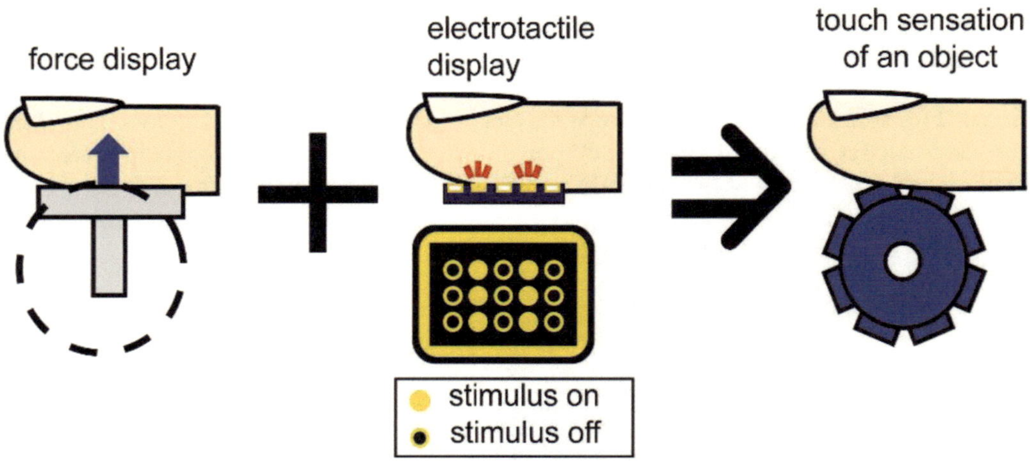

Fig. 3. Touch sensation by integration of electrotactile and force displays.

3. Electrotactile feedback for shape recognition

The electrotactile display may help perceive the shape of an object. Before implementing the integrated haptic display, we evaluated the efficiency of an electrotactile feedback when it is integrated with a force feedback (Sato, et al., 2007a; 2007c).

3.1 Efficiency of electrotactile feedback

First, we evaluated the efficiency of electrotactile feedback for shape recognition. Figure 4 shows the experimental setup. The participants wore a plastic finger case on their fingertip when they touched the object. The electrode plate used for electrotactile feedback was in the finger case. The electrotactile display that we used was the same as that shown in Fig. 2. In this setup, a "real" force sensation was generated by actual contact, and tactile sensation was generated by using the virtual model of the object in a PC. This condition is simulates a "mixed reality" situation.

We prepared three objects with the following characteristics: a flat surface, a curved face, and an edge (Fig. 5). We considered two modes of touching, namely, pushing and tracing (or sliding) as shown in Fig. 5. Experiments were conducted under six conditions as follows:

C1. Pushing with electrotactile feedback
C2. Pushing with force feedback
C3. Pushing with electrotactile and force feedbacks
C4. Tracing with electrotactile feedback
C5. Tracing with force feedback
C6. Tracing with electrotactile and force feedbacks

Under these conditions, we evaluated the accuracy and time taken for shape recognition.

Figure 6 shows the experimental results for all participants. From the results, we confirmed that the correct answer ratio when electrotactile feedback was present was higher than that when it was absent; moreover, the recognition time when electrotactile feedback was present was shorter than that when it was absent. Further, this result was independent of the participant and mode of touching. Therefore, we inferred that the electrotactile feedback improves the efficiency of shape recognition.

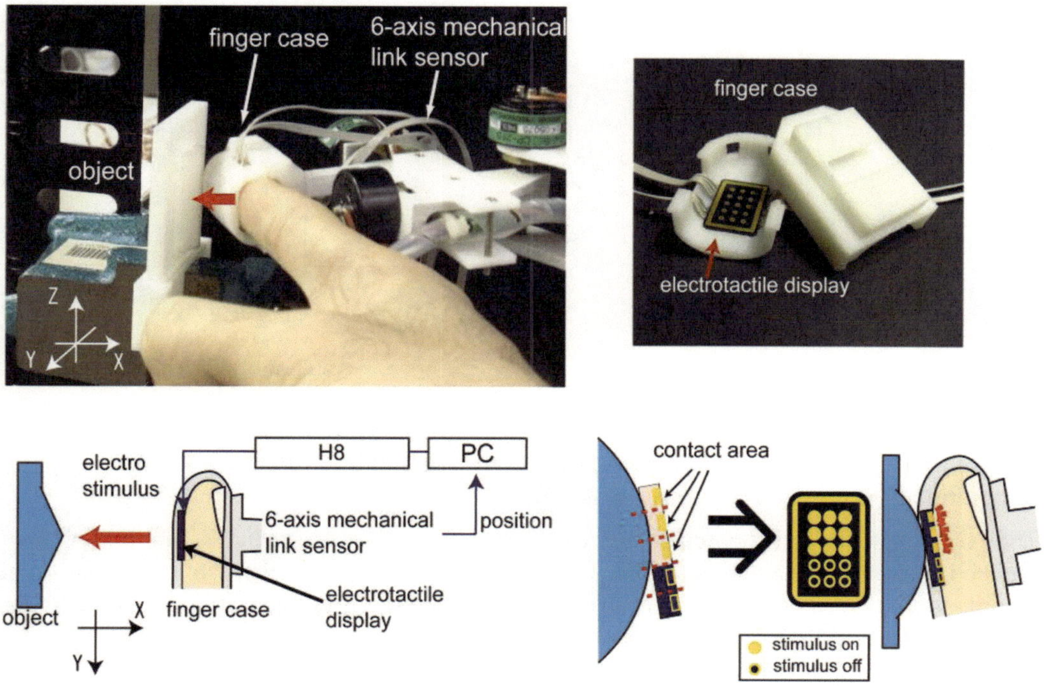

Fig. 4. Experimental environment.

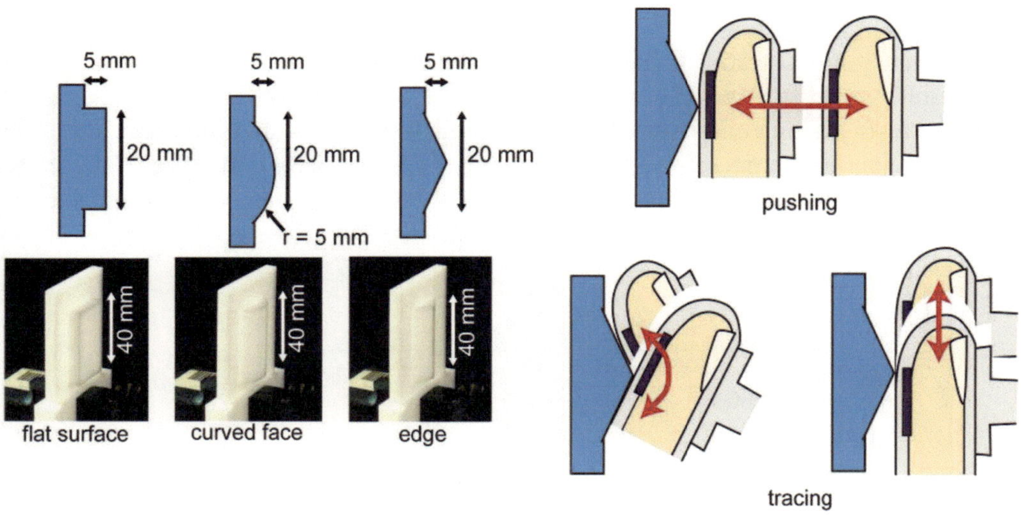

Fig. 5. Objects that participants touched and two mode of touching.

3.2 Importance of electrotactile feedback

For shape recognition, electrotactile feedback is more important than force sensation; a number of shape sensations are generated by the electrotactile stimulus. For example, when the force display generates the sensation of an "object with an edge" while the electrotactile display generates the sensation of a "curved object," a human being would perceive the latter. We investigated the responses of the participants to the force or electrotactile sensations.

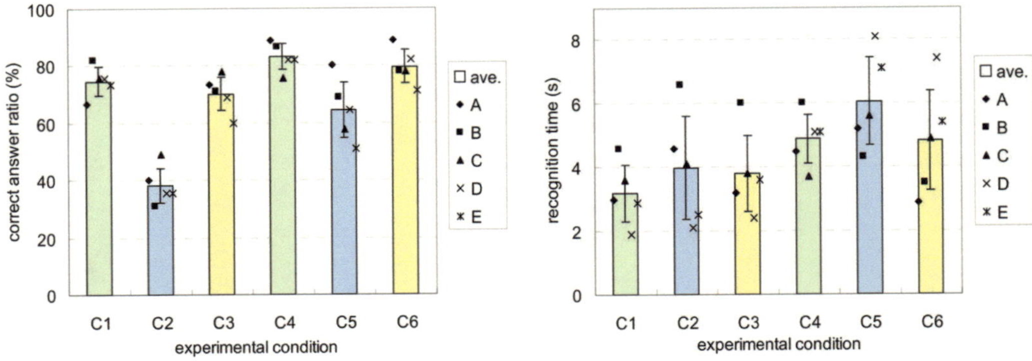

Fig. 6. Results of the shape recognition experiment. The horizontal and vertical axes represent the abovementioned experimental conditions and the correct answer ratio or recognition time, respectively. (Sato, et al., 2007c)

The participants traced the object surface in the manner shown in Fig. 5. The objects they touched were an edge and a curve (Fig. 5). Two stimulation modes were tested for electrical stimulation. The first mode stimulated a "curvature"; the second, an "edge". The experimental conditions were as follows.

C1. Touching curved face with electrotactile feedback of curved face

C2. Touching curved face with electrotactile feedback of edge

C3. Touching edge with electrotactile feedback of curved face

C4. Touching edge with electrotactile feedback of edge

The average response ratio of the "curve" is shown in Fig. 7. In this experiment, the participants tended to respond to an object on the basis of the electrotactile feedback. This result supports the hypothesis that the electrotactile sensation is more important than the force sensation in shape recognition. Therefore, it is suggested that the electrotactile stimulus is efficient in generating the shape sensation. In addition, we suggest that any touch sensation related to a typical object shape can be generated by integrating an electrotactile display with force display.

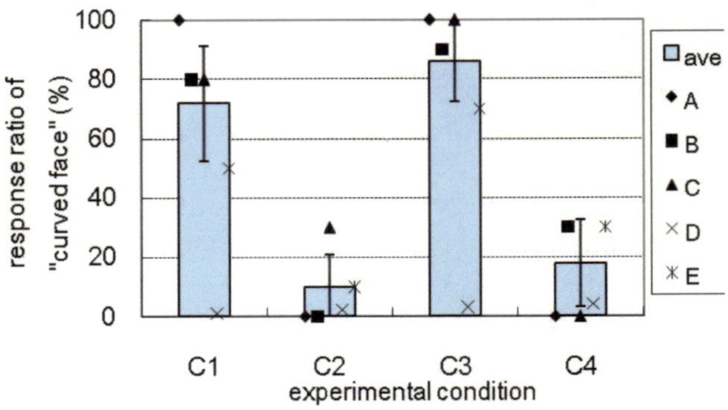

Fig. 7. Experimental result. The horizontal and vertical axes represent the experimental conditions and the response ratio of the "curve," respectively. (Sato, et al., 2007a; 2007c)

4. One-fingered system

We constructed the one-fingered system of the electrotactile and force integration. Then, we evaluated the performance of the integrated system and the efficiency of the integration of electrotactile and force displays for a particular task (Sato, et. al., 2007b; 2007e).

4.1 Integration of electrotactile display with PHANToM

Figure 8 shows the configuration of the one-fingered system. In this system, we used PHANToM Omni (SensAble Tec.) as a force display. It provides a wide workspace and generates sufficient force for one finger. We mounted the electrotactile display on the end-effector of the PHANToM. The users placed the tip of their index finger on the electrotactile display and moved the end-effector of the PHANToM. They could control the cursor in the virtual environment using their fingertips. The fingertip was fixed on the end-effector by rubber bands. The electrotactile display that we used is same as shown in Fig. 2.

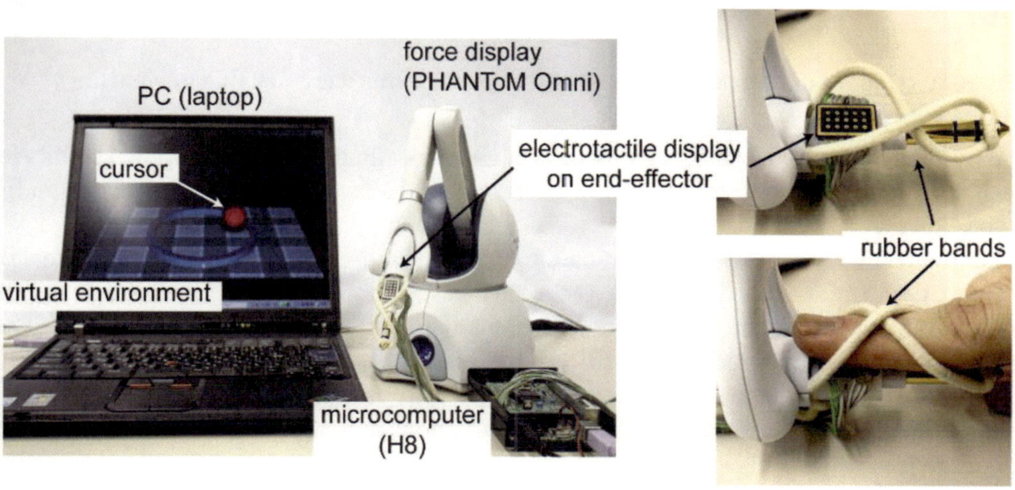

Fig. 8. Overview of the single-fingered system and electrotactile display on the end-effector of PHANToM.

The position data of the user's index finger is captured by the PHANToM and translated to the PC. Then, the position of the cursor in the virtual environment is updated. On the basis of the cursor position, the reflection force and the electric current at the electrode pin are calculated. The reflection force is calculated by using the spring-damper model. Current is passed through the electrodes on the basis of the position of the contact field between the cursor and the virtual object. This implies that the electrostimulus is provided by the electrodes at the position corresponding to the contact position of a finger pad and an object. For example, when the finger pad is in contact with the face of a cube, all electrodes send a current to the finger. When the center of the finger pad touches the edge of the cube, the electrodes located in a line send the current.

4.2 Basic performance of the one-fingered system

We used the constructed system to examine the space resolution of the electrotactile feedback by distance and width discrimination. Subsequently, we evaluated the strength resolution of the electrical stimulus by strength discrimination.

We chose three experimental conditions: 2-line, width, and strength conditions. In each condition, there was a floor, a cursor, and two lines (a standard line and a comparison line) in the virtual environment. We specified two modes of touching the lines—pushing and sliding (Fig. 9).

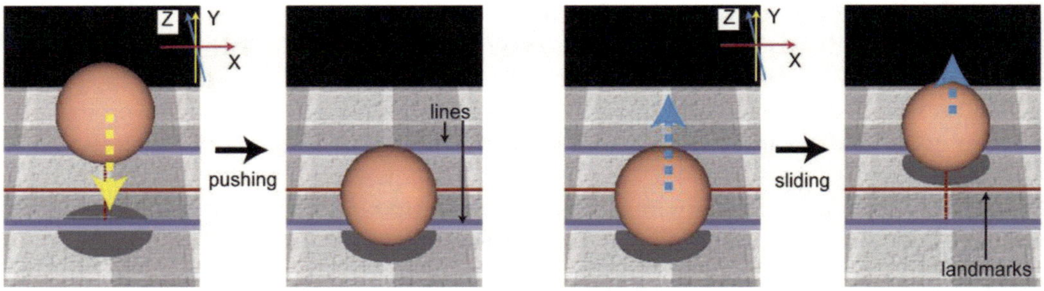

Fig. 9. Two modes of touching lines. (Note that participants were not able to view lines during experiments.)

We conducted each experiments by method of constant stimuli. The experimental results for each setting are shown in Fig. 10. From the results, the effect of the touching modes on the resolution seems to be small.

From the results of the 2-line discrimination, the threshold is observed to be approximately 9.5 mm. On the electrotactile display, the electrical current flows from the electrode only to the adjacent electrodes. Therefore, the discrimination threshold should be around 5.0 to 7.5 mm. However, under practical conditions, the electrical current leaks to the surrounding electrodes. This leakage current results in a wide area of contact sensation. Therefore, we believe that the leakage current will cause complications in identifying whether the lines are identical or not.

The width discrimination threshold for the 7.5 mm line is approximately 2.0 mm. On the basis of the distance between the centers of the electrodes, the width discrimination threshold is considered to range from 0.0 to 2.5 mm. This result is in accordance with the theoretical value. Therefore, we conclude that the abovementioned leakage current does not affect width discrimination.

In the case of strength discrimination, the upper and lower thresholds are approximately 0.12 and 0.06 mA, respectively. These thresholds are considered to be small as compared to the range of the strength of the electrical stimuli that the participants could feel comfortably (1.5 mA). Therefore, we believe that the electrotactile display has a high strength resolution. On the basis of this result, it is possible to implement the presentation of magnitude of the pressures by means of the strength of the electrotactile stimulus.

4.3 Tracing task efficiency

Using the one-fingered system, we evaluated the manipulation efficiency in track tracing task. The participants controlled the cursor and traced a circular path in a virtual environment using the constructed system (Fig. 11). The experiment was conducted under the following four feedback conditions:

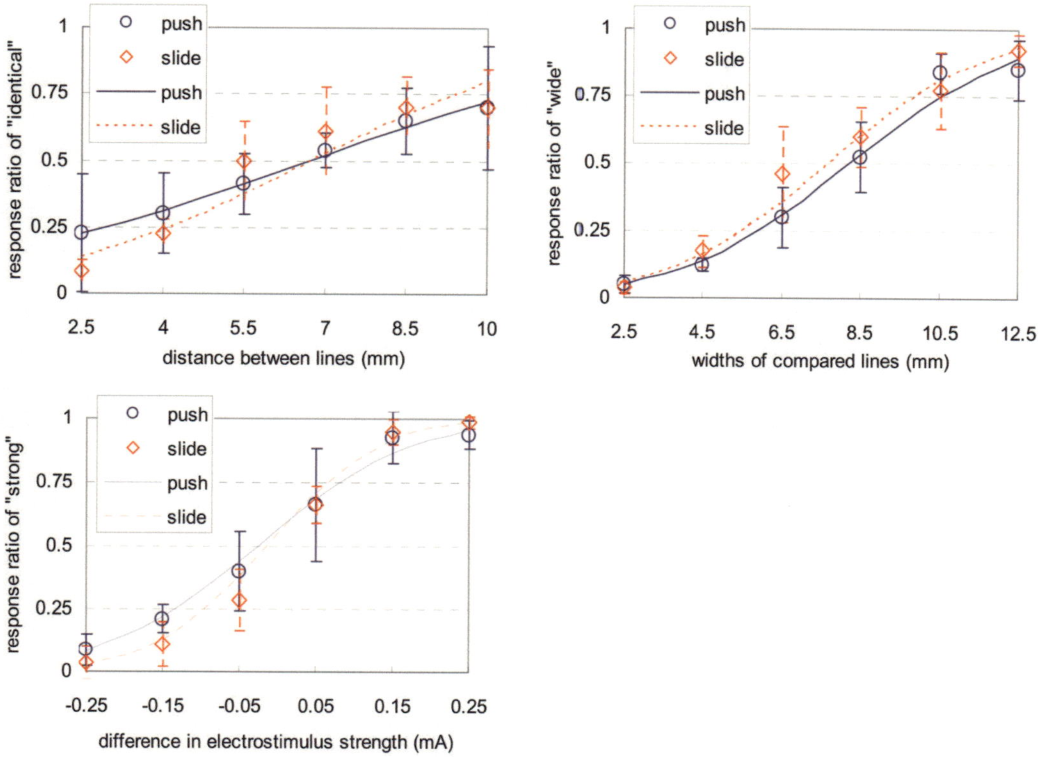

Fig. 10. Results of experiments on 2-line, width, and strength discriminations. The horizontal and verrtical axes represent the reference value of each experiment and represents the response ratio of participants, respectively. (Sato, et al., 2007e)

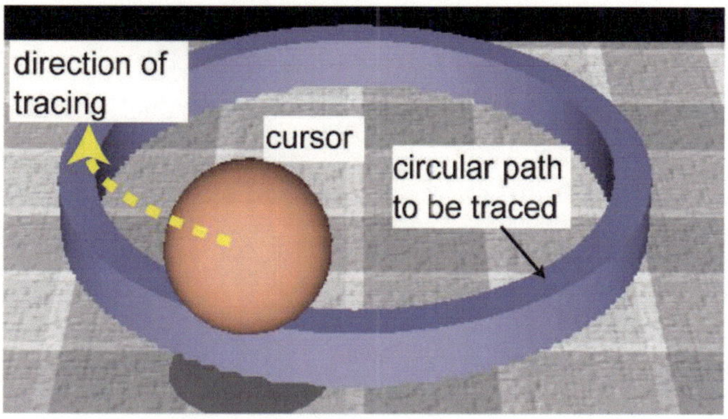

Fig. 11. Overview of tracing a circular path in a virtual environment.

C1. Integration 1: reflection force and position sensation
C2. Integration 2: reflection force and contact sensation
C3. Force: reflection force
C4. Electrotactile: position sensation

The position and contact sensation were generated by the electrotactile display. In C1, a two-dimensional contact position sensation was generated by each electrode of the electrotactile

display. This shows the participant's finger tip where the cursor touches the circular path. In C2, the contact sensation was generated by all the electrodes of the electrotactile display.
Figure 12 shows the result of the evaluation of the track-tracing task. In order to evaluate the accuracy of the tracing task, we assumed the trajectory that traces the center of the path to be the optimal trajectory. Then, we compared the avarage error between the optimal trajectory and the measured trajectory.
The error in C1 is the smallest for all participants. Therefore, we can confirm that the electrotactile and force integration is effective in the case of the track-tracing task. When we compare the errors in C1, C3, and C4, we find that the error in the case in C4 is the largest. This shows that the force feedback is more important than the electrotactile feedback in for stablity in operation. When we compare the errors in C2 and C3, the error in C2 is larger than that in C3 even though more haptic information is generated in C2. This may mean that tonly contact sensation cannot improve the task efficiency. This result confirms the importance of the proposed spatially distributed tactile feedback.

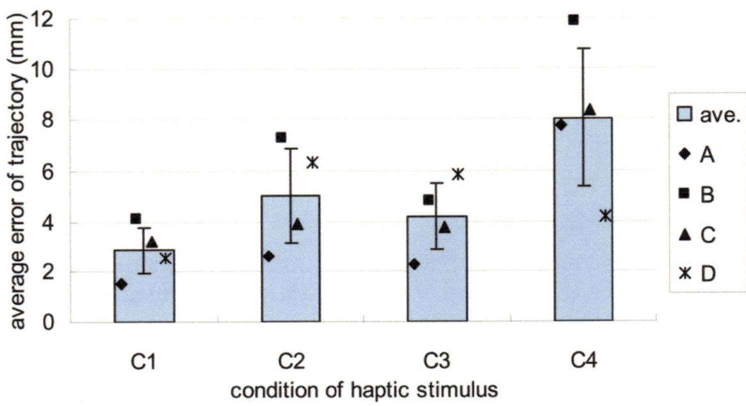

Fig. 12. Result of the evaluation of the track-tracing task. The horizontal and vertical axes represent the haptic condition and the trajectory error, respectively. (Sato, et al., 2007b)

5. Multi-fingered robotic hand system: Haptic Telexistence

By integrating electrotactile and force displays, we constructed a multi-fingered robotic hand master-slave system named Haptic Telexistence.

5.1 Configuration

Our system consists of four devices, namely, a multi-fingered slave hand, a finger-shaped haptic sensor for the slave hand, an exoskeleton encounter-type master hand, and electrotactile display (Fig. 13).
We mounted the electrotactile display on a multi-fingered master hand (Nakagawara, et al., 2005). This hand has two features. One is a compact exoskeleton mechanism called "circuitous joint," which covers the wide workspace of an operator's finger. The other is the encounter-type force feedback. These features help avoid unnecessary contact sensation and enable the unconstrained motion of the operator's fingers. We set the electrotactile display on the tips of each finger mechanism.

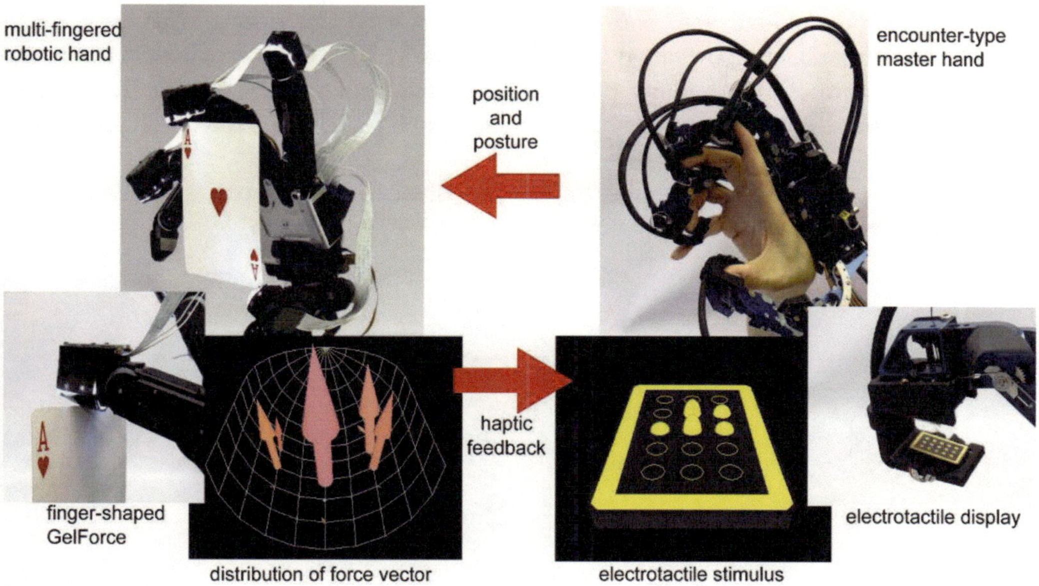

Fig. 13. Configuration of Haptic Telexistence system.

The multi-fingered slave hand (Hoshino & Kawabuchi, 2005) has the following futures. This hand has 15 DOFs — five DOFs for the thumb, one for abduction of other fingers, three for the index finger, and two for the remaining fingers. Each fingertip has an independent DOF, and the index finger and the thumb can be moved in opposite directions. Therefore, a pinching operation by the fingertip is possible. In addition, we developed a finger-shaped haptic sensor (Sato, et al., 2008) using the GelForce technology (Kamiyama, et al., 2005) for this robotic hand. GelForce is a haptic sensor that measures the distribution of both the magnitude and the direction of force.

The master-slave manipulation is realized by bilateral position control of the multi-fingered slave hand and the encounter-type master hand. This control is exercised from the position of the master and slave fingers. The position is calculated using the angle of each finger joint. The refresh rate of the control is 1 kHz. Therefore, we can operate the multi-fingered slave hand smoothly and perceive sufficient force sensation.

When the slave hand touches an object, the finger-shaped GelForce mounted on the slave hand acquires haptic information such as the distribution of the magnitude and the direction of force. Then, this information is transmitted to the master system. The electrotactile display provides a tactile sensation on the basis of this information. Information regarding the distribution of the force is obtained from the pin location which provides electrostimulus. Subsequently, information regarding the magnitude of the force at each position is obtained form the strength of electrostimulus. As a result, we can feel the field, edge, peak, and the movement of an object. By integrating these force and tactile sensations, we can perceive the exact shape and stiffness of the object. This enables highly realistic interactions with remote objects.

5.2 Exhibition of Haptic Telexistence

Figure 14 represents the Haptic Telexistence system designed by us. We exhibited this system in some conferences such as ACM SIGGRAPH 2007 (Sato, et al., 2007d). During the exhibitions, approximately one thousand participants used this system. The participants could feel an object being touched with the finger of slave hand due to the electrotactile and force feedbacks. In addition, many participants pointed out that the Haptic Telexistence system is a useful technology for tele-communication and tele-manipulation in fields such as relesurgery.

In the future, we will evaluate the haptic telexistence system from the viewpoint of efficiency of transmission of haptic information and tele-manipulation.

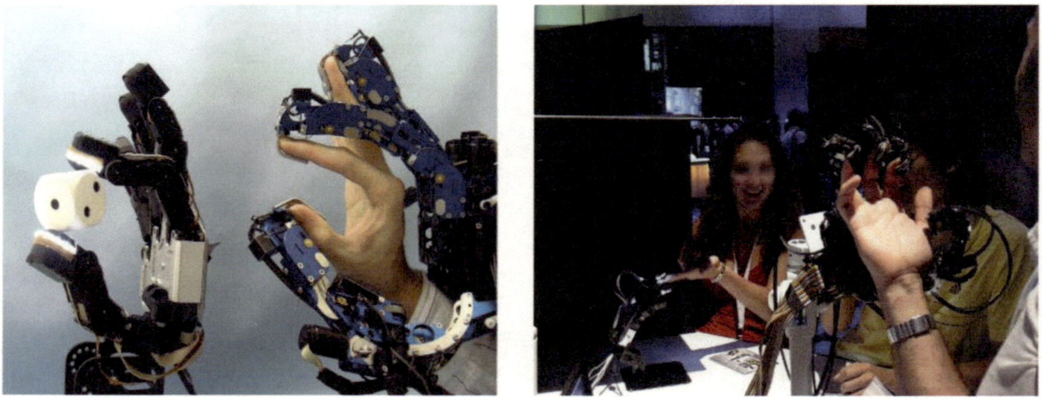

Fig. 14. Haptic Telexistence system and its exhibition at a conference. (Sato, et al., 2007d)

6. Conclusion

In this chapter, we described a robotic system that enables us to interact with a remote human or object. We proposed the integration of electrotactile and force feedback for dexterous tele-manipulation. The electrotactile feedback can provide spatially distributed tactile sensation; therefore, we consider that the integration of electrotactile and force feedback is effective in perceiving the shape of an object and in manipulating it. We have confirmed the effectiveness of the electrotactile feedback and constructed a multi-fingered telexistence system named Haptic Telexistence.

In the future, we plan to provide more object properties such as texture and temperature. Not only will we be able to shake hands with people at remote locations but we will be able to feel the warmth of their hands. In the case of internet shopping, we will be able to check the texture of an article before purchase. We expect that the Haptic Telexistence system will dramatically improve the human interaction with a remote object.

7. Acknowledgement

This study is partly supported by Grant-in-Aid for JSPS Fellows (20·10009).

8. References

Bar-Cohen, Y.; Mavroidis, C.; Bouzit, M.; Pfeiffer, C. & Magruder, D. (2000). Haptic Interfaces, *Chapter in Automation, Miniature Robotics and Sensors for Non-Destructive*

Testing and Evaluation, Y. Bar-Cohen (Ed.), The American Society for Nondestructive Testing, Inc., pp. 461-468

Hoshino, K. & Kawabuchi, Y. (2005). Pinching at finger tips for humanoid robot hand, *Journal of Robotics and Mechatronics*, Vol. 17, No. 6, pp. 655–663

Kajimoto, H.; Kawakami, N.; Maeda, T. & Tachi, S. (2004). Electro-Tactile Display with Tactile Primary Color Approach, *Proceedings of International Conference on Intelligent Robots and Systems*

Kamiyama, K.; Vlack, K.; Mizota, T.; Kajimoto, H.; Kawakami, N. & Tachi, S., 2005, Vision-Based Sensor for Real-Time Measuring of Surface Traction Fields, *journal of IEEE Computer Graphics & Applications Magazine*, Vol. 25, No. 1, pp. 68-75

Kim, Y.; Oakley, I. & Ryu, J. (2006). Combining Point Force Haptic and Pneumatic Tactile Displays, *Proceedings of the EuroHaptics 2006*, pp. 309-316

Methil, N. S.; Shen, Y.; Zhu, D.; Pomeroy, C.A.; Mukherjee, R.; Xi, N. & Mutka, M. (2006). Development of supermedia Interface for Telediagnostics of Breast Pathology, *Proceedings of IEEE International Conference on Robotics and Automation*, pp. 3911-3916

Nakagawara, S.; Kajimoto, H.; Kawakami, N. & Tachi, S. (2005). An Encounter-Type Multi-Fingered Master Hand Using Circuitous Joints, *Proceedings of IEEE International Conference on Robotics and Automation (ICRA2005)*, Barcelona, Spain

Sato, K.; Kajimoto, H.; Kawakami, N. & Tachi, S. (2007a). Improvement of Shape Distinction by Kinesthetic-Tactile Integration, *Proceedings of World Haptics 2007*, pp. 391-396, Tsukuba, Japan

Sato, K.; Kajimoto, H.; Kawakami, N. & Tachi, S. (2007b). Electrotactile and Kinesthetic Integration for Dexterous Manipulation, *Proceedings of JSME Robotics and Mechatoronics Conference*, Akita, Japan (in Japanese)

Sato, K.; Kajimoto, H.; Kawakami, N. & Tachi, S. (2007c). I Display of Shape Sensation by Electrotactile-Kinesthetic Integration, *Journal of Human Interface*, Vol. 9, No. 3, pp. 71- 76 (in Japanese)

Sato, K.; Minamizawa, K.; Kawakami, N. & Tachi, S. (2007d). Haptic Telexistence, *34th Int. Conf. On Computer Graphics and Interactive Techniques (ACM SIGGRAPH 2007)*, San Diego, USA

Sato, K.; Kajimoto, H.; Kawakami, N. & Tachi, S. (2007e). Electrotactile Display for Integration with Kinesthetic Display, *Proceedings of 16th IEEE International Symposium on Robot & Human Interactive Communication (RO-MAN2007)* , pp. 3-8, Jeju, Korea

Sato, K.; Kamiyama, K.; Nii, H.; Kawakami, N. & Tachi, S. (2008). Measurement of Force Vector Field of Robotic Finger using Vision-based Haptic Sensor, *Proceedings of IEEE Intelligent Robotics and Systems (IROS) 2008*, pp. 488-493, Nice, France

Shimoga, K.B. (1993a). A Survey of Perceptual Feedback Issues in Dexterous Telemanipulation: Part 1. Finger Force Feedback, *Proceedings of the IEEE Virtual Reality Annual International Symposium*, pp. 263-270

Shimoga, K.B. (1993b). A Survey of Perceptual Feedback Issues in Dexterous Telemanipulation: Part 2. Finger Touch Feedback, *Proceedings of the IEEE Virtual Reality Annual International Symposium*, pp. 271-279

Tachi, S. & Yasuda, K. (1994). Evaluation experiments of a telexistence manipulation system. *Presence*, Vol. 3, No. 1, pp. 35–44

Wagner, C. R.; Perrin, D. P.; Feller, R. L.; Howe, R. D.; Clatz, O.; Delingette, H. & Ayache, N. (2005). Integrating Tactile and Force Feedback with Finite Element Models, *Proceedings of IEEE International Conference on Robotics and Automation*, pp. 1–10, Barcelona, Spain

12

Predictive Tracking in Vision-based Hand Pose Estimation using Unscented Kalman Filter and Multi-viewpoint Cameras

Albert Causo[1], Kentaro Takemura[1], Jun Takamatsu[1], Tsukasa Ogasawara[1], Etsuko Ueda[2] and Yoshio Matsumoto[3]
[1]*Nara Institute of Science and Technology*
[2]*Nara Sangyo University*
[3]*National Institute of Advanced Industrial Science and Technology*
Japan

1. Introduction

One of the major challenges in human-robot interaction is how to enable the use of unrestricted hand motion as a tool for communication. The direct use of hand as an input tool enables the user to connect to systems more naturally, allowing them to become an integral part of our daily lives. A vision-based approach, using cameras to capture data, supports non-contact and unrestricted movement of the hand. Nonetheless, the high degrees of freedom (DOF) of the hand is an essential issue to tackle in articulated hand motion tracking and pose estimation.

In this paper, we present our vision-based model-based approach, which uses multiple cameras and predictive filtering, to estimate the pose of the hand. We build on the research of Ueda *et al.*, whose work can separately estimate the global pose (wrist position and palm orientation) and the local pose (finger joint angles), but not simultaneously (Ueda et al., 2003). We address the problem through the use of a non-linear filter, Unscented Kalman Filter (UKF), to track the motion and simultaneously estimate the global and local poses of the hand.

The rest of the paper is organized as follows. Section 2 presents the related works and Section 3 discusses the UKF. Section 4 explains the hand pose estimation system and Section 5 details how we use the UKF for tracking and pose estimation. Experimental results and discussions are found in Section 6.

2. Related works

There are two main techniques to a vision-based hand pose estimation: appearance-based and model-based. Appearance-based approach uses two dimensional features such as silhouettes or colors, in order to compare the input image to a database of pose images and determine the hand pose (Athitsos & Sclaroff, 2003; Shimada et al., 2001). However, appearance-based approach is perspective-limited and usually gives solution only to a specific task problem (Pavlovic et al., 1997).

In model-based approach, the hand motion is modeled parametrically, giving a more precise and generic result. It tries to minimize the error between a predefined model of the hand and the observation data.

A full DOF hand has at least 27 parameters composed of 6 global (rotation and translation of the hand) and 21 local (finger joint angles). In a model based approach, the hand is represented as a model that describes its characteristics. The models can be classified in three groups: geometric model, statistical model, and physical based model. Geometric model uses various geometric primitives to represent the physical structure of the hand. Examples of geometric model include truncated quadrics (Stenger et al., 2001) and cardboard (Wu et al., 2001). On the other hand, statistical models define a hand shape as a variation of a mean model shape (Huang & Jeng, 2001). A physical based model considers the effect of various forces on the hand pose. Skeletal model covered with B-spline surface (Kuch & Huang, 1994), quadric surface (Ueda et al., 2003), or voxels (Causo et al., 2009) are examples of a physical based model.

Erol *et al.* classified hand pose estimation and motion tracking methods as either single-hypothesis tracking or multiple hypotheses tracking (Erol et al., 2007). In the former, the matching error between the model and the observation data is minimized by a best fit search. This technique includes optimization based methods like Gauss Newton (Rehg & Kanade, 1994), Genetic Algorithm (Lien & Huang, 1998), or Stochastic Gradient Descent (Bray et al., 2004) and physical-force models that uses force (Ueda et al., 2003), Unscented Kalman Filter (UKF) (Stenger et al., 2001; Causo et al., 2008) or Iterative Closest Point (ICP) algorithm (Delamarre & Faugeras, 1999).

Multiple hypotheses tracking, wherein multiple pose estimates are considered at each time frame, tries to address the issues of single hypothesis tracking such as the presence of singularity or spurious local minima. This includes tree-based search (Thayananthan et al., 2003), template matching (Shimada et al., 2001), and particle filtering (Lin et al., 2002).

Hand motion tracking is not a linear problem, but predictive tracking solutions for non-linear systems are available including Extended Kalman Filter (EKF), UKF, Gaussian sum filter, particle filter, and grid-based methods. Extended Kalman Filter is a straight-forward adaptation of the Kalman Filter to non-linear systems. Shimada *et al.* used EKF to estimate the pose of the hand and refine the 3D shape model even when using only a monocular camera and without any depth information (Shimada et al., 1998). A modified EKF through constraint fusion was used by Azoz *et al.* to localize and track an articulated arm (Azoz et al., 1998). Another extension of the Kalman Filter is the UKF (Julier & Uhlmann, 1997) which Stenger *et al.* used to track the motion of the hand modelled as truncated quadrics (Stenger et al., 2001).

Gumpp *et al.* used particle filtering (PF) to track the hand motion of the user in order to control a 20 DOF robot hand (Gumpp et al., 2006). Lin *et al.* parametrized the hand configuration space to be able to use a lower number of particles and consequently speeded up the computation (Lin et al., 2002). Thayananthan *et al.* and Stenger *et al.* both used grid-based filtering to search for the representative pose by traversing the tree nodes with high probabilities (Thayananthan et al., 2003; Stenger et al., 2004). They were able to do a fast search because the tree nodes' probabilities are updated during tracking and they skip the children of the nodes with small probabilities.

3. The Unscented Kalman Filter (UKF)

Unscented Kalman Filter belongs to the Kalman Filter (KF) family. It is a recursive estimator that uses information from the previous time frame in addition to the current observation measurement to make an estimate of the current state. Unlike the KF though, EKF and UKF are designed for non-linear systems. Extended Kalman Filter (EKF) is the more commonly used technique between the two. However, it requires the computation of Jacobian matrices, which is non-trivial in most cases.

In contrast, UKF uses unscented transformation method, which calculates the statistics of a random variable that undergoes non-linear transformation (Julier & Uhlmann, 1997). It is accurate up to the second order and requires fewer samples compared to a similar particle filter. Xiong *et al.* studied the performance of UKF under certain conditions and showed that it performs robustly in general tracking applications of non-linear systems (Xiong et al., 2006).

Figure 1 shows the overview of the UKF process, which is composed of two main parts, similar to the KF. First is the time-update, wherein the initial state estimate is computed by selecting sigma points and solving for its mean and covariance. The observation is also propagated in this step and its mean and covariance are also calculated. The second part is the measurement update. The Kalman gain and cross-covariance of the propagated state and the propagated observation are calculated and used to update the state and its covariance. The computational details are discussed next.

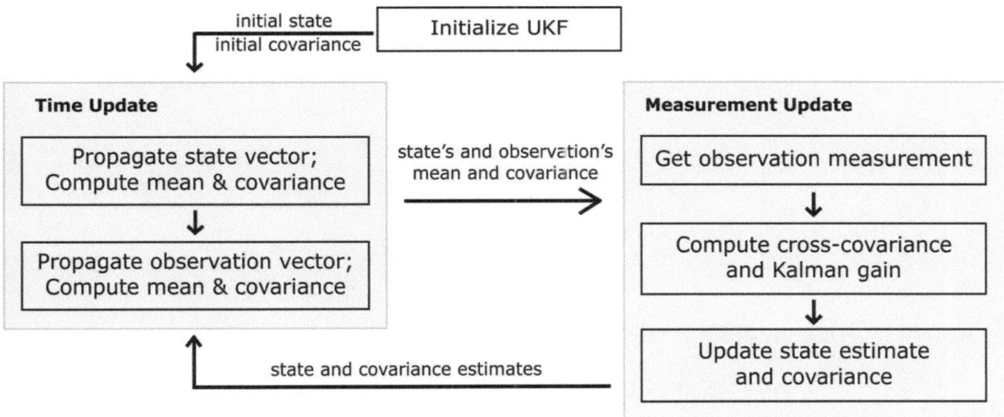

Fig. 1. The Unscented Kalman Filter (UKF) process. This is very similar to the Kalman Filter, except that the initial state estimate (under the Time Update box) is obtained from the sigma (particle) propagation.

For a given tracking problem, consider the state dynamics,

$$\mathbf{X_k} = f(\mathbf{X_{k-1}}, \mathbf{R_k}) \tag{1}$$

where:
f is the system dynamics,
$\mathbf{X_k}$ is the state vector of size n at time k, and
$\mathbf{R_k}$ is the state noise covariance.

UKF makes an initial estimate of the state vector, by selecting sigma points through Equation 2:

$$\mathbf{X_k^i} = \begin{cases} \mathbf{X_k^0} = \bar{\mathbf{X}}_{\mathbf{k-1}} & \\ \mathbf{X_k^i} = \bar{\mathbf{X}}_{\mathbf{k-1}} - (\phi)_i & i = 1,\ldots,n \\ \mathbf{X_k^i} = \bar{\mathbf{X}}_{\mathbf{k-1}} + (\phi)_{i-n} & i = n+1,\ldots,2n \end{cases} \tag{2}$$

where:
ϕ is the i_{th} column of $\sqrt{(n+\lambda)\mathbf{P_{k-1}}}$,
$\mathbf{P_{k-1}}$ is the covariance estimate from the previous iteration,
$\bar{\mathbf{X}}_{\mathbf{k-1}}$ is the state estimate from the previous iteration, and
λ is the scaling parameter.
$2n$+1 sigma points are selected to approximate the posterior mean and covariance of the state vector. The selection of the sigma points is deterministic and is set by adjusting the scaling parameter λ:

$$\lambda = \alpha^2(n+\kappa) - n \tag{3}$$

where:
α determines the distribution of the points around the mean and is set to a small positive value, and
κ is a secondary parameter set to 0 or $3 - n$.
Equation 1 is applied to $\mathbf{X_k^i}$ to obtain the propagated state vector $\hat{\mathbf{X}}_{\mathbf{k}}^{\mathbf{i}}$:

$$\hat{\mathbf{X}}_{\mathbf{k}}^{\mathbf{i}} = f(\mathbf{X_k^i}, \mathbf{R_k^i}) \tag{4}$$

The mean $\bar{\hat{\mathbf{X}}}_{\mathbf{k}}$ and the covariance $\hat{\mathbf{P}}_{\mathbf{k}}$ of the propagated sigma points are computed:

$$\bar{\hat{\mathbf{X}}}_{\mathbf{k}} = \sum_{i=0}^{2n} W_i \hat{\mathbf{X}}_{\mathbf{k}}^{\mathbf{i}} \tag{5}$$

$$\hat{\mathbf{P}}_{\mathbf{k}} = \sum_{i=0}^{2n} W_i \left[\hat{\mathbf{X}}_{\mathbf{k}}^{\mathbf{i}} - \bar{\hat{\mathbf{X}}}_{\mathbf{k}}\right] \left[\hat{\mathbf{X}}_{\mathbf{k}}^{\mathbf{i}} - \bar{\hat{\mathbf{X}}}_{\mathbf{k}}\right]^T \tag{6}$$

The weight W_i is computed according to the following:

$$\begin{aligned} W_0 &= \{\lambda/(n+\lambda)\} + (1-\alpha^2+\beta) \\ W_i &= 1/\{2(n+\lambda)\} \qquad\qquad i = 1,2,...,2n. \end{aligned} \tag{7}$$

β is used to include information about the distribution of the state variable. It is found to be optimal at $\beta = 2$ for a Gaussian distribution.
Likewise, the observation vector is propagated using the propagated sigma points:

$$\hat{\mathbf{Y}}_{\mathbf{k}} = h(\hat{\mathbf{X}}_{\mathbf{k}}, \mathbf{S_k}) \longrightarrow \hat{\mathbf{Y}}_{\mathbf{k}} \approx \left\{\hat{\mathbf{Y}}_{\mathbf{k}}^{\mathbf{0}}, \hat{\mathbf{Y}}_{\mathbf{k}}^{\mathbf{1}}, \hat{\mathbf{Y}}_{\mathbf{k}}^{\mathbf{2}}, \ldots, \hat{\mathbf{Y}}_{\mathbf{k}}^{\mathbf{2N}}\right\} \tag{8}$$

where:
h describes the nonlinear observation function,
$\hat{\mathbf{Y}}_{\mathbf{k}}$ is the propagated observation vector,
$\hat{\mathbf{X}}_{\mathbf{k}}$ is the propagated state vector, and
$\mathbf{S_k}$ is the measurement noise covariance.
Then $\bar{\hat{\mathbf{Y}}}_{\mathbf{k}}$, the mean of the propagated observation vector, and its covariance $\hat{\mathbf{P}}_{\mathbf{yy_k}}$ are calculated using the same weights defined in Equation 7:

$$\bar{\hat{\mathbf{Y}}}_{\mathbf{k}} = \sum_{i=0}^{2n} W_i \hat{\mathbf{Y}}_{\mathbf{k}}^{\mathbf{i}} \tag{9}$$

$$\hat{\mathbf{P}}_{\mathbf{yy_k}} = \sum_{i=0}^{2n} W_i \left[\hat{\mathbf{Y}}_{\mathbf{k}}^{\mathbf{i}} - \bar{\hat{\mathbf{Y}}}_{\mathbf{k}}\right] \left[\hat{\mathbf{Y}}_{\mathbf{k}}^{\mathbf{i}} - \bar{\hat{\mathbf{Y}}}_{\mathbf{k}}\right]^T \tag{10}$$

Up until this point is the time update block of Fig. 1.

The succeeding steps are part of the measurement update. The observation vector $\mathbf{Y_k}$ is obtained from sensor measurements. Then the cross-covariance of the state and the observation vectors, $\hat{\mathbf{P}}_{\mathbf{xy_k}}$, is calculated in order to derive the Kalman gain $\mathbf{K_k}$.

$$\mathbf{P}_{\mathbf{xy_k}} = \sum_{i=0}^{2n} W_i \left[\hat{\mathbf{X}}_{\mathbf{k}}^{\mathbf{i}} - \bar{\hat{\mathbf{X}}}_{\mathbf{k}}\right] \left[\hat{\mathbf{Y}}_{\mathbf{k}}^{\mathbf{i}} - \bar{\hat{\mathbf{Y}}}_{\mathbf{k}}\right]^T \tag{11}$$

$$\mathbf{K_k} = \mathbf{P}_{\mathbf{xy_k}} \hat{\mathbf{P}}_{\mathbf{yy_k}}^{-1} \tag{12}$$

Finally, the state and covariance estimates are updated:

$$\bar{\mathbf{X}}_{\mathbf{k}} = \bar{\hat{\mathbf{X}}}_{\mathbf{k}} + \mathbf{K_k}(\mathbf{Y_k} - \bar{\hat{\mathbf{Y}}}_{\mathbf{k}}) \tag{13}$$

$$\mathbf{P_k} = \hat{\mathbf{P}}_{\mathbf{k}} - \mathbf{K_k} \hat{\mathbf{P}}_{\mathbf{yy_k}} \mathbf{K_k}^T \tag{14}$$

where $\bar{\mathbf{X}}_{\mathbf{k}}$ is the state estimate, and $\mathbf{P_k}$ is its covariance at time k. These values become the input to the next iteration, i.e., $\bar{\mathbf{X}}_{\mathbf{k}}$ becomes $\bar{\mathbf{X}}_{\mathbf{k-1}}$ and $\mathbf{P_k}$ becomes $\mathbf{P_{k-1}}$. Then the whole process repeats again.

Upon initialization of the filter, $\bar{\mathbf{X}}_{\mathbf{k-1}}$ and $\mathbf{P_{k-1}}$ in Equations 1 and 2 are set to some initial values and become $\bar{\mathbf{X}}_{\mathbf{0}}$ and $\mathbf{P_0}$, respectively. The scaling parameter values are adjusted heuristically.

For further discussion and details on the implementation of UKF, consult Julier & Uhlmann (Julier & Uhlmann, 1997) and Wan & Van der Merwe (Wan & van der Merwe, 2000).

4. Hand pose estimation using multi-viewpoint cameras

The vision-based hand pose estimation system takes its input from multiple cameras, which are positioned so that they see with the least amount of occlusion (Fig. 2).

Fig. 2. Multiple viewpoint camera system.

The system is model-based as it uses a skeletal model of the hand (Fig.3). It represents the hand as a set of five manipulators with a common base point at the wrist. Each finger is a manipulator with several links and joints. The metacarpophalangeal (MCP) and carpometacarpal (CMC) joints have two DOFs each to account for flexion-expansion and abduction-adduction motions. The rest of the joints has one DOF each. The thumb has a special configuration. It has only 4 joints and 4 links, for a total of 5 DOFs. The wrist, which accounts for the global pose, has six DOFs for translation and rotation. The model has a skin composed of quadric surfaces, which will be referred to as surface model throughout this paper. The surface model represents the underlying skeletal configuration of the links and the joints. In summary, the skeletal model has 19 joints, 31 DOFs, 24 links and a total of 744 surface quadrics.

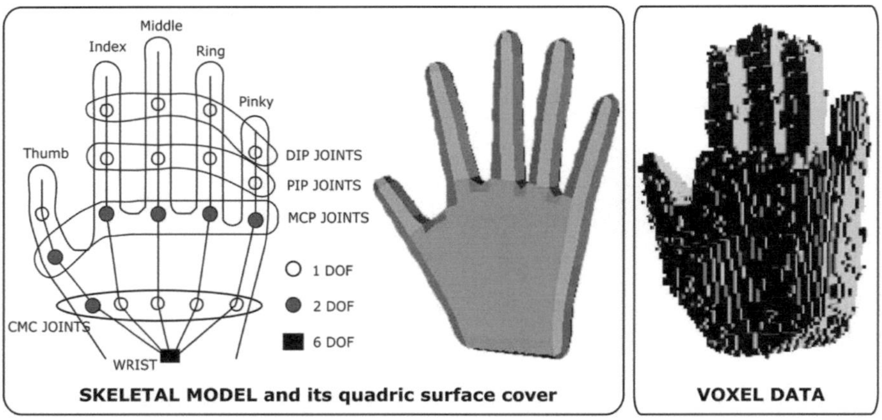

Fig. 3. The hand models. The skeletal model is covered with quadric surfaces. The voxel data is derived from the silhouettes of the input images.

The observation data of the system is the images from the cameras converted to voxel data by shape-from-silhouette technique (Szeliski, 1993). In other words, the voxel data represents the current hand pose as seen by the cameras.

Ueda *et al.*, minimized the error between the voxel data and the skeletal model using virtual force (Ueda et al., 2003). Although their technique has the advantages of being simple and fast, it cannot estimate the global and local poses simultaneously. It also has difficulty in recovering from erroneous estimation.

In our proposed approach, we estimated the global and the local parameters simultaneously using UKF. Instead of being limited to finger movements, it will also allow the palm's rotation and the wrist's translation to be estimated. This enables the hand pose estimation system to accept a more dynamic hand motion as input.

5. UKF in hand pose estimation

We present in this section how we used UKF to estimate the hand pose. We chose UKF over EKF or particle filter because of its simple implementation, fewer number of particles needed, and accuracy of up to the second order (Julier & Uhlmann, 1997). Moreover, for our system, the relationship between the observation data and the hand pose is non-linear.

Figure 4 illustrates the process of using UKF to estimate the skeletal pose of the hand using the voxel data as input. The step by step explanation is as follows:

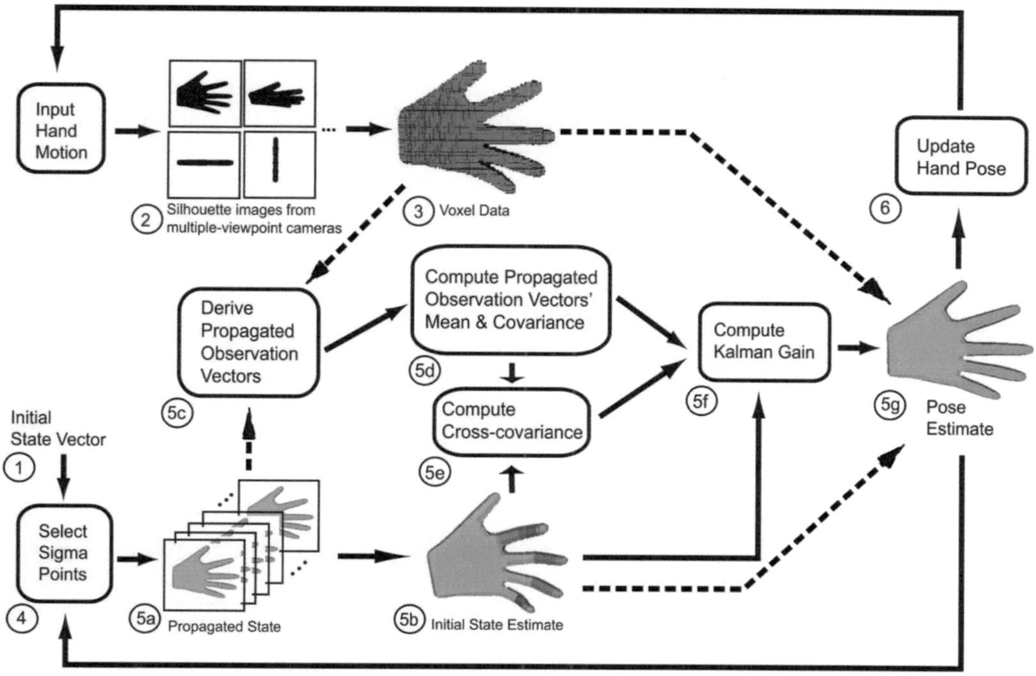

Fig. 4. Details of the proposed method. Dashed lines indicate comparison of models in order to obtain error measurements.

1. The state vector is set to $\bar{X}_{k-1}$ while the state covariance is set to $\mathbf{P}_{k-1}$. During initialization, the state is set to zero ($\mathbf{X}_0$) while the state covariance is set to some value ($\mathbf{P}_0$).
2. The color image inputs from the multiple cameras are converted to silhouettes.
3. Using shape-from-silhouette approach, the silhouette images are converted to voxel data.
4. Sigma points are selected using $\bar{X}_{k-1}$ and $\mathbf{P}_{k-1}$.
5. Hand pose is estimated using UKF:
   a. Apply Equation 1, the state dynamics equation, to the sigma points X^i_k. This gives the propagated state vectors $\hat{X}^i_k$, illustrated as variations of hand poses.
   b. Calculate the mean value of the propagated state vectors $\bar{\hat{X}}_k$ and its covariance $\hat{P}_k$. The $\bar{\hat{X}}_k$ is the filter's initial state estimate.
   c. Propagate the observation vector $\hat{Y}^i_k$ by computing the error between the propagated state $\hat{X}^i_k$ and the voxel data.
   d. Calculate $\bar{\hat{Y}}_k$, the mean value of the propagated observation vectors, and its covariance $\hat{\mathbf{P}}_{yy_k}$.
   e. Calculate the cross covariance $\hat{\mathbf{P}}_{xy_k}$.
   f. Compute Kalman gain $\mathbf{K}_k$ using Equation 12.
   g. Compute state estimate $\bar{X}_k$ and its covariance $\mathbf{P}_k$. The hand pose estimate is defined by $\bar{X}_k$.
6. Update the hand pose. $\bar{X}_k$ and $\mathbf{P}_k$ become the next iteration's $\bar{X}_{k-1}$ and $\mathbf{P}_{k-1}$, respectively.
7. The process repeats from Step 2 for the next iteration.

5.1 The state dynamics and composition of the state vector

A key factor in using a predictive filter is using the correct state dynamics. For the hand pose estimation, we used a second order dynamics or constant acceleration model, which describes the change in the state vector over time. It also captures the nature of the hand motion better than a constant velocity model does.

In Equation (15), $\mathbf{X_k}$ is the state vector at time $\mathbf{k}$, Δt is the time interval between frames, $\mathbf{V_k}$ is the noise covariance of the state vector, and $\mathbf{I}$ is the identity matrix. The state noise covariance accounts for all the disturbances not accounted for by the dynamics; it was determined heuristically in the experiments. The uncertainties of the dynamics are modeled to be independent for the position, velocity, and acceleration components.

$$\mathbf{X_k} = \begin{bmatrix} \mathbf{I} & \Delta t & \frac{1}{2}\Delta t^2 \\ 0 & \mathbf{I} & \Delta t \\ 0 & 0 & \mathbf{I} \end{bmatrix} [\mathbf{X_{k-1}}] + [\mathbf{V_k}] \tag{15}$$

The state vector $\mathbf{X}$ is composed of both global (rotation and translation) and local (finger joint angles) pose parameters and their respective first and second order derivatives (velocity and acceleration). In Equation 16, θ_n is either a global or local parameter, $\dot{\theta}_n$ is its velocity, and $\ddot{\theta}_n$ its acceleration.

$$\mathbf{X} = \left[\theta_1, \theta_2, \ldots, \theta_n, \dot{\theta}_1, \dot{\theta}_2, \ldots, \dot{\theta}_n, \ddot{\theta}_1, \ddot{\theta}_2, \ldots, \ddot{\theta}_n\right]^T \tag{16}$$

5.2 The observation function and composition of the observation vector

The observation function of our system is the non-linear process of obtaining the observation vector given a hand pose configuration. A given state vector, $\mathbf{X}$, specifies a combination of finger joint angles and wrist data that can be represented by the hand model as a particular pose. After converting the state vector to a hand model's pose, we obtain the error between the voxel data (Step 3 of Fig. 4) and the hand model. We designed the observation vector to contain the error measurement between the voxel data (the observed hand pose) and the hand model (the surface model).

The observation vector is composed of the geometric distance measured between the voxel data and the hand surface model. The distance is computed by checking whether each quadric Q_i of the surface model is located inside or outside of the voxel data. If it is outside, the Manhattan distance d_i between the center of the quadric and the nearest voxel is measured. If it is inside, d_i is set to zero. It is repeated for all the quadrics of the surface model. Figure 5 illustrates the process.

The distance values are stacked to form the observation vector $\mathbf{Y}$:

$$\mathbf{Y} = [\ldots, d_{i-1}, d_i, d_{i+1}, \ldots]^T \tag{17}$$

In order to lessen the computation time, the size of the observation vector can be decreased. In our experiments, instead of obtaining distance measurements from all quadrics (744 in total), we only sampled 140 quadrics. It made the computation time more manageable.

Additionally, at every time step, we always take distance measurements from the same set of quadrics. We are able to follow the motion of the finger links from one time frame to

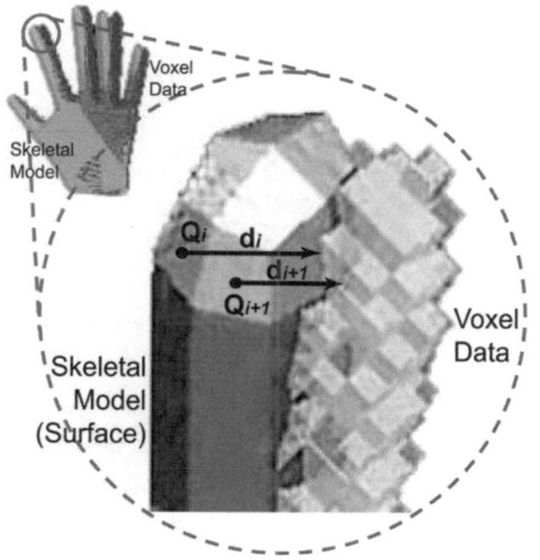

Fig. 5. The calculation of geometric error between surface model and voxel data.

another by keeping track of the changes in the distance values of the selected quadrics. That way, we can say that a sense of direction is encoded in the observation vector. Thus, the observation vector contains the magnitude of change of the finger link's motion, as well as its general sense of direction.

To interpret the observation vector, a zero error between the model and the observation (i.e., $\mathbf{Y} = 0$) implies that the hand model is completely inside the voxel data. Some value in the observation vector indicates that the fingers have moved in certain direction.

Finally, in Equation 13, $\mathbf{Y}_k$ is always set to zero, based on two things. First, comparing the voxel data to itself and computing distance measurements will just yield zero. Second, from the perspective of the filter, $\mathbf{Y}_k = \mathbf{0}$ can be interpreted to mean that the observation (sensor) measurement is not completely reliable and that we have to correct it through the $\mathbf{K}_k(\mathbf{Y}_k - \bar{\hat{\mathbf{Y}}}_k)$ term of Equation 13.

5.3 Initialization and filter tuning

Filter fine tuning and proper parameter initialization are important tasks when incorporating a predictive filter to a motion tracking solution. As mentioned above, the state vector is set to zero value ($\mathbf{X}_0$) at the initial step. The zero values assigned to the state parameters mean that the hand model is at its initial pose. The hand is said to be at initial pose when the palm is flat open and the fingers are extending away from the palm. Likewise, the state covariance matrix's diagonal is set to some value ($\mathbf{P}_0$).

The fine-tuning parameters of λ (see Equation 3) were also determined heuristically. For example, α was set to a small value between 1 and $1x10^{-4}$, κ was set to $(3 - n)$, and β was set to 2. Likewise, selection of the noise covariances $\mathbf{R}_k$ and $\mathbf{S}_k$ is also critical.

6. Experimental results and discussion

For all the experiments we have done, we used eight cameras in order to get finer voxel data. Additionally, the voxel resolution used was 2×2×2[mm] per octant or voxel unit. We also estimated a total of 15 hand pose parameters: 3 global and 12 local. The global parameters are roll, pitch and yaw. The local parameters are the 2 DOFs of the MCP and 1 DOF of the PIP. The following constraint gives the value of the DIP joint angle relative to the PIP:

$$\theta_{DIP} = \frac{2}{3}\theta_{PIP}. \tag{18}$$

The proposed method was tested on several hand motions. Various hand motion data were obtained using a dataglove. These data, which we considered as the ground truth for all our experiments, were then used to create virtual versions of the different hand motions. These virtual motions were used as input to the pose estimation system and tracked. Proper initialization of the hand model and the voxel data (i.e., they must overlap initially) is necessary for filter convergence. Fortunately, the use of simulated motion eliminated this issue.

Figures 6, 7, and 8 show a hand motion that has been tracked successfully. The motion (Motion A) is that of a hand whose wrist is rotating and twisting, while the fingers (with the exception of the thumb) are simultaneously closing slowly. This motion involves three global and 12 local parameters. The wrist's roll, pitch, and yaw (see Fig. 6) and the four fingers' PIP (1 DOF) and MCP (2 DOFs) were estimated with good accuracy. Figure 7 shows only the MCP's expansion-flexion data (left column) and the PIP (right column).

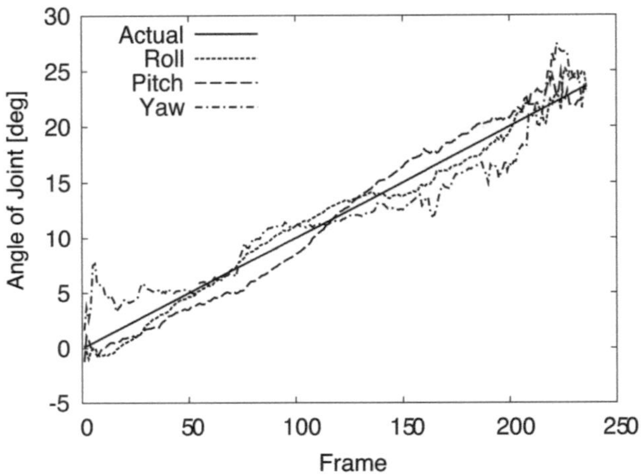

Fig. 6. Global estimation result when the fingers are closing simultaneously. Solid line is the ground truth values (actual); broken lines are the estimate (roll, pitch, yaw).

For Fig. 6 and Fig. 7, the black solid line is the ground truth value while the dotted and dashed lines are the estimate values. For all the fingers, the filter initially shows estimation errors by as much as 10 degrees, although it eventually converges to the desired value. The filter also gets lost but tries to get back on track. This can be seen as a noisy estimation in the pinky's MCP joint estimation (Fig.7 left side, top graph). We had to implement range constraints on the finger motion to ensure that awkward poses, for example fingers bending backward too much, do not happen. This can be seen as a plateau on the pinky's PIP estimation graph (Fig.7 right side, top graph).

Snapshots of the motion described above are shown in Fig.8. The top row is the virtually-generated motion and the bottom row is the result of the pose estimation. The numbers above each column of image correspond to the points in Fig. 7 when the images were taken. The local motion manifests in the images as the closing and opening of the fingers, while the global motion shows as the twisting of the wrist and palm.

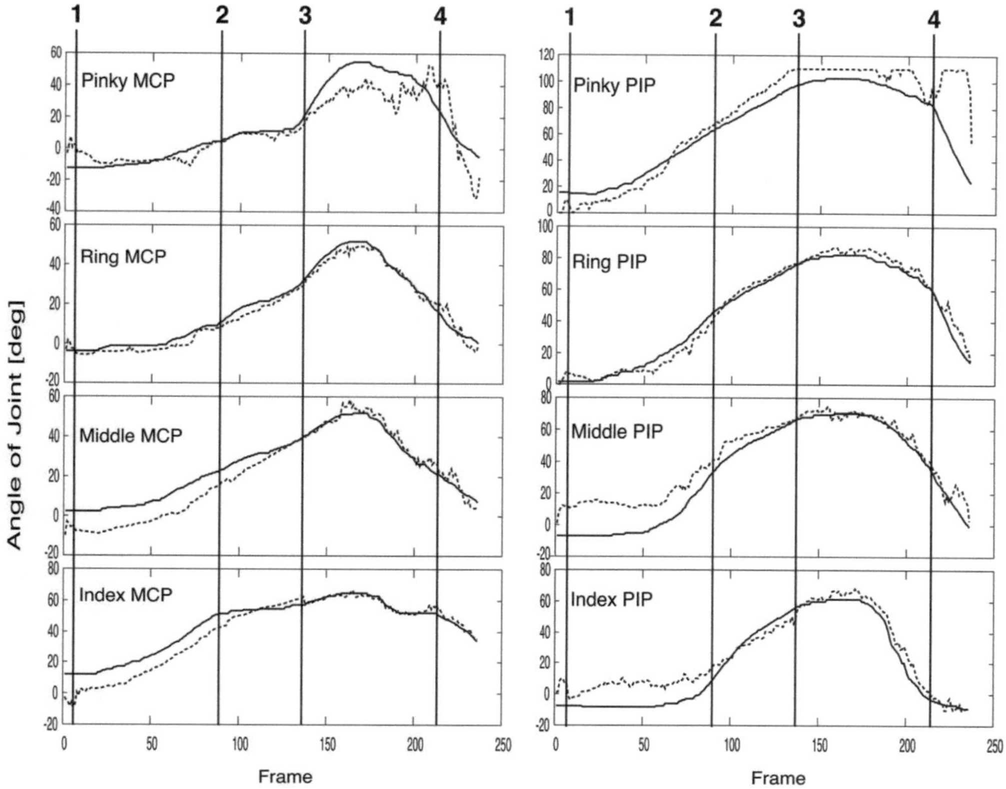

Fig. 7. MCP and PIP estimation results for fingers closing simultaneously while the wrist is rotating. Solid lines are the ground truth values; the dotted lines are the pose estimation result. The numbered vertical lines show when the snapshots in Fig. 8 were taken.

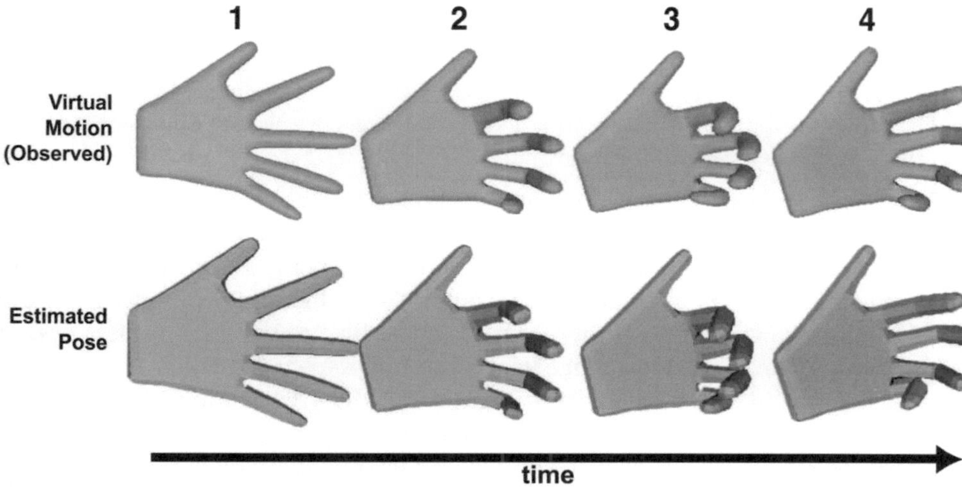

Fig. 8. Snapshots of estimation result. The numbers above each image column correspond to the points in Fig. 7 when the snapshots were taken. The motion is for a rotating wrist while the fingers are closing simultaneously.

Two more motions were tested to demonstrate the flexibility of the system. Snapshots of the estimation results are shown in Fig.9 (Motion B) and Fig.10 (Motion C). For both motions, the wrist is rotating and twisting due to roll, pitch, and yaw motions. In Fig.9, the hand is moving two fingers at a time. In Fig.10, the fingers successively bending towards the palm one by one, starting from the pinky toward the index finger and then opening in the reverse order.

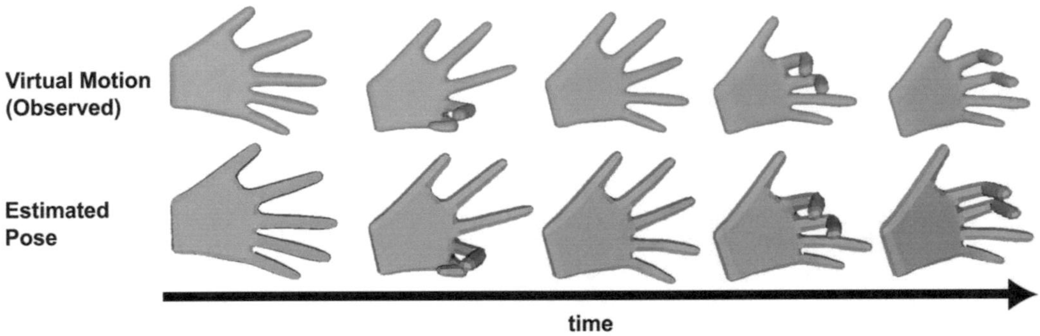

Fig. 9. Snapshots of the observed hand motion and their corresponding estimated hand poses. The motion is that of a hand rotating while the fingers are closing two at a time.

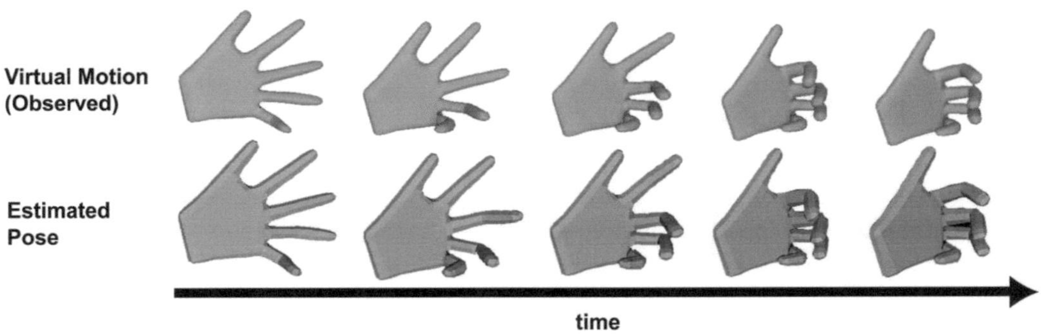

Fig. 10. Snapshots of the observed hand motion and their corresponding estimated hand poses. The motion is that of a hand rotating while the fingers are closing one at a time starting from the pinky going to the index.

To compare the accuracy of our estimation results, Fig.11 shows the average of absolute errors for all the joints estimated. The absolute errors range from 0.20 to 3.40 degrees per joint for every iteration. However, the actual change of angle per iteration of any joint, based on the ground truth data, is less than 1 degree only. We can interpret this range of absolute error as an indication of the filter's effort to converge to the ground truth value. Physically speaking, even a three degree motion of a joint is not easy to perceive due to the presence of the muscle and the skin covering the finger bones. Thus, the converging behavior is noticeable in the graphs of Fig.6 and 7 but imperceptible in the snapshots of Fig.8.

Furthermore, we compared our results with the original model-fitting approach's in (Ueda et al., 2003); a predictive filtering versus model-fitting comparison. Fig.12 establishes the robustness of using the UKF against using the virtual force based model-fitting. The figure shows the estimation result of both methods for the Index PIP joint. Both methods try to converge to the true value, but a closer look shows that the model-fitting has more difficulty in doing so. Between frames 100 to 200, the Index PIP is expanding and flexing (i.e., bending and stretching), and the UKF is able to track this movement quite well. The filter's estimation results fluctuate as it tries to converge to the true value yet manages to recover from the fluctuations. On the other hand, it takes some time for the model-fitting approach to recover from its over-estimation and overshoots its estimates. In short, the proposed method showed better error recovery than the model-fitting method.

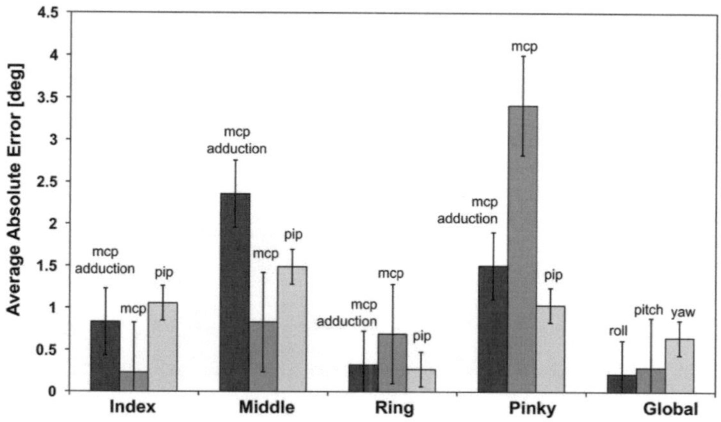

Fig. 11. Average absolute error of each DOF. The motion is that of a hand rotating while the fingers are closing simultaneously.

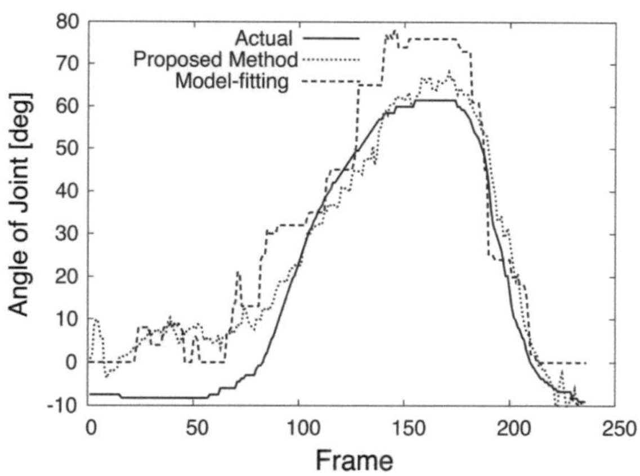

Fig. 12. Comparison of Index PIP estimation results of the original model fitting approach and the proposed method.

There are several issues to address when implementing a predictive filter in hand pose estimation. First is the composition of the state and observation vectors, more importantly, its size. In our experiments, the dimension of the state vector was 45: the 15 hand parameters (3 global + 12 local) and their respective first and second order derivatives while the observation vector's was 140. The size of the observation vector was adjusted until the optimum size is attained. A trade off between size and computation speed is needed here. If the observation vector is too small, there would not be enough information for the filter to process, but too big a size and the computation time increases considerably.

For the state vector, the size is largely determined by the dynamics model of the system. Since we chose a constant acceleration dynamics, we had to incorporate the first and second derivatives of the state variables in the state vector. Fortunately, inclusion of known hand constraints can help lessen the dimensions of the state vector. For example, we used the coupling constraint between the PIP and the DIP (Equation 18), thus shrinking the state vector by nine parameters.
The second important consideration in UKF is the noise covariance of the state (Equation 1) and the observation (Equation 8) vectors. The stability and convergence of the filter depend on the accurate choice of covariances (Xiong et al., 2006). In our case, all the covariances, listed in Table 1 were determined heuristically. The same noise covariances were applied to all the motions discussed here. Likewise, noise covariances for the observation measurement were also determined heuristically. We used different noise covariances for the different motions (see Table 2).

State Parameter	Covariance Value
θ	0.1
$\dot{\theta}$	0.01
$\ddot{\theta}$	0.001

Table 1. Covariance values used for the state vector.

Hand Motion	Covariance Value
Motion A (Fig.8)	0.001
Motion B (Fig.9)	0.1
Motion C (Fig.10)	0.1

Table 2. Covariance values used for the observation vector.

Lastly, the filter's computation speed is another important consideration. As mentioned before, for the UKF, computation speed depends largely on the size of the state vector and the observation vector. Minimizing either or both can result in faster computations, which in turn leads to a more stable and accurate filtering. Modifications to UKF, or its equivalent methods, to further lessen the number of sigma particles from $2n + 1$ have already been reported in the literature. For example, Julier *et al.* used only $n + 1$ number of particles (Julier & Uhlmann, 2002). La Viola compared the performance of EKF and UKF in head tracking and found that using quaternions to encode the joint angles resulted to better estimation, even by just using EKF (La Viola, 1996).
In our experiments, the computation speed of the filter is around 0.87 seconds for every iteration or roughly 1Hz. However, the usual frame capture speed of cameras is around 30Hz. Thus, there is a need to speed up the proposed method.

7. Conclusion and future work

We introduced a predictive filter, Unscented Kalman Filter, to a vision-based model-based system in order to estimate the global and local poses of the hand simultaneously. The UKF minimizes error between the hand model and the voxel data and computes the initial pose estimate by propagating $2n + 1$ sigma particles. We were able to show estimation results for up to 3 global and 12 local pose parameters in different motions and demonstrate better error recovery than a previous pose estimation technique. The results presented in this paper used virtually generated motion obtained from actual hand motion to verify our method.

Our future work includes the implementation of the proposed method in a real camera system and the use of a calibrated hand model. Moreover, an adaptation of the original UKF technique to the hand dynamics is necessary in order to speed up the computation and improve the accuracy and over-all stability of the filtering process.

8. References

Athitsos, V. & Sclaroff, S.J. (2003). Estimating 3D Hand Pose from a Cluttered Image, *Proc. of the IEEE Computer Society Conf. on Computer Vision and Pattern Recognition*, Vol. 2, pp. 432-442, Madison,WI, USA, Jun 2003

Azoz, Y.; Devi, L. & Sharma, R. (1998). Tracking Hand Dynamics in Unconstrained Environments, *Proc. of Third Int. Conf. on Automatic Face & Gesture Recognition*, pp. 274-279, Nara, Japan, Apr 1998

Bray, M.; Koller-Meir, E., Müller, P., Gool, L.V. & Schraudolph, N.N. (2004). 3D Hand Tracking by Rapid Stochastic Gradient Descent using a Skinning Model, *Proc. of the First European Conf. on Visual Media Production*, pp. 59-68, London, 2004

Causo, A.; Ueda, E., Kurita, Y., Matsumoto, Y. & Ogasawara, T. (2008). Model-based Hand Pose Estimation using Multiple View-point Images and Unscented Kalman Filter, *Proc. of the Seventeenth International Symposium Robot and Human Interactive Communication (RO-MAN 2008)*, pp. 291-296, Munich, Germany, Aug 2008

Causo, A.; Matsuo,M., Ueda, E., Takemura, K., Matsumoto, Y., Takamatsu, J. & Ogasawara, T. (2009). Hand Pose Estimation using Voxel-based Individualized Hand Model. *Proc. of the 2009 IEEE/ASME Int. Conf. on Advanced Intelligent Mechatronics*. pp. 451-456. Singapore, Jul 2009

Delamarre, Q. & Faugeras, O. (1999). 3D Articulated Models and Multi-view Tracking with Silhouettes, *Proc. of the Seventh IEEE Int. Conf. on Computer Vision*, Vol. 99, pp. 716-721, Kerkyra, Greece, Sep 1999

Erol, A.; Bebis, G., Nicolescu M., Boyle, R.D.& Twombly, X. (2007). Vision-based Hand Motion Estimation: A Review, *Comput. Vis. Image Underst.*, Vol. 108, No. 1-2 (Oct 2007) pages (52-73)

Gumpp, T.; Azad, P., Welke, K., Oztop, E., Dillmann, R. & Cheng, G. (2006). Unconstrained Real-time Markerless Hand Tracking for Humanoid Interaction, *Proc. of Sixth IEEE/RAS Int. Conf. on Humanoid Robots*, pp. 88-93, Genova, Italy, Dec 2006

Huang, C.L. & Jeng, S.H. (2001). A Model-based Hand Gesture Recognition System, *Machine Vision and Applications*, Vol. 12, No. 5 (Mar 2001) pages 243-258

Julier, S.J. & Uhlmann, J.K. (1997). A New Extension of the Kalman Filter to Nonlinear Systems, *Proc. of Conf. on Signal Processing, Sensor Fusion, and Target Recognition*, pp. 182-193, Orlando, FL, 21-24 Apr 1997

Julier, S.J. & Uhlmann, J.K. (2002). Reduced Sigma Point Filters for the Propagation of Means and Covariances through Non-linear Transformations, *Proc. of 2003 American Control Conf.*, pp. 887-892, Anchorage, AK, USA, 8-10 May 2002

Kuch, J.J. & Huang, T.S. (1994). Vision-based Hand Modeling and Tracking: A Hand Model, *Proc. of Twenty-Eighth Asilomar Conf. on Signal, Systems and Computers*, pp. 1251-1256, 31 Oct - 2 Nov 1994

La Viola Jr., J.J. (1996). A Comparison of Unscented and Extended Kalman Filtering for Estimating Quaternion Motion, *Proc. of American Control Conf.*, Vol. 3, pp. 2435-2440, Denver, CO, USA, Jun 2003

Lien, C.C. & Huang, C.L. (1998). Model-based articulated hand motion tracking for gesture recognition. *Image and Vision Computing*, Vol. 16, No. 2, (Feb 1998) page numbers (121-134)

Lin, J.; Wu, Y. & Huang, T.S. (2002). Capturing Hand Motion in Image Sequences, *Proc. of IEEE Workshop on Motion and Video Computing*. Orlando, FL, pp. 99-104, Dec 2002

Pavlovic, V.; Sharma, R. & Huang, T. (1997). Visual Interpretation of Hand Gestures for Human-Computer Interaction: A Review. *IEEE Transactions on Pattern Analysis and Machine Intelligence*, Vol. 19, No. 7, (July 1997) page numbers (677-695)

Rehg, J.M. & Kanade, T. (1994). DigitEyes: Vision-based Hand Tracking for Human-Computer Interaction, *Proc. of IEEE Workshop on Motion of Non-Rigid And Articulate Objects*, pp. 16-22, Austin, TX, USA, Nov 1994

Shimada, N.; Shirai, Y., Kuno, J., & Miura, J. (1998). Hand Gesture Estimation and Model Refinement using Monocular Camera - Ambiguity Limitation by Inequality Constraint, *Proc. of the Third IEEE Int. Conf. on Face and Gesture Recognition*, pp. 268-273, Nara, Japan, Apr 1998

Shimada, N.; Kimura, K. & Shirai, Y. (2001). Real-time 3D Hand Posture Estimation based on 2-D Appearance Retrieval using Monocular Camera, *Proc. of IEEE ICCV Workshop on Recognition, Analysis, Tracking of Faces and Gestures in Real-Time Systems*, pp. 23-30, Vancouver, Canada, Jul 2001

Stenger, B.; Mendonca, P.R.S. & Cipolla, R. (2001). Model based 3D Tracking of an Articulated Hand, *Proc. of the IEEE Computer Society Conf. on Computer Vision and Pattern Recognition*, Vol. 2, pp. 310-315, Hawaii, USA, Dec 2001

Stenger, B.; Thayananthan, A., Torr, P. & Cipolla, R. (2004). Hand Pose Estimation using Hierarchical Detection. *Lecture Notes in Computer Science*, No. 3058, (2004) page numbers (105-116)

Szeliski, R. (1993). Rapid Octree Construction from Image Sequences. *CVGIP: Image Understanding*, Vol. 58, No. 1, (Jul 1993) page numbers (23-32)

Thayananthan, A.; Stenger, B., Torr, P.H.S. & Cipolla, R. (2003). Learning a Kinematic Prior for Tree-based Filtering, *Proc. of British Machine Vision Conf.*, Vol. 2, pp. 589-598, Norwich, UK, Sep 2003

Ueda, E.; Matsumoto, Y., Imai, M. & Ogasawara, T. (2003). A Hand-Pose Estimation for Vision based Human Interfaces. *IEEE Transactions Industrial Electronics*, Vol. 50, No. 4, (Aug 2003) page numbers (676-684)

Utsumi, A. & Ohya, J. (1999). Multiple-hand Gesture Tracking using Multiple Cameras, *Proc. of the IEEE Computer Society Conf. on Computer Vision and Pattern Recognition*, Vol. 1, pp. 473-478, Ft. Collins, CO, USA, Jun 1999

Wan, E. & van der Merwe, R. (2000). The Unscented Kalman Filter for Nonlinear Estimation, *Proc. of the IEEE Adaptive Systems for Signal Processing, Communications, and Control Symp.*, pp. 153-158, Oct 2000

Wu, Y.; Lin, J.Y. & Huang, T.S. (2001). Capturing Natural Hand Articulation, *Proc. of the Eighth IEEE Int. Conf. on Computer Vision*, Vol. 2, pp. 426-432, Kerkyra, Greece, Sep 2001

Xiong, K.; Zhang, H.Y.& Chan, C.W. (2006). Performance Evaluation of UKF-based Nonlinear Filtering. *Automatica*, Vol. 42, No. 2, (Feb 2006) page numbers (261-270)

13

Real Time Facial Feature Points Tracking with Pyramidal Lucas-Kanade Algorithm

F. Abdat, C. Maaoui and A. Pruski
Laboratoire d'Automatique humaine et de Sciences Comportementales, Université de Metz France

1. Intoduction

Facial expression tracking is a fundamental problem in computer vision due to its important role in a variety of applications including facial expression recognition, classification, and detection of emotional states, among others H. Xiaolei (2004). Research on face tracking has been intensified due to its wide range of applications in psychological facial expression analysis and human computer interaction. Recent advances in face video processing and compression have made face-to-face communication be practical in real world applications. However, higher bandwidth is still highly demanded due to the increasing intensive communication. And after decades, robust and realistic real time face tracking still poses a big challenge. The difficulty lies in a number of issues including the real time face feature tracking under a variety of imaging conditions (e.g., skin color, pose change, self-occlusion and multiple non-rigid features deformation) K. Ki-Sang (2007).

Our study aims to develop an automatic facial expression recognition system. This system analysis the movement of the eyebrows, lips and eyes from video sequences, to determine whether a person is happy, sad, disgust or fear.

In this paper, we concentrate our work on facial feature tracking. Our real time facial features tracking system is outlined in figure 1, which is constituted of two important modules:

1. Extract features in facial image, using a geometrical model and gradient projection Abdat et al. (2008).
2. Facial feature points tracking with optical flow (pyramidal Lucas-Kanade algorithm) Bouguet (2000).

The organization of this paper is as follows: in section 2, we will present a face detection algorithm with HAAR-like features. Facial feature points extraction with a geometrical model and gradient projection will be described in section 3. The tracking of facial feature points with Pyramidal Lucas-Kanade will be presented in section 4. Finally the concluding remarks will be given in section 5.

2. Face detection

Face detection is the first step in our facial expression recognition system, which consist to delimit the face area with a rectangle. For this, we have used a modified Viola & Jones's face detector based on the Haar-like features Viola & Jones (2001).

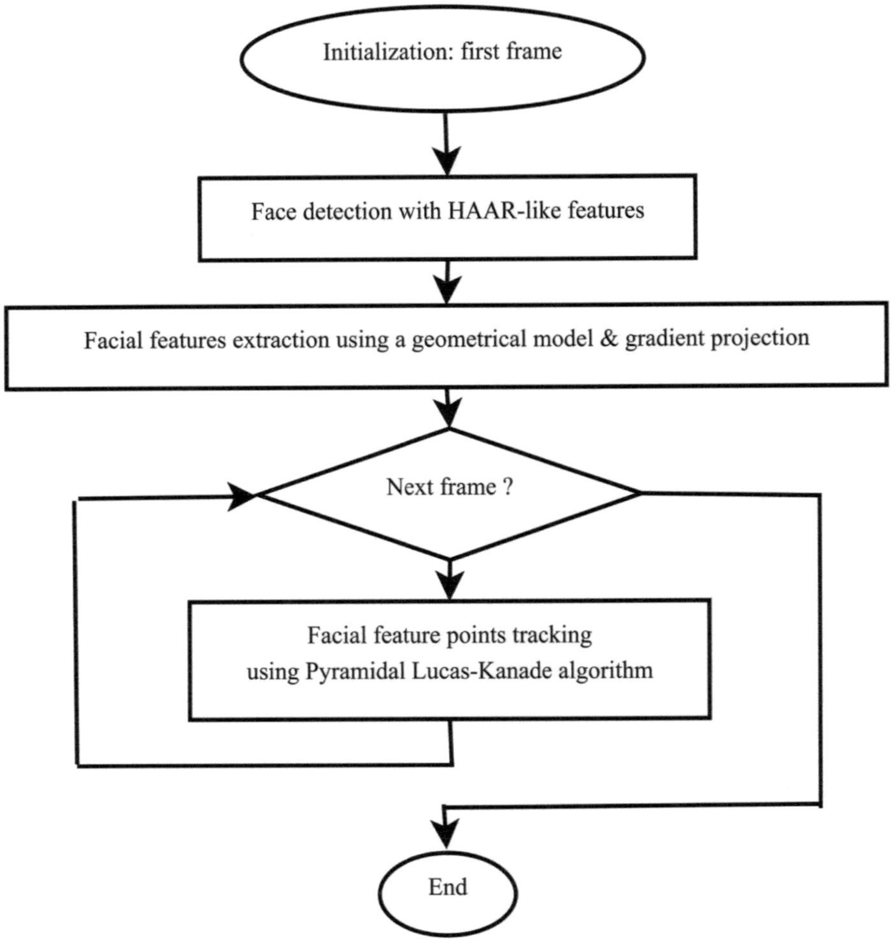

Fig. 1. Real time facial feature points tracking system.

A statistical model of the face is trained. This model is made of a cascade of boosted tree classifiers. The cascade is trained on face and non-face examples of fixed size 24x24. Face detection is done using a retinal approach. A 24x24 sliding window scans the image and each sub-image is classified as face or non-face. To deal with face size the cascade is scaled with a factor of 1.2 by scaling the coordinates of all rectangles of Haar-like features.

2.1 Haar-like features

The pixel value inform us only about luminance and color of a given point. It is therefore more interest to find a detectors based on more global characteristics of the object. This is the case of HAAR descriptors, where the functions allow the knowledge of the contrasts difference between several rectangular regions in image. They encode the existing contrasts in a face and their spatial relationships.
Figure 2 represents the shapes of the used features. Actually, hundreds of features are used as these shapes are applied at different position in the 24x24 retina; a feature is defined by its shape (including its size depending on a scale factor that define the expected face size) and its location.

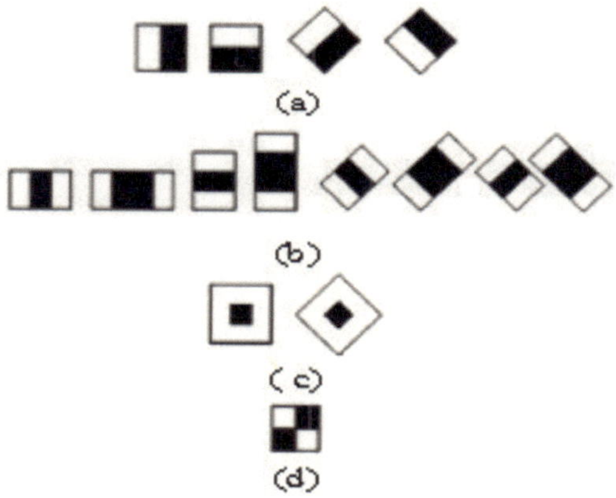

Fig. 2. Haar-like feature extended set.

A feature's value is the weighted sum of pixels over the whole area added to the weighted sum over the dark rectangle R. Belaroussi & Milgram (2006). Absolute value of black area weight is inversely proportional to its area as shown in Figure 3.

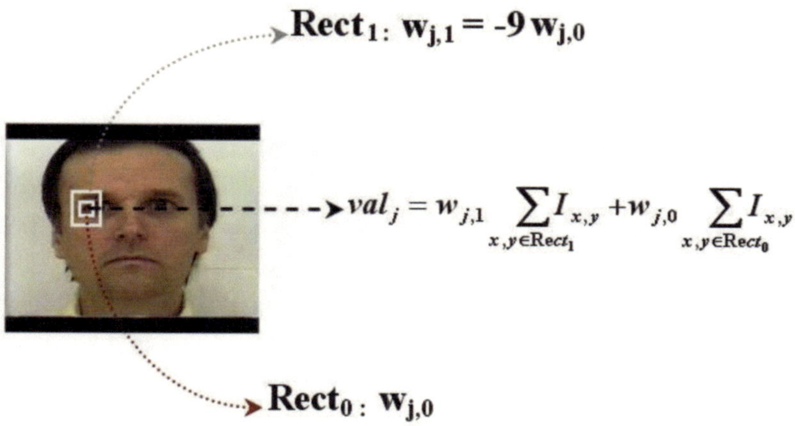

Fig. 3. Shape and location in the search window of the feature j.

2.2 Cascade classifier

A simple decision tree classifier, referred to as "weak" classifier, processes the feature value. A complex classifier

$$F_k = sign(\sum_{i=1}^{n}(c_i f_i)) \quad (1)$$

is iteratively computed as a weighted sum of weak classifiers using a boosting procedure. At each iteration, a weak classifier parameters are trained and a weight c_j is assigned to the weak classifier relatively to its error on the training set. The trained weak classifier is then added to the sum and the training samples weights are updated in order to emphasize the misclassified ones. Finally, an intentional cascade is implemented: it is a cascade of boosted classifiers with increasing complexity. As shown in Figure 4, the simplest classifiers comes first and is intended to reject majority of sub-window before calling more complex classifiers P. Viola & M. Jones (2001).

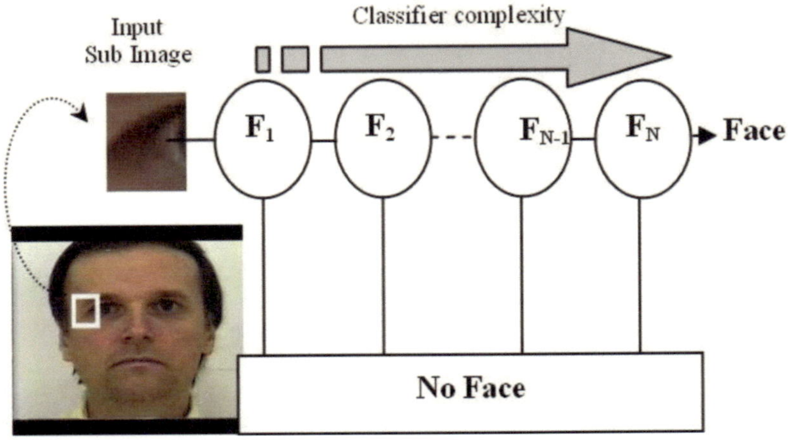

Fig. 4. Cascade of boosted classifiers.

The real-time implementation of this detector using our database, shows that the detector is fast (~10 frames per second) and robust to illumination conditions (Figure 5). However, the detector work hardly when face pose is too slanted. Figure 6 illustrates the limitation of this detector where the bowed face was not detected.

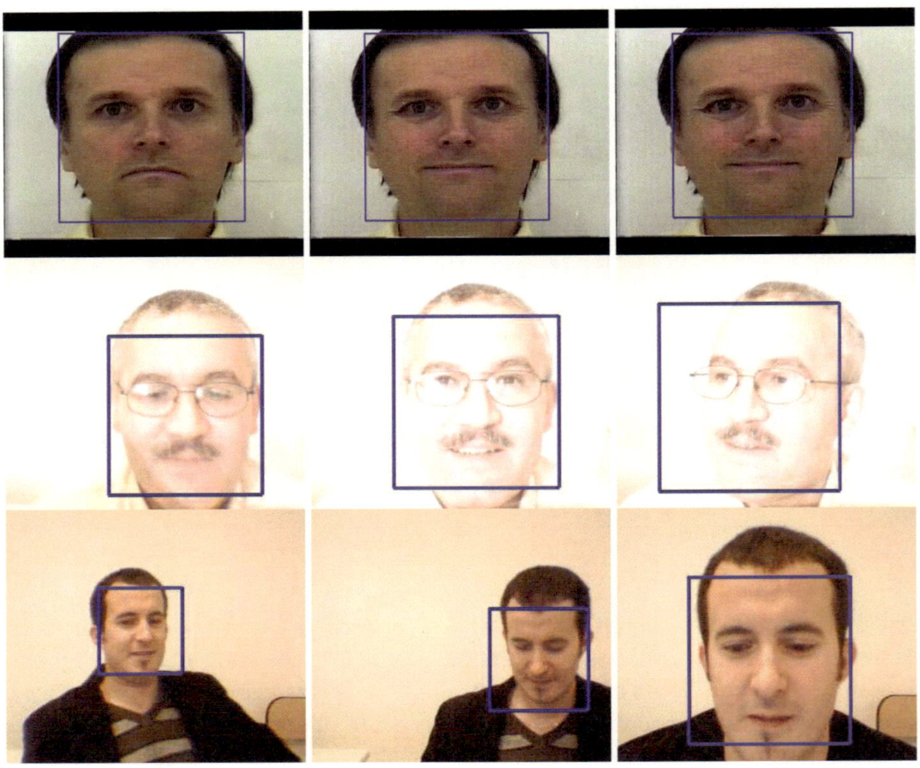

Fig. 5. Face detector.

Fig. 6. Limits of the face detector.

3. Facial feature extraction

After face detection in the first frame, the next step is to extract necessary information about the facial expression presented in the image sequence. When facial muscles contract, the transformation of the corresponding skin areas attached to the muscles produces changes in the appearance of facial features and results in a certain type of visual effect. The movements of facial points (eyebrows, eyes, and mouth) have a strong relation to the information about the shown facial expression. Therefore, many approaches greatly depend on the tracking of permanent facial features (eyebrows, eyes, mouth, and furrows that have become permanent with age) and/or transitional facial features (facial lines and furrows that are not present at a neutral state). In fact, the extraction of facial features is a very challenging task. Facial features cannot always be obtained reliably because of the quality of images, illumination, and some other disturbing factors. Furthermore, it usually takes a lot of computations to extract precise facial features.

3.1 Facial features localization

For facial features localization using the geometric face model, we have used the following stages as in Abdat et al. (2008):

1. Eyes axis location is determined by the maximum of the projection curve which has a high gradient. First we calculate the gradient of the image I:

$$\nabla I_x = \frac{\delta I}{\delta x}\vec{i} \tag{2}$$

   ∇I_x corresponds to the differences in the x (column) direction. The spacing between points in each direction is assumed to be one. Computing the absolute gradient value in each line given by:

$$HI_x(x) = \sum_{y=1}^{n} \nabla I_x(x,y) \tag{3}$$

   Then, we find the maximum value which corresponds to the line contains eyes. This line corresponds to many transitions: skin to sclera, sclera to iris, iris to pupil and the same thing for the other side (high gradient).

2. Median axis location is a vertical line which devises the frontal face in two equal sides. In other words, it is the line passed by the nose. To determine the median axis, we take the median of the bounding box of the face.

3. Mouth axis location is determined as the same way of eyes axis. For the localization of this axis, we look for the maximum value of the projection curve in the low part of the bounding box from eye axis.

Once the eyes and mouth axis are located, we use the geometric face model Shih & Chuang (2004) which suppose that:

- The vertical distance between two-eyes and the center of mouth is D.
- The vertical distance between two-eyes and the center of the nostrils is 0.6D.
- The width of the mouth is D.
- The width of nose is 0.8D.
- The vertical distance between eyes and eyebrows is 0.4D.

Figure 7 shows the results of the facial feature localization for a video sequence and for a real time acquisition. The eyes and the mouth are well located by rectangular windows.

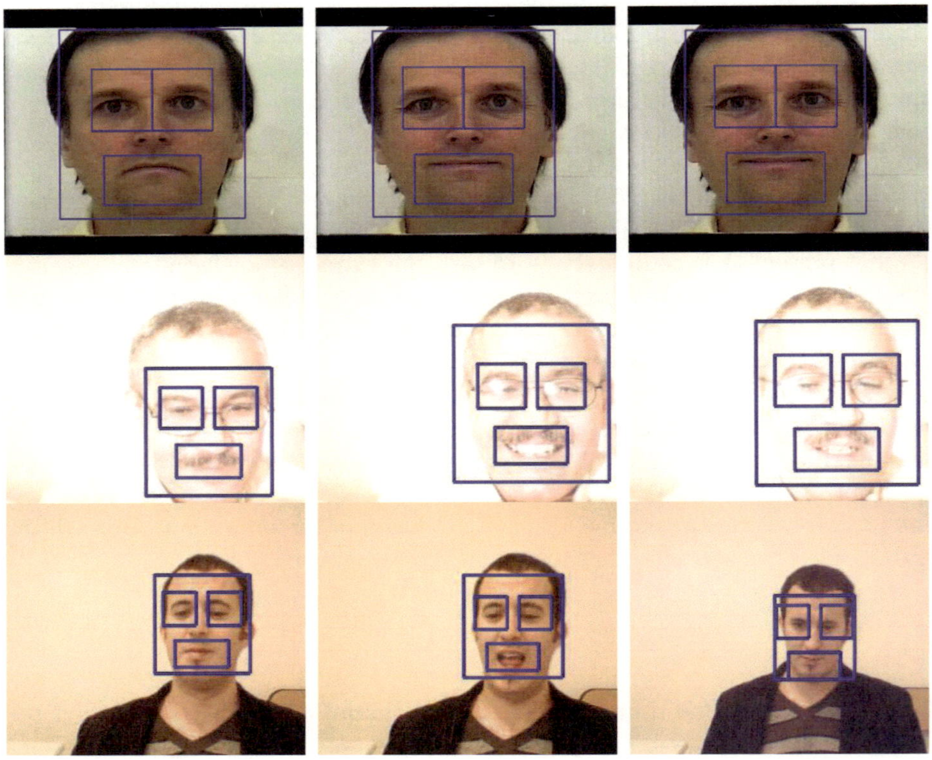

Fig. 7. Facial feature localization

3.2 Facial features points selection

The detected rectangles in the previous step do not give accurate information on facial features. To describe the movement of these features, we detected interest points. As first step, we used the uniform distribution, which consists on sampling the points of the rectangles in the directions of x and y with a step one-fifth $\frac{1}{5}$ of the rectangles size.

Figure 8 illustrates three refined rectangles, while the feature points are uniformly distributed in each rectangle. This selection of feature points is used in Shih & Chuang (2004). After this extraction step, the facial feature points will be tracked using an algorithm of optical flow which is pyramidal Lucas-Kanade tracker.

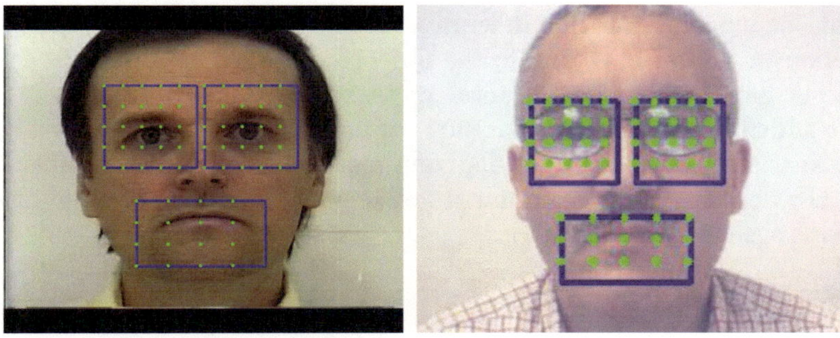

Fig. 8. Uniform distribution selection from the bounding box of facial feature.

4. Optical flow tracking

Optical flow is defined as an apparent motion of image brightness. Let $I(x,y,t)$ be the image brightness that changes in time to provide an image sequence. Two main assumptions can be made Su & Hsieh (2007):

1. Brightness $I(x,y,t)$ smoothly depends on coordinates x, y in greater part of the image.
2. Brightness of every point of a moving or static object does not change in time.

Let some object in the image, or some point of an object, move and after time dt the object displacement is (dx,dy). Using Taylor series for brightness $I(x,y,t)$, we obtain:

$$I(x+dx,y+dy,t+dt) = I(x,y,t) + \frac{\delta I}{\delta x}dx + \frac{\delta I}{\delta y}dy + \frac{\delta I}{\delta t}dt + \ldots \tag{4}$$

where "..." are higher order terms.
Then, according to assumption 2:

$$I(x+dx,y+dy,t+dt) = I(x,y,t) \tag{5}$$

and

$$\frac{\delta I}{\delta x}dx + \frac{\delta I}{\delta y}dy + \frac{\delta I}{\delta t}dt + \ldots = 0 \tag{6}$$

Dividing equation 6 by dt gives:

$$\frac{\delta I}{\delta t} = \frac{\delta I}{\delta x}\frac{\delta x}{\delta t} + \frac{\delta I}{\delta y}\frac{\delta y}{\delta t} \tag{7}$$

Usually, equation 7 called optical flow constraint equation, where:

$$\tfrac{\delta x}{\delta t} = u \text{ et } \tfrac{\delta y}{\delta t} = v$$

are components of optical flow field $\vec{U}$ in x and y coordinates respectively.
Calculate optical flow returns to calculate for each point in the image the following equation:

$$\frac{\delta I}{\delta t} = u \cdot \frac{\delta I}{\delta x} + v \cdot \frac{\delta I}{\delta y} \tag{8}$$

However, the equation 8 can not determine with a single way the optical flow. The indetermination of optical flow due to the absence of global constraint in the precedent equations, only gradients which are local measures are taken into account. Lucas and Kanade have added new constraints to ensure the uniqueness of the solution. The method of Lucas and Kanade consists to find $\vec{U}$ applying a calculation of least squares to minimize constraint. They define a pre-neighborliness, and they optimize $\vec{U}$ in order to give a solution of the following system for n points:

$$\begin{bmatrix} \frac{\delta I}{\delta x}(p_1) & \frac{\delta I}{\delta y}(p_1) \\ \cdot & \cdot \\ \cdot & \cdot \\ \cdot & \cdot \\ \frac{\delta I}{\delta x}(p_i) & \frac{\delta I}{\delta y}(p_i) \\ \cdot & \cdot \\ \cdot & \cdot \\ \cdot & \cdot \\ \frac{\delta I}{\delta x}(p_n) & \frac{\delta I}{\delta y}(p_n) \end{bmatrix} \cdot \begin{bmatrix} u \\ v \end{bmatrix} = \begin{bmatrix} -\frac{\delta I}{\delta t}(p_1) \\ \cdot \\ \cdot \\ \cdot \\ -\frac{\delta I}{\delta t}(p_i) \\ \cdot \\ \cdot \\ \cdot \\ -\frac{\delta I}{\delta t}(p_i) \end{bmatrix} \quad (9)$$

4.1 Discussion

After feature points extraction using the uniform distribution, we have used pyramidal Lucas-Kanade algorithm to track those points as shown in Figure 9. This algorithm has less computation. So, it is adapted for real time application. A motion, caused by a real moving-face, should be highly correlated in space and time domains. In other words, a moving-face in a video sequence should be seen as the conjunction of several smoothed and coherent observations over time.

Tracking a set of interest points is based on valuation techniques of movement between two consecutive images. To obtain a reliable tracking, it is important that these issues be discriminating in the image. For example, a point in the midst of a region of a uniform image may not be identified precisely because all the neighboring pixels are similar. Also, an interest point is normally a point which has a position in the image with strong bi-directional changes. The points tracking consist to identify a set of N interest points in order to model the interest region, and compute a location of each item according to calculations of optical flow.

Figure 9, shows an example of points tracking. These points are selected using the uniform distribution. It can be noted that from the second image, points began to disperse in arbitrary manner diverging from the correct position.

With the uniform distribution, we have got a bad results because these points haven't a strong bidirectional variation. In order to resolve this problem, we search for the strong points in the image, for this reason, we have look for good features to track of Shi & Tomasi (1994).

4.2 Good features to track of Shi and Thomasi:

In order to compare the obtained results using uniform distribution, we have used the method of Shi and Thomasi for interest points extraction. This method is based on the general assumption that the luminance intensity does not change for image acquisition.

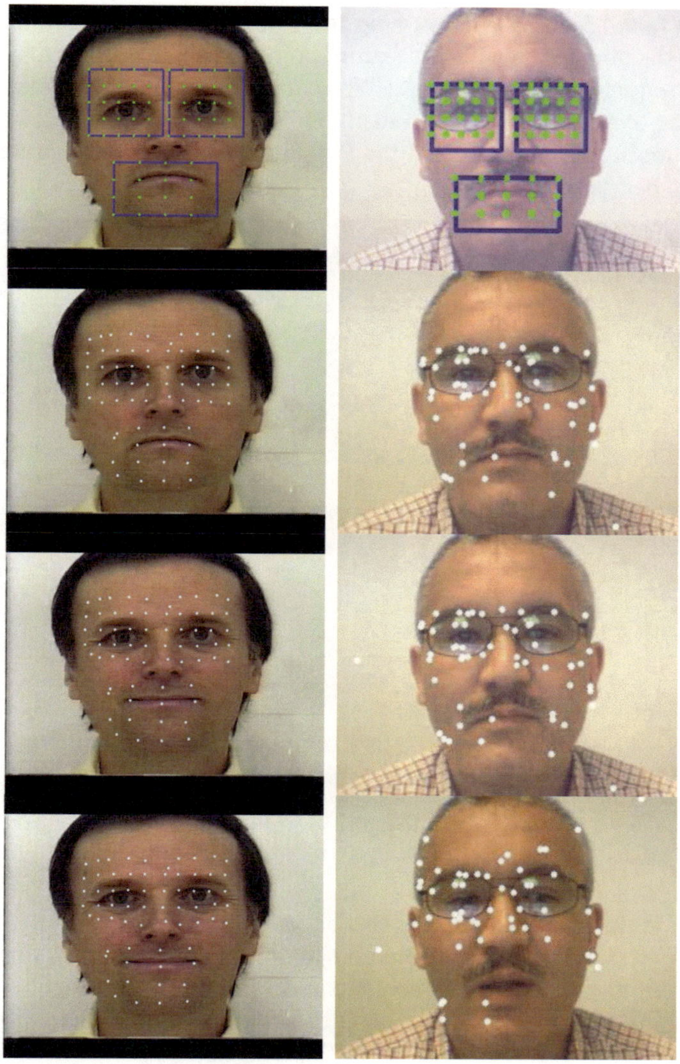

Fig. 9. Tracking of the uniform distribution for video sequence.

To select interest points, a neighbourhood N of nxn pixels is selected around each pixel in the image. The derivatives Dx and Dy are calculated with a Sobel operator for all pixels in the block N. For each pixel the minimum eigenvalue λ is calculated for matrix A where

$$A = \begin{bmatrix} \sum D_{x_{i,j}}^2 & \sum D_{x_{i,j}} \sum D_{y_{i,j}} \\ \sum D_{x_{i,j}} \sum D_{y_{i,j}} & \sum D_{y_{i,j}}^2 \end{bmatrix} \quad (10)$$

and Σ is performed over the neighborhood of N. The pixels with the highest values of λ are then selected by thresholding.

The next step is rejecting the corners with the minimal eigenvalue less than some threshold. Finally, a test is made, all the found corners are distanced enough from one another by getting two strongest features and checking that the distance between the points is satisfactory. If not, the point is rejected. For further details see Shi & Tomasi (1994).

4.3 Detection of facial feature points using the Shi and Thomasi method:

Figure 10 shows the obtained results for feature points detection with the method of Shi and Thomasi (video sequence, real time acquisition) applied to the whole image. We can see a good tracking for these points in the remaining of the sequence, unlike the first method (uniform distribution), which prove that the Pyramidal Lucas-Kanade Feature Tracker need a strong points to be tracked.

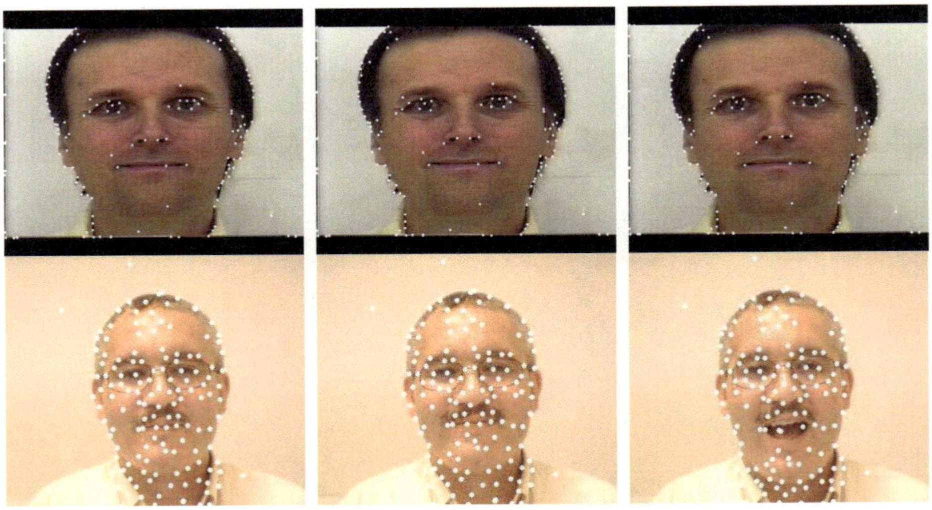

Fig. 10. Extraction of feature points in the first frame and feature points tracking using pyramidal Lucas-Kanade feature tracker in the remaining of the sequence.

The method of Shi and Thomasi ensures good detection of points that have strong gradient. This good detection leads to a good tracking of these points.

4.4 Detection of facial feature points in the bounding box:

In the previous section, we have presented a detection of interest points in the face, however, we need only the points which surround facial features such as eyes, eyebrows and mouth. For this reason, we will reject all the pixels outside the rectangle. Figure 11 shows interest region, which will be used for the detection of points with Shi and Thomasi method.

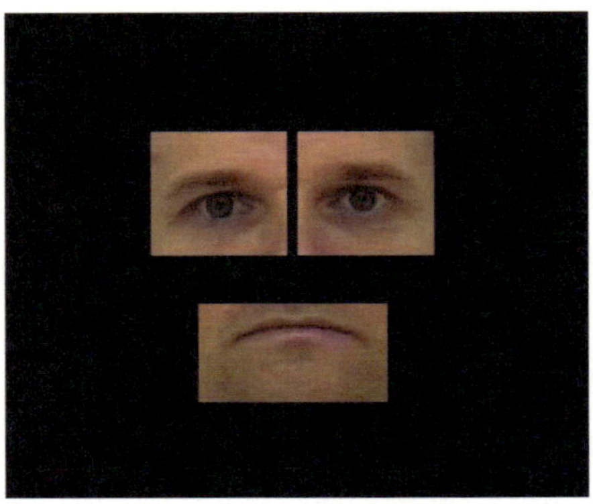

Fig. 11. The interest region feature point extraction in the first frame.

Figure 12 shows an example of points tracking in the bounding box. The tracking is very well; the first image presents the detection of points in the bounding boxes which delimit the facial features using Shi and Thomasi method. These detected points correspond to pixels with strong gradient. The following images present the 1st, 2nd, 22th and 46th frames in the first video sequence and the 1st, 2nd, 51th and 67th for the second sequence.
Our system is implemented in VC.NET on a pentium IV with $2GHz$ under windows XP. The table 1 presents the elapsed time for each step in our system. The size of the frame is 576 $*$ 720 and the video sequence format is $AVI - I420$.
For the first frame, the elapsed time for rectangle localization is $0.281S$ and the elapsed time for point detection is $1.40S$. For the remaining of the sequence, we can track the detected points for $0.031s$.

Operation	Time (S)
Rectangles localization	0.281
Point detection	0.140
Tracking	0.031

Table 1. Elapsed time for each step.

5. Conclusion and future works

In this paper, we have presented a face tracking algorithm in real time camera input environment, in order to use it in the facial expression recognition system. To detect the face in the image, we have used a modified face detector based on the Haar-like features. This face detector is fast and robust to illumination condition but hardly work when face pose is too slanted. For feature points extraction, we have used the algorithm of Shi and Thomasi to extract feature points. This method gives good results. To track the facial feature points, Pyramidal Lucas-Kanade Feature Tracker KLT algorithm is used. We have got a bad results with a uniform distribution of feature points which explain that this algorithm need a strong points. However, using detected points with the algorithm of Shi and Thomasi, we have got good results in video sequence and in real time acquisition. The obtained results indicate that the proposed algorithmcan accurately extract facial features points. The future work will include extracting feature points with some conditions to limit the number of feature points in bounding box and choose only the points which describe well the shape of the facial feature. This work will be used for real time facial expression recognition application.

6. References

Abdat, F., Maaoui, C. & Pruski, A. (2008). Real facial feature points tracking with pyramidal lucas-kanade algorithm, *IEEE RO-MAN08, The 17th International Symposium on Robot and Human Interactive Communication, Germany*.

Bouguet, J. (2000). Pyramidal implementation of the lucas kanade feature tracker, *Intel Corporation* Microprocessor Research Labs.

H. Xiaolei, Z. Song, W. Y. M. D.-S. D. (2004). A hierarchical framework for high resolution facial expression tracking, *3rd IEEE Workshop on articulated and non rigid motion ANM 2004*.

K. Ki-Sang, J. Dae-Sik, C. H.-I. (2007). Real time face tracking with pyramidal lucas-kanade feature tracker, *Computational science and its applications ICCSA 2007* 4705: 1074–1082.

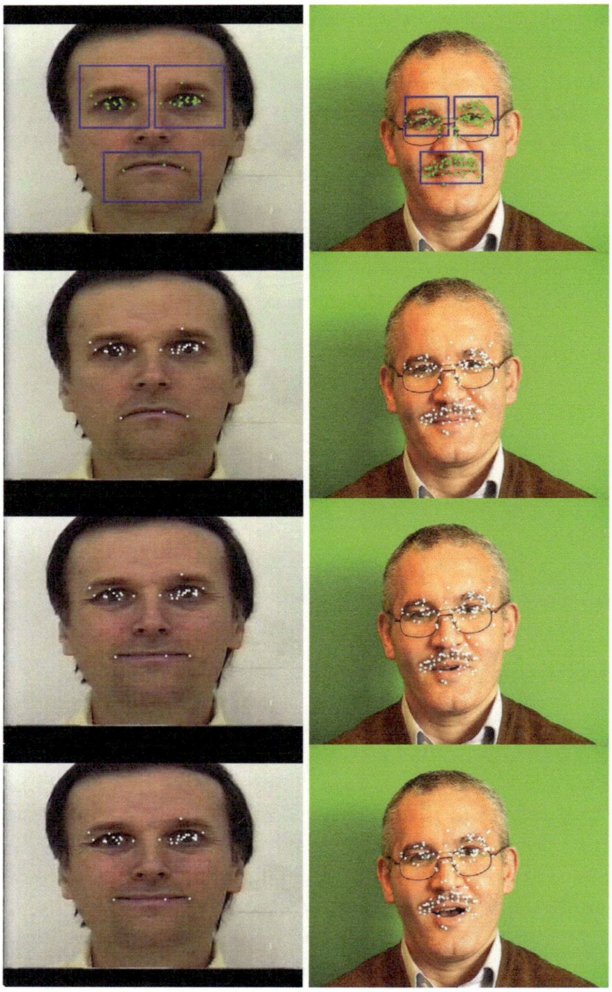

Fig. 12. Extraction of feature points in the bounding box for the first frame and Feature points tracking using pyramidal Lucas-Kanade in the remaining of the sequence.

P.Viola & M.Jones (2001). Rapid object detection using a boosted cascade of simple features, *Conference on CVPR 2001.*

R.Belaroussi & Milgram, M. (2006). Face tracking and facial features detection with a webcam, *CVMP 2006.*

Shi, J. & Tomasi, C. (1994). Good features to track, *IEEE Conf. Computer Vision and Pattern Recognition Seattle* CVPR'94.

Shih, F. & Chuang, C. (2004). Automatic extraction of head and face boundaries and facial features, *Information Sciences* 158: 117–130.

Su, M. & Hsieh, Y. (2007). A simple approach to facial expression recognition, *Proceeding WSEAS 2007 Australia.*

Viola, P. & Jones,M. (2001). Robust real-time object detection, 2nd international workshop on statistical and computational theories of vision - modeling, learning, computing, and sampling Vancouver, Canada.

14

Improving Human-Robot Interaction through Interface Evolution

Brenden Keyes[1], Mark Micire[2], Jill L. Drury[1] and Holly A. Yanco[2]
[1]*The MITRE Corporation, Bedford, MA,*
[2]*University of Massachusetts Lowell, Lowell, MA,*
USA

1. Introduction

In remote robot operations, the human operator(s) and robot(s) are working in different locations that are not within line of sight of each other. In this situation, the human's knowledge of the robot's surroundings, location, activities and status is gathered solely through the interface. Depending on the work context, having a good understanding of the robot's state can be critical. Insufficient knowledge in an urban search and rescue (USAR) situation, for example, may result in the operator driving the robot into a shaky support beam, causing a secondary collapse. While the robot's sensors and autonomy modes should help avoid collisions, in some cases the human must direct the robots' operation. If the operator does not have good awareness of the robot's state, the robot can be more of a detriment to the task than a benefit.

The human's comprehension of the robot's state and environment is known as situation awareness (SA). Endsley developed the most generally accepted definition for SA: "The perception of elements in the environment within a volume of time and space [Level 1 SA], the comprehension of their meaning [Level 2 SA] and the projection of their status in the near future [Level 3 SA]" (Endsley, 1988). Drury, Scholtz, and Yanco (2003) redefined this definition of situation awareness to make it more specific to robot operations, breaking it into five categories: human-robot awareness (the human's understanding of the robot), human-human awareness, robot-human awareness (the robot's information about the human), robot-robot awareness, and the humans' overall mission awareness. In this chapter, we focus on two of the five types of awareness that relate to a case in which one human operator is working with one robot: human-robot awareness and the human's overall mission awareness. Adams (2007) discusses the implications for human-unmanned vehicle SA at each of the three levels of SA (perception, comprehension, and projection).

In Drury, Keyes, and Yanco (2007), human-robot awareness is further decomposed into five types to aid in assessing the operator's understanding of the robot: location awareness, activity awareness, surroundings awareness, status awareness and overall mission awareness (LASSO). The two types that are primarily addressed in this chapter are location awareness and surroundings awareness. Location awareness is the operator's knowledge of where the robot is situated on a larger scale (e.g., knowing where the robot is from where it started or that it is at a certain point on a map). Surroundings awareness is the knowledge the operator has of the robot's circumstances in a local sense, such as when there is an

obstacle two feet away from the right side of the robot or that the area directly behind the robot is completely clear.
Awareness is arguably the most important factor in completing a remote robot task effectively. Unfortunately, it is a challenge to design interfaces to provide good awareness. For example, in three studies that examined thirteen separate USAR interfaces (Yanco et al., 2002; Scholtz et al., 2004; Yanco & Drury, 2006), there were a number of critical incidents resulting from poor awareness. For instance, an operator moved a video camera off-center to conduct victim identification. After allowing the robot to drive autonomously out of a tight area, the operator forgot that the camera was off-center, resulting in poor robot operation, including collisions and operator confusion (Yanco et al., 2004).
This chapter presents lessons learned from the evolution of our human-robot interaction (HRI) design for improved awareness in remote robot operations, including new design guidelines. This chapter brings together for the first time the four different versions of the interface created during the evolution of our system, along with the motivation for each step in the design evolution. For each version of the interface, we present the results of user testing and discuss how those results influenced our next version of the interface. Note that the final version of the interface discussed in this chapter (Version 4) was designed for multi-touch interaction, and the study we conducted on this version establishes a performance baseline that has not been previously documented.
The next section presents summaries of some of the previous interfaces that have influenced our design approaches, followed by our design and testing methodology in Section 3. Section 4 describes briefly the robot hardware that was controlled by the various interfaces. After a section presenting our general interface design approach, the next four sections describe the four generations of the evolving interface. Finally, we present conclusions and plans for future work.

2. Related work

We did not design in a vacuum: there have been numerous attempts in the past decade to design remote robot interfaces for safety-critical tasks. Remote robot interfaces can be partitioned into two categories: map-centric and video-centric (Yanco et al., 2007). A map-centric interface is an interface in which the map is the most predominant feature in the interface and most of the frequently used information is clustered on or near the map. Similarly, in a video-centric interface, the video window is the most predominant feature with the most important information located on or around the video screen.
Only a few interfaces are discussed here due to space limitations; for a survey of robot interfaces that were used in three years of the AAAI Robot Rescue competition, see Yanco and Drury (2006).

2.1 Map-centric interfaces

It can be argued that map-centric interfaces are better suited for operating remote robot teams than video-centric interfaces due to the inherent location awareness that a map-centric interface can provide. The relationship of each robot in the team to other robots as well as its position in the search area can be seen in the map. However, it is less clear that map-centric interfaces are better for use with a single robot. If the robot does not have adequate sensing capabilities, it may not be possible to create maps having sufficient accuracy. Also, due to an emphasis on location awareness at the expense of surroundings awareness, it can be difficult to effectively provide operators with a good understanding of the area immediately around the robot.

One example of a map-centric interface, developed by the MITRE Corporation, involved using up to three robots to build a global map of the area covered by the robots. Most of the upper portion of the display was a map that gradually updated as ranging information was combined from the robots. The interface also had the ability to switch operator driving controls among the three robots. Small video windows from the robots appeared under the map. The main problems with this interface were the small size of the video screens as well as the slow updates (Drury et al., 2003).

Brigham Young University and the Idaho National Laboratory (INL) also designed a map-centric interface. The INL interface has been tested and modified numerous times, originally starting as a video-centric interface before changing to a map-centric interface (Nielsen et al., 2007; Nielsen & Goodrich, 2006; Nielsen et al., 2004). This interface combines 3D map information using blue blocks to represent walls or obstacles with a red robot avatar indicating its position on the map. The video window is displayed in the current pan-tilt position with respect to the robot avatar, indicating the orientation of the robot with respect to where the camera is currently pointing. If the map is not generated correctly due to moving objects in the environment, faulty sensors or other factors, however, the operator can become confused regarding the true state of the environment. We have witnessed cases in which the INL robot slipped and its map generation from that point on shifted to an offset from reality, with the consequence that the operator became disoriented regarding the robot's position.

Because of these drawbacks for remote robot operations (overreliance on potentially inaccurate maps and a smaller video displays due to larger maps), we found inspiration for our interface designs in video-centric interfaces.[1]

2.2 Video-centric interfaces

Video-centric interfaces are by far the most common type of interface used with remote robots. Operators rely heavily on the video feed from the robot and tend to ignore any other sensor reading the interface may provide (e.g., see Yanco & Drury, 2004). Many commercially available robots have video-centric interfaces (e.g., iRobot's Packbot and Foster Miller's Talon).

ARGOS from Brno University of Technology is an excellent example of a video-centric interface (Zalud, 2006). It provides a full screen video interface with a "heads up" display (HUD) that presents a map, a pan/tilt indicator and also a distance visualization widget that displays the detections from the laser sensor on the front of the robot. What makes this interface unique is its use of virtual reality goggles. These goggles not only display the full interface, but the robot also pans and tilts the camera based on where the operator is looking, making scanning an area as easy as turning your head in the direction you want to look. It also eliminates issues with forgetting that the camera is not centered.

The CASTER interface developed at the University of New South Wales (Kadous et al., 2006) also provides a full screen video interface but incorporates a different arrangement of small sensor feeds and status readouts placed around the edges.

---

[1] Readers who are interested in map-based interfaces in collocated operations may find the guidelines and heuristics in Lee et al. (2007) to be helpful.

Researchers at Swarthmore College (Maxwell et al., 2004) have designed a video-centric interface that includes a main panel showing the view of the video camera. It has the unique feature of overlaying green bars on the video which show 0.5 meter distances projected onto the ground plane. The interface also has pan-tilt-zoom indicators on the top and left of the video screen, and it displays the current sonar and infrared distance data to the right of the video window.

Inspired by these video-centric systems, we have incorporated into our interface a large video feed in the central portion of the interface and a close coupling between pan-tilt indicators and the video presentation.

3. Methodology

3.1 Design methodology

We started with an initial design based upon a list of guidelines recommended by Yanco, Drury and Scholtz (2004) and Scholtz et al. (2004). The guidelines state that a USAR interface should include:

- A map of where the robot has been.
- Fused sensor information to lower the cognitive load on the user.
- Support for multiple robots in a single display (in the case of a multi-robot system).
- Minimal use of multiple windows.
- Spatial information about the robot in the environment.
- Help in deciding which level of autonomy is most useful.
- A frame of reference to determine position of the robot relative to its environment.
- Indicators of robot health/state, including which camera is being used, the position(s) of camera(s), traction information and pitch/roll indicators.
- A view of the robots' body so operators can inspect for damage or entangled obstacles.

We also kept in mind the following design heuristics, which we adapted from Nielsen (1993):

- Provide consistency; especially consistency between robot behavior and what the operator has been led to believe based on the interface.
- Provide feedback.
- Use a clear and simple design.
- Ensure the interface helps to prevent, and recover from, errors made by the operator or the robot.
- Follow real-world conventions, e.g., for how error messages are presented in other applications.
- Provide a forgiving interface, allowing for reversible actions on the part of the operator or the robot as much as possible.
- Ensure that the interface makes it obvious what actions are available at any given point.
- Enable efficient operation.

Finally, we designed to support the operator's awareness of the robot in five dimensions:

- Enable an understanding of the robot's location in the environment.
- Facilitate the operator's knowledge of the robot's activities.
- Provide to the operator an understanding of the robot's immediate surroundings.
- Enable the operator to understand the robot's status.
- Facilitate an understanding of the overall mission and the moment-by-moment progress towards completing the mission.

We realized that we were not likely to achieve an optimal design during the first attempt, so we planned for an iterative cycle of design and evaluation.

3.2 SA measurement techniques

Because it is important to characterize and quantify awareness as a means to evaluate the interfaces, we discuss SA measurement techniques here. Hjelmfelt and Pokrant (1998) state that experimental methods for measuring SA fall into three categories:

1. Subjective: participants rate their own SA
2. Implicit performance: Experimenters measure task performance, assuming that a participant's performance correlates with SA and that improved SA will lead to improved performance
3. Explicit performance: Experimenters directly probe the participant's SA by asking questions during short suspensions of the task

For these studies, we elected to use mainly implicit measures to associate task outcomes with implied SA; in particular, we focused on task completion time and number of collisions. A faster completion time as well as fewer collisions implies better SA. We also performed an explicit measure at the end of some studies, in which the user was asked to complete a secondary task that required awareness, such as: return the robot to a particular landmark that was previously visited. We used post-task questions that asked for participants' subjective assessment of their performance. We did not place great weight on the subjective assessments, however. Even if participants reported that they had performed well, their assessments were not necessarily accurate. In the past, we had observed many instances in which participants reported that the robot had not collided with obstacles when they had actually experienced collisions that caused damage to the arena (e.g., see Yanco et al., 2004).

3.3 General testing methodology

For all of our evaluations, we used similar test arenas that were based upon the National Institute of Standards and Technology (NIST) USAR arena (Jacoff et al., 2000; Jacoff et al., 2001; Jacoff et al., 2002). Each study used multiple arena orientations and robot starting positions, which were permuted to eliminate learning effects. In all the studies, except for the one that was performed on Version 3 of the interface, the users had a set time limit to complete their task. In most cases, participants were told that a disaster had occurred and that the participant had a particular number of minutes to search for and locate as many victims as possible. The time limit was between 15 and 25 minutes depending on the study.

We used an "over-the-shoulder" camera that recorded the user's interaction with the interface controls as well as the user's think-aloud comments (Ericsson & Simon, 1980). Think-aloud is a protocol in which the participants verbally express their thoughts while performing the task assigned to them. They are asked to express their thoughts on what they are looking at, what they are thinking, why they are performing certain actions and what they are currently feeling. This allows the experimenters to establish the reasoning behind participants' actions. When all the runs ended, the experimenter interviewed the participant. Participants were asked to rate their own performance, to answer a few questions about their experience, and to provide any additional comments they would like to make.

During the run, a camera operator and a person recording the robot's path on a paper map followed the robot through arenas to create a record of the robot's progress through the test

course. The map creator recorded the time and location on the map of critical incidents such as collisions with obstacles. The map and video data were used for post-test analysis to determine the number of critical incidents and to cross-check data validity.
We analyzed this data to determine performance measures, which are implicit measures of the quality of the user interaction provided to users. As described above, we inferred awareness based on these performance measures. We recorded the number of collisions that occurred with the environment, because an operator with good surroundings awareness should hit fewer obstacles than an operator with poor surroundings awareness. We also analyzed the percentage of the arena covered or the time to complete the task, depending on the study. Operators with good location awareness should not unknowingly backtrack over places they have already been, and thus should be able to cover more area in the same amount of time than an operator with poor awareness, who might unknowingly traverse the same area multiple times. Similarly, we expected study participants with good awareness to complete the task more quickly than users with poor awareness, who may be confused and need additional time to determine a course of action. Participants' think-aloud comments were another important implicit measure of awareness. These comments provided valuable insight into whether or not a participant was confused or correctly recognized a landmark. For example, users would often admit to a loss of location awareness by saying "I am totally lost," or "I don't know if I've been here before," (speaking as a "virtual extension" of the robot).

4. Robot hardware

Our system's platform is an iRobot ATRV-JR. It is 77cm long, 55cm high and 64cm wide. It is a four-wheeled, all-terrain research platform that can turn in place due to its differential (tank-like) steering. The robot has 26 sonar sensors that encompass the full 360 degrees around the robot as well as a SICK laser range finder that covers the front 180 degrees of the robot. It has two pan/tilt/zoom cameras, one forward-facing and one rear-facing. To help with dark conditions in USAR situations, we added an adjustable lighting system to the robot.
The robot system has four autonomy modes: teleoperation, safe, shared, and escape, based upon Bruemmer et al. (2002). In the teleoperation mode, the operator makes all decisions regarding the robot's movement. In safe mode, the operator still directs the robot, but the robot uses its distance sensors to prevent the operator from driving into obstacles. Shared mode is a semi-autonomous navigation mode that combines the user's commands with sensor inputs to promote safe driving. Escape mode is the only fully autonomous mode on the system and is designed to drive the robot towards the most open space.

5. General interface design

Our interface was designed to address many of the issues that emerged in previous studies. The interface also presents easily readable distance information close to the main video so that the user is more likely to see and make use of it. The system also provides access to a rear camera and automatic direction reversal as explained below.
The main video panel is the hub of the interface. As Yanco and Drury (2004) state, users rely heavily on the main video screen and rarely notice other important information presented on the interface. Therefore, we located the most important information on or around the

main video screen so that the operator would have a better chance of noticing it. The main video screen was designed to be as large as possible so users can better perceive the visual information provided by the cameras. Further, we overlaid a small cross on the screen to indicate the direction in which the camera is pointing. These crosshairs were inspired by the initial design of the Brno robot system (Zalud, 2006).

In the prior studies discussed by Yanco, Drury and Scholtz (2004), we observed that more than 40% of robot collisions with the environment were on the rear of the robot. We believe a lack of sensing caused many of these rear collisions, so we added a rear-looking camera. Since the rear-looking camera would only be consulted occasionally, we mirrored the video feed and placed it in a similar location to a rear-view mirror in a car.

To further reduce rear collisions, we implemented an Automatic Direction Reversal (ADR) system. When ADR is in use, the interface switches the video displays such that the rear view is expanded in the larger window. In addition, the drive commands automatically remap so that forward becomes reverse and reverse becomes forward. The command remapping allows an operator to spontaneously reverse the direction of the robot in place.

The interface also includes a map panel, which displays a map of the robot's environment and the robot's current position and orientation within the environment. As the robot moves throughout the space, it generates a map using the distance information received by its sensors using a Simultaneous Localization and Mapping (SLAM) algorithm. The placement of this panel changed throughout the evolution of the interface, but to ensure it is easily accessible to users, it has always remained at the same horizontal level as the video screen.

Throughout the evolution of our interface, the distance panel has been the main focus of development. It is a key provider of awareness of all locations out of the robot's current camera view. The distance panel displays current distance sensor readings to the user. The presentation of this panel has differed widely during the course of its progression and will be discussed more thoroughly in the next sections.

The autonomy mode panel has remained the same in all of our interface versions; it allows for mode selection and displays the current mode. The status panel provides all status information about the robot, including the battery level, the robot's maximum speed and if the lighting system is on or off.

6. Version 1

6.1 Interface description

The first version of the interface consisted of many of the panels described above in Section 5 and is shown in the top row of Table 1. The large video panel is towards the left center of the screen. The rear-view camera panel is located above the video panel to mimic the placement of a car's rear-view mirror. Bordering the main video screen are color-coded bars indicating the current values returned by the distance sensors. In addition to the color cues, multiple bars were filled in, with more bars meaning a closer object, to aid people with color deficiencies. Directly below the video screen is the mode panel. The illustration in Table 1 indicates that the robot is in the teleoperation mode. Directly to the right of the main video screen is the map panel. On the top-right of the interface is the status panel.

6.2 Evaluation description

We designed a study to determine if adding the rear-facing camera would improve awareness (Keyes et al., 2006). We created three variations of the interface, which we refer

Illustration	Version Description
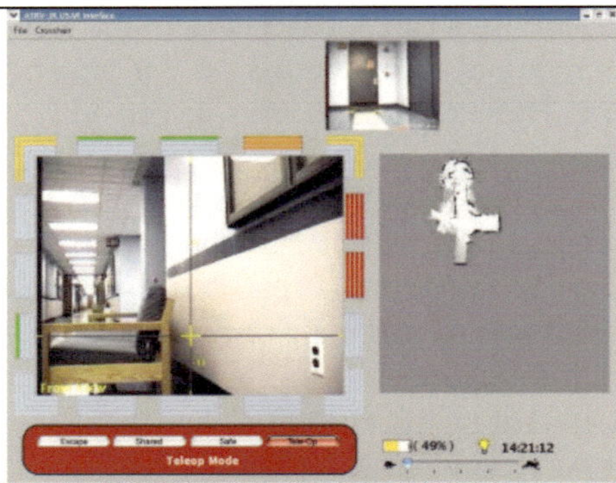	Version 1 consisted of the main video screen as well as the rear view camera, map, mode and status panels. The distance panel, placed around the main video screen, displays how close an obstacle is to the robot by filling in the colored bars. The interface was controlled via keyboard and joystick.
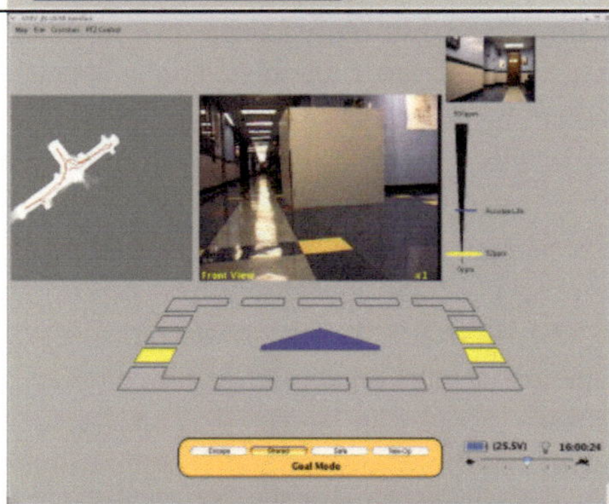	Version 2 moves the video panel to the center of the screen. The distance panel is placed in a perspective view below the video screen, and turns from grey to yellow to red as objects get closer to the robot. It also rotates in response to the panning of the camera. Zoom mode (not shown here, but later in Figure 1), is displayed over the map panel when it is toggled on. This version was controlled via keyboard and joystick.
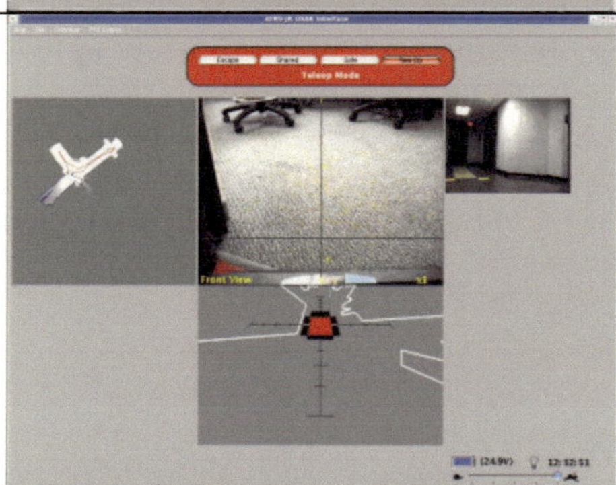	Version 3 replaces the distance panel with a zoom mode inspired panel. Instead of colored boxes, lines are drawn around a scale model of the robot based on sensor information. This version was controlled via keyboard and joystick.
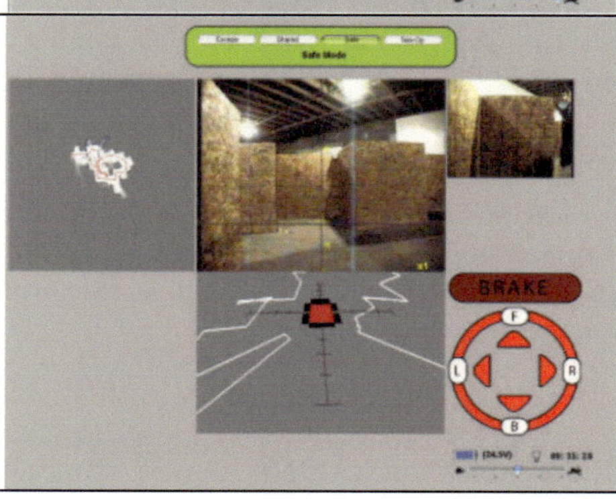	Version 4 keeps the visual presentation the same as Version 3 while replacing the input method with multi-touch gesture activation. The virtual joystick in the lower right-hand corner provided the rotation and translation of the robot along with brake control. The visual feedback from Version 3, such as speed control and mode selection became interactive sliders and buttons in this interface.

Table 1. Summary of the Interface Versions

to as Interfaces A, B, and C. Interface A consisted of the main video panel, distance panel, pan-tilt indicator, mode bar and status panel. For this interface, the participants only had access to the front camera's video stream. Interface B displayed of all the same panels as Interface A, but the user could switch the main video panel to display the rear camera's video feed, triggering ADR mode. Interface C added the rear view camera panel and also had ADR mode, providing the users with the full Version 1 interface. Nineteen people participated, ranging in age from 18 to 50, with 11 men and 8 women. Using a within-subjects design, each participant operated the robot through the three different arena configurations using a different interface each time, with the order of the interface use and arena configurations being randomized.

6.3 Evaluation results

As expected, participants who had access to the rear camera had greater awareness than participants who did not. Using a two-tailed paired *t*-test, we found a significant difference in the number of collisions that occurred between the different interfaces. Participants made significantly more collisions when using Interface A (no rear-looking camera) than Interface C (both front- and rear-looking cameras displayed simultaneously) ($M_A = 5.4$ collisions, $SD_A = 3.2$, $M_C = 3.6$, $SD_C = 2.7$, $p < 0.02$).

Participants also made significantly more collisions when using Interface A than Interface B (front and rear cameras both available but not displayed simultaneously) ($M_A = 5.4$ collisions, $SD_A = 3.2$, $M_B = 3.9$, $SD_B = 2.7$, $p < 0.04$). These results indicate that awareness regarding the rear of the robot is improved by having access to the rear camera, even if the rear camera is not constantly being displayed. We did not find any significant difference in the time it took to complete the task.

There was only one user in this study who did not use the rear camera at all. The other eighteen participants made at least one camera switch when using Interface B. For Interface C, three of eighteen participants did not switch camera modes. One user stated that it was not necessary to switch camera modes because both cameras were being displayed already. Another user discussed being reluctant to switch views because it caused confusion when trying to keep track of the robot's current environment.

Five of the nineteen participants stated that they preferred to use only the front camera because they were able to pan the camera down to see the front bumper of the robot. The front of the robot has a larger bumper than the back of the robot, so the front camera is the only camera that can see the robot chassis. We found that the five users who had the strategy of looking at the bumper to localize the robot in the environment had fewer collisions ($M = 8.0$ collisions, $SD = 4.1$) than the other fourteen participants ($M = 14.7$ collisions, $SD = 6.6$).

We found that most of the collisions between the robot and the arena occurred on the robot's tires. Seventy-five percent of all the front collisions that occurred with the robot involved the robot's tires. These tires lie just outside the visible area and widen the robot by about five inches on each side. Despite warnings by the experimenter, users acted under the assumption that the boundaries of the video reflected the boundaries of the robot. It is also important to note that 71% of the total collisions in the study occurred on the tires. Because the tires make up almost the entire left and right sides of the robot, this result is unsurprising. The use of two cameras helped to improve situation awareness with respect to the front and rear of the robot, but users still lacked SA with respect to the sides of the robot.

Fifteen of the nineteen participants (79%) preferred the interface with two camera displays. Three of the participants preferred the interface with two cameras that could be switched in a single video window. Two of these participants had little computer experience, which suggests that they might have been overwhelmed by the two video windows. The final participant expressed no preference between the two interfaces with two cameras but did prefer these two to the single camera case. No participant preferred the single camera case.

Two of the users in this study found the distance panel to be unintuitive. They thought the bars on top of the video window corresponded to distance sensors pointing directly up from the robot and the bars on the bottom represented distance sensors that were pointing down from the bottom of the robot. We also noted that due to the number of colors displayed by the bars, as well as the fact that different numbers of bars were filled, it was difficult for users to keep track of what was important. Often the display panel appeared to be blinking due to the high frequency with which distance values were changing. This resulted in the undesirable situation in which users started to ignore the panel altogether. While the addition of the rear camera helped improve SA significantly, the distance panel was not particularly helpful to prevent collisions on the side of the robot.

7. Version 2

Based upon the results of the previous study, particularly with respect to the lack of surroundings awareness relating to the sides of the robot, the focus of this design iteration was to improve the distance panel. Version 2 of the interface is the second image from the top in Table 1.

7.1 Interface description

The range data was moved from around the video window to directly below it. We altered the look and feel of the distance panel by changing from the colored bars to simple colored boxes that used only three colors (gray, yellow and red) to prevent the distance panel from constantly blinking and changing colors. In general, when remotely operating the robot, users only care about obstacles in close proximity, so using many additional colors to represent faraway objects was not helpful. Thus, in the new distance panel, a box would turn yellow if there was an obstacle within one meter of the robot and turn red if an obstacle was within 0.5 meters of the robot.

The last major change to the distance panel was the use of a 3D, or perspective, view. This 3D view allows the operator to easily tell that the "top" boxes represent forward-facing sensors on the robot. We believe this view also helps create a better mental model of the space due to the depth the 3D view provides, thus improving awareness around the sides of the robot. Also, because this panel was in 3D, it was possible to rotate the view as the user panned the camera. This rotation allows the distance boxes to line up with the objects the user is currently seeing in the video window. The 3D view also doubles as a pan indicator to let the user know if the robot's camera is panned to the left or right.

This version of the interface also included new mapping software, PMap from USC, which added additional functionality, such as the ability to display the robot's path through the environment (Howard, 2009).

One feature that resulted from the use of PMap was a panel that we termed "zoom mode." This feature, which can be seen in Figure 1 on the left, represents a zoomed-in view of the map. It takes the raw laser data obtained in front of the robot and draws a line connecting the sensor readings together. The smaller rectangle on the bottom of this panel represents the robot. As long as the sensor's lines do not touch or cross the robot rectangle, the robot is not in contact with anything. This sensor view gives highly accurate, readily visible cues regarding whether the robot is close to an object or not. Our goal was to develop an approach to make it easier to visualize the environment than the information from Version 1's colored boxes by requiring the operator make fewer mental translations. However, due to the PMap implementation, the zoom mode and the map display panel were mutually exclusive (only one could be used at a time).

The video screen was moved from the left side to the center of the screen. This shift was mainly due to the fact that the new distance panel was larger, and, with the rotation feature, was not fully visible on the screen. Placing it in the center allowed for the full 3D view to be displayed at all times. The map was moved to the right side of the video.

7.2 Evaluation description

During Version 2's evaluation, we studied the differences between our video-centric interface and INL's map-centric interface (Yanco et al., 2007). Similar to the evaluation for Version 1, we designed a within-subjects study, counterbalancing whether participants began with the Version 2 interface or INL's interface. We also varied the robot's starting point into the arena to avoid introducing a learning effect. Seven men and one woman participated, ranging in age from 25 to 60. All had experience in search and rescue.

7.3 Evaluation results

Here we concentrate on the lessons learned about the Version 2 interface rather than the comparison between the Version 2 and INL interfaces. We found that users liked the new distance panel as well as the zoom mode capability. Although the colored boxes on the new distance panel resulted in better performance than the previous colored bar approach, the new distance panel design did not result in a large performance improvement. The main problem was that the use of only two colors, yellow and red, was too simple. When in a tight area, which is often the case in a USAR environment, the robot may not have 0.5 meters on either side of it; this was the case during the experiment. If a distance box was red, the user knew that an obstacle was within 0.5 meters but did not know exactly how close it was. When the robot is in a very confined area, 0.5 meters is a large distance. While the interface could have been tuned so that the boxes only turned red at 0.1 meters or even 0.05 meters, the basic problem would remain. The colored boxes are not informative enough, and that uncertainty causes the user to distrust the system.

The zoom mode feature helped to address the uncertainty caused by not knowing how far the robot is situated from an obstacle. Using the lines provides a concrete idea of how close an obstacle is to the robot without overwhelming the user. It is also extremely accurate, which can help produce a better mental model of the environment. The operator does not have to extrapolate what the area might look like based on colored boxes, thus reducing cognitive load. The user can also see the flow of the obstacles with respect to the robot's movements. The zoom mode feature also helps to give the user a more accurate idea regarding the layout of obstacles.

8. Version 3

The results of the Version 2 user study demonstrated the utility of the zoom mode feature (shown on the left in Figure 1). The distance panel was still problematic and so became the focus of our next iteration.

8.1 Interface description

The Version 2 distance panel was removed and replaced by a distance panel based on the zoom mode feature (the Version 3 interface is the third image down in Table 1). The zoom mode was extended to encompass the entire circumference of the robot. The view of the front part of the robot would be based on laser data, whereas the views of the left, right, and rear of the robot would use sonar data. We also added tick marks at 0.25 meter increments to the lines to indicate distance. This panel was again placed directly under the main video display. Unlike the previous top-down zoom mode, this new panel also had the ability to provide a perspective view (the top down and perspective view distance panels are shown in the center and right of Figure 1). Results from the previous study indicated that users liked having the ability to go from a 2D map to a 3D map, so we felt users would appreciate this toggle ability here as well. Also, as with the previous distance panel, this panel also rotated with as the user panned the camera.

8.2 Evaluation description

We conducted an evaluation of this version of the interface. This new study consisted of 18 users, 12 men and 6 women. They varied in age from 26 to 39, with varying professions. None of them were USAR experts. The main purpose of this study was to compare the Version 2 distance panel (referred to as Interface D here) with the new Version 3 distance panel (referred to as Interface F here). For experimentation purposes, we also included a modified version of the distance panel from Interface D that overlaid the distance values in meters on the colored boxes (referred to as Interface E here) to give users exact distance information.

This test differed slightly from our previous studies. For this study, the user was only tasked to go through an arena and back again. Unlike all of the previous arenas, there was only one path for the user to take. When the user reached the end of the path, he was asked to turn around and come back out the way he had come in. Participants were not searching for victims: they were only asked to maneuver through the course. The courses in this study were much narrower than ones from previous studies. In some cases, there were only three centimeters of clearance on either side of the robot. This was done to fully exploit the weaknesses of the distance panels on each interface to determine which facilitated the best performance. If the arenas were wide open and easy to navigate, it would have been more difficult to discern differences between the interfaces.

We also forced the participants to use only teleoperation mode, which was not done in the previous studies, to remove the confounder of autonomy. Because we were studying the effects of the distance panel, if we allowed the robot to take some initiative, such as stopping itself, it may have prevented many of the collisions from happening and thus skewed the results.

For this study, we hypothesized that Interface D would perform the worst due to the lack of information that it provides. We believed that Interface D would result in the most collisions due to the lack of specificity of the distance information. However, we believed that Interface D would result in relatively fast time-on-task since the user would eventually come to ignore this unhelpful distance information, and it always takes time to perceive information in an interface.

We hypothesized that Interface E would result in fewer collisions than Interface D due to the exactness of the data presented. However, because the user would have to interpret the numerical data, we felt Interface E would lead to longer run times than both Interfaces D and F. We felt users would perform the best using Interface F due to how easy it is to visually process the information. We hypothesized that Interface F would yield fewer collisions and faster run times than either of the other interfaces. It is very easy to interpret, thus it is extremely easy to recognize if an obstacle is close to the robot without having to expend mental effort calculating distances. Due to the constantly changing numerical values on Interface E, we believed that users might experience cognitive overload and misinterpret the values. With Interface F, we felt this would not be an issue. Even though Interface F's data presentation is still technically not as precise as Interface E's, the fact that the user can instantly know if obstacles are close or not would provide much better surroundings awareness.

We did not want the user to get lost in the arena, so the courses were designed to have exactly one possible path. Because we were primarily interested in which interface resulted in the fastest run times, we did not want the results to be skewed by the users becoming lost in the arena. If a user was confused as to the direction in which to proceed, the test administrator indicated verbally the correct direction. (This was the only information the test administrator gave the operators while the runs were in progress.)

Once again, participants used all three versions of the interface in a within-subjects study design and the order of the interface use was randomized. We collected and analyzed the time it took each participant to finish the task.

8.3 Evaluation results

When comparing time on task, our initial hypotheses held true for most of the test cases. Using two-tailed paired *t*-tests (df = 17), we found significant differences between the interfaces in the amount of time on task. Interface D (the distance panel with colored boxes) was significantly faster than Interface E (the distance panel with colored boxes and distance values) (M_A = 508 seconds, SD_A = 283.6, M_B = 635, SD_B = 409.1, $p = 0.02$). Interface F (the "Zoom mode"-inspired panel) was also significantly faster than Interface E (M_B = 635 seconds, SD_B = 409.1, M_C = 495, SD_C = 217.8, $p = 0.031$). These results indicate that Interface F was the fastest while Interface E was by far the slowest. We believe this difference is due to likely cognitive loads induced by the two distance panels. Interface E requires many mental calculations to yield important results, whereas no mental number calculations are needed to use Interfaces D or F. We suspect that Interface D performed similarly to Interface F in part because there were no calculations to be done and in part because it provided only vague information. For most of the run, the boxes displayed on Interface D were all red, so users tended to ignore them.

When comparing the number of collisions that occurred, the data supported our initial hypotheses. The number of collisions experienced using Interface D versus E was not significant (M_A = 8.78 collisions, SD_A = 3.72, M_B = 7.61, SD_B = 3.11, $p = 0.14$). However, both of these interfaces resulted in significantly more collisions than Interface F (M_C = 6.00 collisions, SD_C = 3.07; $p = 0.007$ and $p = 0.041$ for Interfaces D and E, respectively).

Overall, this study provided very conclusive results. We found the data closely matched our initial hypotheses. Interface F had significantly fewer collisions and yielded significantly faster run times than both Interfaces D and E. The total number of collisions that stemmed from this experiment was much larger than the number of collisions seen in our previous studies. As was previously stated, the arenas in this experiment were extremely narrow and operators were only allowed to be in teleoperation mode, so a larger number of total collisions were expected.

One factor that may limit generalization is that this experiment differed more than the studies that were carried out with the other interface versions. In this study, participants traversed a path and then returned along the same path. They did not have to search for victims or traverse a maze, which are challenges that the previous studies presented. As a result, participants may have been more apt to concentrate on the distance panel more than they would have otherwise because there was no threat of missing a victim or important landmark in the video. However, this study still shows that the zoom mode inspired distance panel of Interface F is superior to the previous one. As future work, we would like to conduct a study on Version 3 utilizing tasks more similar to those in the studies of Versions 1 and 2.

With respect to the new distance panel, the majority of the users (11 of 18) preferred Interface F, while six of the eighteen participants preferred Interface E. Some commented that having the exact numbers were a huge benefit and indicated that if somehow the numbers could be shown along with the lines from Interface F, they would like it better. Only one user selected Interface D as being the best, stating that he/she liked how it was less cluttered, but also that he/she was more used to the system by the third run (during which he/she used Interface D). We note that had this participant used a different interface on the final run, he/she may have chosen that interface as the favorite instead.

Three users did, however, say they liked Interface F the least. All three commented that the lines kept changing their distance, which made it hard to track. The sides and rear of the robot use sonar sensors to detect distance. Sonar sensors are inherently unstable and fluctuate a great deal. There is an averaging algorithm being performed as the robot collects the distance readings to try to minimize this fluctuation, but because Interface F is easy to interpret, every shift is noticed. With Interface D, the box will most likely stay the same color, or in Interface E, users may not notice fluctuating numbers as much if they are not looking directly at the panel. We believe this result is related to the quality of the sensor data, rather than the quality of the interface, because if there were laser sensors on the sides and rear of the robot, instead of the sonar sensors, these fluttering lines would not occur. The movement of the lines as the robot moves through the environment would be much more fluid. Fourteen of the eighteen uses disliked Interface D the most and one user disliked Interface E the most.

About half the users preferred having Interface F in its perspective view, while the other half preferred it in the top-down view, which suggests that the ability to switch between views should be preserved in future versions of the interface. Most users generally chose which view they preferred at the beginning of the run and continued to use it throughout the study. Several participants, however, did change the panel's view at various times during the run. Generally these users would put the panel in the top-down view when the front of the robot was very close to an obstacle.

9. Version 4

9.1 Interface description

For this version of the interface, we investigated the impact of a multi-touch interaction device on robot control.

The last few years have shown a great deal of interest in multi-touch tabletop and screen display research. Hardware solutions such as the Mitsubishi DiamondTouch (Dietz & Leigh, 2001), Microsoft Surface, and Touchtable by TouchTable, Inc., have been in low volume production for some time now. Increases in processor and graphics co-processor speeds have allowed for innovative software solutions that now rival the responsiveness of exclusively hardware solutions (Han, 2005).

By removing the joystick, mouse, or keyboard from the interaction, we increase the degree of direct manipulation by removing a layer of interface abstraction (Shneiderman, 1983). In the case of human-robot interaction, this should allow users to more directly interact with the robot and affect its behavior. To our knowledge, this study represents the first use of a multi-touch table with a physical agent. Many unexpected events occur when a system contains a moving, semi-autonomous physical object that is affecting the world. As such, we must determine if multi-touch interaction decreases the performance of systems in the real, dynamic, and noisy world.

To enable a baseline comparison that isolated the effect of the different interaction modality (traditional display device and joystick versus multi-touch device), we needed to ensure that the multi-touch interface was visually identical to Version 3 of the interface. The goal was to duplicate all of the functionality without creating any confounding issues in presentation or arrangement of display elements. Each of the discrete interaction elements is shown in Figure 2 and was previously described in Section 5. Besides the drive control panel, we made no visible changes to the interface.

Despite this attention to visual similarity between Versions 3 and 4, the DiamondTouch was immediately able to provide more functionality than the joystick could alone. For example,

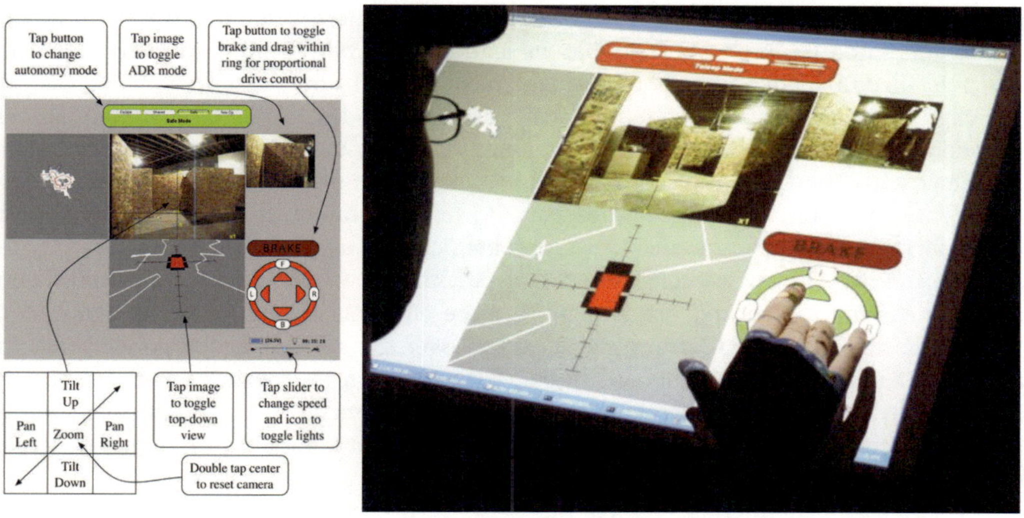

Fig. 2. Screenshot and guide to gestures (left) allowed the user (right) to activate interface features and autonomy modes.

the autonomy mode selection was offloaded to the keyboard in the joystick interface due to a limited number of buttons. In the case of the DiamondTouch, the buttons that were already displayed were used for this purpose. This "free functionality" was also true for the distance panel, speed control, and light control.

9.2 Evaluation description

Since the two interfaces make use of the same graphical elements and provide the same functionality, we hypothesized that performance using the two interfaces would be comparable. As in the previous studies, we used a within-subject design with all participants using both the Version 3 and Version 4 interfaces. Participants consisted of six trained search and rescue personnel (four men and two women).

We assessed the positive, or constructive, aspects of performance based on measuring the number of victims found and the amount of new or unique territory the robot covered while traversing the arena. These measurements are related because it is difficult to find additional victims if the operator is not successful in maneuvering the robot into previously unexplored areas. Collisions constituted a measure of destructive performance.

9.3 Evaluation results

Participants explored an average of 376 square feet and found an average of 5 victims when using the joystick-based interface (SD = 90.4 and SD = 2.3, respectively). The DiamondTouch interface shows remarkably similar results: participants directed robots to 373.3 square feet of territory and found 5.7 victims (SD = 107.4, SD = 2.9, respectively). Thus, there is no significant difference in the constructive performance of the two interfaces.

Paired, two-tailed *t*-tests (df = 5) indicated that there were no significant differences with respect to the numbers of collisions participants made using each interface ($M_{Joystick} = 1.54$ collisions, $SD_{Joystick} = 4.22$, $M_{Touch} = 1.92$, $SD_{Touch} = 2.8$). Thus we confirmed that there was no difference in constructive or destructive performance when using the two interfaces.

Now that we know that performance is not degraded by the act of porting the interface to the DiamondTouch table, we can begin optimizing the design for use with multi-touch interaction based on incorporating what we learned from participants' subjective feedback and a detailed understanding of how they interacted with the interface.

To capture participants' preferences, we asked them six Semantic Differential-scale questions. Using a scale of one to five, we asked how they would rate each interface along six dimensions: hindered in performing the task/helped in performing the task, difficult to learn/easy to learn, difficult to use/easy to use, irritating to use/pleasant to use, uncomfortable to use/comfortable to use, and inefficient to use/efficient to use.

Prior to the experiment we conjectured that participants would find the DiamondTouch interface easier to learn and use and to be more efficient. The rationale for the ease of learning is that the controls are more dispersed over the table and incorporated into the areas that they relate to, as opposed to being clustered on the joystick where users must remember what motions and buttons are used for what functions. The predictions for ease of use and efficiency result from our postulation that an interface with a higher degree of direct manipulation will be easier and faster to use.

The DiamondTouch interface scored the same or higher on average in all categories, although four of these categories evidenced no statistically significant difference. We found weak significance using a paired, 1-tailed *t*-test (df = 5) for ease of learning ($M_{Joystick} = 4.7$,

$SD_{Joystick} = 0.5$, $M_{Touch} = 5.0$, $SD_{Touch} = 0.0$, $p = 0.088$,) and efficiency ($M_{Joystick} = 3.3$, $SD_{Joystick} = 1.2$, $M_{Touch} = 4.33$, $SD_{Touch} = 0.8$, $p = 0.055$) and assert that it is likely we would have attained true significance with a greater number of participants.

We believe that the scores given the DiamondTouch interface with respect to its ease of use suffered because of several implementation problems. Sometimes the robot did not receive the "recenter camera" command despite the fact that the participants were using the correct gesture to send that command, requiring the participants to frequently repeat the recentering gesture. At other times, the participants attempted to send that command by tapping on the very edge of the region in which that command could be activated, so sometimes the gesture was effective and at other times it failed, and it was difficult and frustrating for the participants to understand why the failures occurred. Also, it was not always clear to participants how to form the optimal gestures to direct the robot's movement.

Because differences in Semantic Differential-scale scores for ease of learning and pleasantness to use were on the edge of significance, we looked for other supporting or disconfirming evidence. We noted that participants asked questions about how to activate functions during the runs, which we interpreted as indication that the participants were still learning the interface controls despite having been given standardized training. Accordingly, we investigated the number of questions they asked about each system during the runs as well as the number of times they showed uncertainty in finding a particular function such as a different autonomy mode. We found that five of the six participants asked a total of eight questions about the joystick interface and one participant asked two questions about the DiamondTouch interface ($p = 0.072$, df = 5 for paired, 1-tailed t-test). This result, while again being on the edge of significance due to the small sample size, tends to support the contention that the DiamondTouch interface is easier to learn than the joystick interface.[2]

10. Conclusions and future work

Through our iterative design and testing process, we succeeded in providing a useful surroundings awareness panel that displays accurate data to the user in an easy-to-interpret manner. In the testing for Version 3, the current distance panel was proven to provide faster run times, with fewer collisions than the previous two versions.

Our results support the usefulness of the guidelines we followed in the creation of the interface. For example, we fused sensor information to lower the cognitive load on the user. Having the laser and sonar sensor values being displayed in the same distance panel provided users a single interface through which to access distance information. Through an iterative process, we gradually improved the distance panel. This panel rotates when the operator pans the camera, which allows the user to line up the obstacle they see in the video with where it is represented in the distance panel, to help reduce cognitive load.

The distance panel also functions as a camera pan indicator. To provide redundant cueing in a location where operators will be naturally focused much of the time, crosshairs are overlaid on the video screen to show the current pan/tilt position of the main camera.

---

[2] For an analysis of the gestures participants used to direct the robot using the Version 4 multi-touch interface, see Micire et al. (2009).

Additionally, we provide indicators of robot health and state, as well as include information on which camera is currently in the main display. Finally, we have shown that the ability to see the robot's chassis improves surroundings awareness. This finding provides strong support for the guideline that states the operator should have the ability to inspect the robot's body for damage or entangled obstacles.

Through this iterative design and testing process, we have also added the following guidelines to enhance the list of previously-reported guidelines:

- Important information should be presented on or very close to the video screen. Users primarily pay attention to the video screen, so keeping important information on or near it makes it more noticeable.
- If the robot system has more than one camera, a second camera should be mounted facing the rear of the robot to provide enhanced awareness of the robot's surroundings.
- If the robot system has more than one camera, the system should include an ADR mode to improve awareness and reduce the number of collisions that occur while the robot is backing up.

Once we completed three iterations of the interface design, we investigated alternative interaction methods. A joystick interface limits the user to a relatively small set of interaction possibilities. The multi-touch surface is quite different, allowing for numerous interaction methods using a large set of gestures on a 2D plane. However, the flexibility of the interface also presents a problem for the designer, who must carefully choose control methods that give clear affordances and appropriate feedback to the user. Users are accustomed to haptic feedback, such as spring-loaded buttons and gimbals, and auditory feedback, such as clicks, even from a non-force-feedback joystick controller.

Nevertheless, the results show promise for the multi-touch interface since little optimization was actually performed during the porting process. In fact, we know that several of the interaction methods that survived the porting process are sub-optimal, and yet performance was not degraded. This research thus provides a good baseline. We are confident that more can be done to enrich the user experience because we no longer are limited to the constraints of the number of degrees of freedom of a joystick. Because this is a software system, it is easier to iteratively tailor the interaction approach using a multi-touch table than when using a joystick. This feature strikes a beneficial middle ground between a software and hardware solution for interaction functionality.

Our future work will focus on the lessons learned from this experiment, particularly designing new versions of the interface that are optimized for the multi-touch display. We will explore direct map manipulation and "point to send the robot here" commands that can provide easier navigation.

11. Acknowledgments

Michael Baker, Robert Casey, Andrew Chanler, and Philip Thoren contributed to the development of the robot system. The authors wish to thank the experiment participants. We would also like to thank Harold Bufford, Munjal Desai and Kate Tsui for their data gathering assistance and Kristen Stubbs and Kate Tsui for their editing assistance. Thanks to Mitsubishi Electric Research Laboratories for the extended use of a DiamondTouch tabletop. And finally, thanks to Adam Jacoff, Elena Messina, and Ann Virts at NIST for the use of their facilities for and support throughout the testing. This research was supported in part by the National Science Foundation (IIS-0415224, IIS-0308186, IIS-0546309) and the National Institute of Standards and Technology (70NANB3H1116).

12. References

J. A. Adams. Unmanned vehicle situation awareness: a path forward, *Proceedings of the 2007 Human Systems Integration Symposium*, 2007.

D. Bruemmer, D. Dudenhoeffer, and J. Marble. Dynamic autonomy for urban search and rescue. In *Proceedings of the 2002 AAAI Mobile Robot Workshop*, Edmonton, Canada, August 2002.

P. Dietz and D. Leigh. DiamondTouch: a multi-user touch technology. In *Proceedings of the ACM Symposium on User Interface Software and Technology*, pages 219–226, November 2001.

J. Drury, B. Keyes, and H. Yanco. LASSOing HRI: analyzing situation awareness in map-centric and video-centric interfaces. In *Proceedings of the ACM SIGCHI/SIGART Conference on Human-Robot Interaction*, pages 279–286, 2007.

J. Drury, L. Riek, A. Christiansen, Z. Eyler-Walker, A. Maggi, and D. Smith. Evaluating Human-Robot Interaction in a Search-and-Rescue Context. In *Proceedings of the Performance Metrics for Intelligent System (PerMIS) Workshop*, 2003.

J. Drury, J. Scholtz, and H. Yanco. Awareness in human-robot interactions. In *Proceedings of the IEEE International Conference on Systems, Man and Cybernetics*, volume 1, 2003.

M. R. Endsley. Design and evaluation for situation awareness enhancement. In *Proceedings of Human Factors Society 32nd Annual Meeting*, Santa Monica, CA, 1988.

K. A. Ericsson and H. A. Simon. Verbal reports as data. *Psychological Review*, 87:215–251, 1980.

J. Y. Han. Low-cost multi-touch sensing through frustrated total internal reflection. In *Proceedings of the ACM Symposium on User Interface Software and Technology*, 2005.

A. Hjelmfelt and M. Pokrant. Coherent Tactical Picture. CNA RM, pages 97–129, 1998.

A. Howard. Simple Mapping Utilities, http://www-robotics.usc.edu/~ahoward/pmap/index.html Accessed on Oct. 15, 2009.

A. Jacoff, E. Messina, and J. Evans. A standard test course for urban search and rescue robots. In *Proceedings of the Performance Metrics for Intelligent Systems (PerMIS) Workshop*. National Institute of Science and Technology, 2000.

A. Jacoff, E. Messina, and J. Evans. A reference test course for autonomous mobile robots. In *Proceedings of the SPIE-AeroSense Conference*. International Society for Optical Engineering, 2001.

A. Jacoff, E. Messina, and J. Evans. Performance evaluation of autonomous mobile robots. *Industrial Robot: An International Journal*, 29(3):259–267, 2002.

M. W. Kadous, R. Ka-Man Sheh, and C. Sammut. Effective user interface design for rescue robotics. In *Proceedings of the ACM/IEEE Conference on Human-Robot Interaction*, 2006.

B. Keyes, R. Casey, H. Yanco, B. Maxwell, and Y. Georgiev. Camera placement and multi-camera fusion for remote robot operation. In *Proceedings of the IEEE International Workshop on Safety, Security and Rescue Robotics*, 2006.

W. Lee, H. Ryu, G. Yang, H. Kim, Y. Park, and S. Bang. Design guidelines for map-based human-robot interfaces: A collocated workspace perspective. *International Journal Of Industrial Ergonomics*, 37(7), 589—604, July 2007.

B. Maxwell, N. Ward, and F. Heckel. A configurable interface and architecture for robot rescue. In *Proceedings of the 2004 AAAI Mobile Robotics Workshop*, San Jose, CA, 2004.

M. Micire, J. L. Drury, B. Keyes, and H. A. Yanco. Multi-touch interaction for robot control. In *Proceedings of the Intelligent User Interfaces 2009 Conference*, Sanibel Island, FL, February 2009.

C. Nielsen and M. Goodrich. Comparing the usefulness of video and map information in navigation tasks. In *Proceedings of the 1st ACM SIGCHI/SIGART Conference on Human-Robot Interaction*, pages 95–101. ACM New York, NY, USA, 2006.

C. W. Nielsen, M. A. Goodrich, and B. Ricks. Ecological interfaces for improving mobile robot teleoperation. *IEEE Transactions on Robotics and Automation*. Vol 23, No 5, pp. 927-941, October 2007.

C. Nielsen, B. Ricks, M. Goodrich, D. Bruemmer, D. Few, and M. Few. Snapshots for semantic maps. In *Proceedings of the IEEE International Conference on Systems, Man and Cybernetics*, volume 3, 2004.

J. Nielsen. *Usability Engineering*. San Diego: Academic Press, 1993.

J. Scholtz, J. Young, J. L. Drury, and H. A. Yanco. Evaluation of human-robot interaction awareness in search and rescue. In *Proceedings of the IEEE International Conference on Robotics and Automation (ICRA)*, New Orleans, April 2004.

B. Shneiderman. Direct manipulation: a step beyond programming languages. *IEEE Computer*, 16:57–69, 1983.

H. A. Yanco and J. L. Drury. A taxonomy for human-robot interaction. In *Proceedings of the AAAI Fall Symposium on Human-Robot Interaction*, 2002.

H. A. Yanco and J. L. Drury. Rescuing interfaces: a multi-year study of human-robot interaction at the AAAI Robot Rescue Competition. *Autonomous Robots Journal*, Special Issue on the AAAI Mobile Robot Competition and Exhibition, Dec. 2006.

H. A. Yanco, J. L. Drury, and J. Scholtz. Beyond usability evaluation: analysis of human-robot interaction at a major robotics competition. *Human-Computer Interaction*, 19(1&2):117–149, 2004.

H. A. Yanco and J. L. Drury. "Where am I?": acquiring situation awareness using a remote robot platform. In *Proceedings of the IEEE Conference on Systems, Man and Cybernetics*, October 2004.

H. A. Yanco, B. Keyes, J. L. Drury, C. W. Nielsen, D. A. Few, and D. J. Bruemmer. Evolving interface design for robot search tasks. *Journal of Field Robotics*, 24:779–799, August/September 2007.

L. Zalud. ARGOS – System for heterogeneous mobile robot teleoperation. In *Proceedings of the IEEE/RSJ International Conference on Intelligent Robots and Systems*, pages 211–216, 2006.

Skill Based Approach with Human-Robot Interaction

15

Safe Cooperation between Human Operators and Visually Controlled Industrial Manipulators

J. A. Corrales, G. J. Garcia, F. A. Candelas, J. Pomares and F. Torres
University of Alicante
Spain

1. Introduction

The development of industrial tasks where human operators and robots collaborate in order to perform a common goal is a challenging research topic on present robotics. Nowadays, industrial robots are isolated in fenced workspaces where human cannot enter in order to avoid collisions. However, this configuration misses the opportunity of developing more flexible industrial tasks where humans and robots work together. This collaboration takes advantage of the complementary features of both entities. On the one hand, industrial robots are able to perform repetitive tasks which can be exhausting and monotonous for human operators. On the other hand, humans are able to perform specialized tasks which require intelligence or dexterity. Thereby, industrial tasks can be improved substantially by making humans and robots collaborate in the same workspace. The main goal of the work described in this chapter is the development of a human-robot interaction system which enables this collaboration and guarantees the safety of the human. This system is composed of two subsystems: the human tracking system and the robot control system.

The human tracking system deals with the precise real-time localization of the human operator in the industrial environment. It is composed of two systems: an inertial motion capture system and an Ultra-WideBand localization system. On the one hand, the inertial motion capture system is composed of 18 IMUs (Inertial Measurement Units) which are attached to the body of the human operator. These sensors obtain precise orientation measurements of the limbs of the body of the operator and thus full-body movements of the operator are tracked. On the other hand, the Ultra-WideBand localization system obtains a precise measurement of the global position of the human operator in the environment. The measurements of both systems are combined with a Kalman filter algorithm.

Thereby, all the limbs of the body of the human operator are positioned precisely in the environment while the human-robot interaction task is performed. These measurements are applied over a skeleton which represents the basic structure of the human body. However, this representation by itself does not take into account the actual dimensions of the surface of the body because each limb is modelled as a line. The bones of this skeleton have been covered with bounding volumes whose dimensions match approximately the size of the corresponding human limbs. The implemented bounding volumes have been organized in a three-level hierarchy in order to reduce the computational cost of the distance computation.

The robot control system is based on visual servoing. Most current industrial robots are controlled only with kinematic information without permitting the interaction of the robot with its environment. Nevertheless, the robot needs sensors in order to adapt its behaviour depending on the changes in the environment when human-robot interaction tasks are performed. In this work, the robot has a camera mounted at its end-effector to control its movements according to a visual servoing method. The main objective of visual servoing is to minimize the error between the image obtained at the first pose of the robot and the image obtained at the target position of the robot. Visual servoing is adequate to control the robot in situations where external or not planned objects enter the robot workspace, avoiding a possible collision. When the robot has to perform a planned path in a 3D space limited by other objects, classic image based visual servoing fails to track this planned path. Previous research on visual path tracking has tried to solve this problem by making use of visual servoing to follow a desired image path previously sampled in time. These systems can be modified in order to obtain a human safety algorithm. In this way, when a human is dangerously close to the robot, the path tracking must be stopped. However, after the danger of collision has disappeared, previous image trajectory tracking methods based on visual servoing fail to return to the initial path because they are time-dependent. Therefore, a time-independent behaviour of the control system is crucial to develop interactions with the workspace. This time-independent behaviour makes the robot continue on the same path from the same point that it was tracking before the detection of the human. The method described in this chapter guarantees the correct tracking in the 3D space at a constant desired velocity.

As shown before, a safety behaviour which stops the normal path tracking of the robot is performed when the robot and the human are too close. This safety behaviour has been implemented through a multi-threaded software architecture in order to share information between both systems. Thereby, the localization measurements obtained by the human tracking system are processed by the robot control system to compute the minimum human-robot distance and determine if the safety behaviour must be activated.

This chapter is organized as follows. Section 2 describes the human tracking system which is used to localize precisely all the limbs of the human operator who collaborates with the robotic manipulator. Section 3 presents an introduction to visual servoing and shows how the robot is controlled by a time-independent path tracker based on it. Section 4 describes how the safety behaviour which avoids any collision between the human and the robot is implemented. This section presents the hierarchy of bounding volumes which is used to compute the human-robot distance and modify the movements of the robot accordingly. Section 5 enumerates the results obtained in the application of all these techniques in a real human-robot interaction task where a fridge is disassembled. Finally, the last section presents the conclusions of the developed research.

2. Human tracking system

2.1 Components of the human tracking system

The human operator who interacts with robotic manipulators has to be localized precisely in the industrial workplace because of two main reasons. On the one hand, the knowledge of the human location enables the development of safety behaviours which avoid any risk for the physical integrity of the human while their interaction takes place. On the other hand, the localization of the human operator can also be taken into account to modify the movements of the robot accordingly. The movements of the robot are adapted to the human's behaviour and thus more flexible human-robot interaction tasks can be implemented.

The necessary precision of the human localization system depends on the type of interaction task and the distance between the human and the robot. For instance, in interaction tasks where the human operator and the robot collaborate on the assembly of big sized products, a global localization system which only obtains a general position of the human and the robot is sufficient because they work far from each other. Nevertheless, in most interaction tasks, the robot and the human need to collaborate in close distances and a localization of their limbs and links is required. The position of each link of the robotic manipulator can be easily obtained from the robot controller through forward kinematics. However, an additional motion capture system is necessary to register the movements of all the limbs of the human operator.

In this chapter, an inertial motion capture suit (*GypsyGyro-18* from *Animazoo*) has been used to localize precisely all the limbs of the human operator. This system has several advantages over other motion capture technologies (Welch & Foxlin, 2002): it does not suffer from magnetic interferences (unlike magnetic systems), optical occlusions do not affect its precision (unlike vision systems) and it is comfortable to wear (unlike mechanical systems). The main component of this inertial system is a lycra suit with 18 IMUs (Inertial Measurement Units) which is worn by the human operator. Each IMU is composed of 3 MEMS (Micro-Electro-Mechanical Systems) gyroscopes, 3 accelerometers and 3 magnetometers. The measurements from these 9 sensors are combined by a complementary Kalman filter (Foxlin, 1996) in order to obtain the orientation (relative rotation angles: roll, pitch and yaw) of the limb to which the IMU is attached. This inertial motion capture system not only registers the relative orientations of all the limbs of the human's body but it also computes an estimation of the global displacement of the human through a foot-step extrapolation algorithm. Nevertheless, this algorithm sometimes does not detect correctly when a new step takes place and this fact involves the accumulation of a small global position error (drift) which becomes excessive after several steps (Corrales et al., 2008).

An additional global localization system based on UWB signals (*Ubisense v.1* from *Ubisense*) has been used to solve this problem and correct the error accumulated by the inertial motion capture system in the global positioning of the human. This localization system is based on two different devices: four sensors which are installed at fixed positions of the industrial workplace and a small tag which is worn by the human operator. This small tag sends UWB pulses which are processed by the four sensors in order to compute an estimation of the global position of the tag in the environment by implementing a combination of AoA (Angle of Arrival) and TDoA (Time-Difference of Arrival) techniques. Thereby, this UWB system obtains precise estimates of the global position of the human operator. The global position measurements of both systems are combined through the fusion algorithm described in the following section.

2.2 Fusion algorithm

The two systems (inertial motion capture system and UWB system) which compose the developed human tracking system have complementary features which show the suitability of their combination. On the one hand, the inertial motion capture system registers precise relative limbs orientations with a high sampling rate (30 - 120Hz). However, the global position estimated by this system is prone to accumulate drift. On the other hand, the UWB localization system calculates a more precise global position of the human operator but with a considerably lower sampling rate (5 - 10Hz). Furthermore, the measurements of the UWB system can be easily related to fixed objects in the environment because they are represented in a static coordinate system while the measurements of the inertial motion capture system are represented in a dynamic coordinate system which is established every time the system is initialized. All these complementary features have been taken into account in order to develop a fusion algorithm which estimates precisely the position of the human operator from the position measurements of both tracking systems (Corrales et al., 2008). The reader can see Table 1 for a detailed description of the implemented fusion algorithm. The limbs orientation measurements from the inertial motion capture system are not processed by this algorithm because they are very precise and do not need any correction process.

01: Initialize ${}^{U}\mathbf{T}_{G}$ with the first two measurements: $\mathbf{p}_1^G$, $\mathbf{p}_2^U$
02: *for each* measurement $\mathbf{p}_t$
03: *if* $\mathbf{p}_t$ is from the inertial motion capture system *G*
04: *if* $\mathbf{p}_{t-1}$ is from the UWB localization system *U*
05: Recalculate ${}^{U}\mathbf{T}_{G}$ with $\mathbf{p}_t$ and $\mathbf{x}_{t-1}$
06: *end if*
07: Transform $\mathbf{p}_t$ from *G* to *U* with ${}^{U}\mathbf{T}_{G}$
08: $\mathbf{x}_t$ =KalmanFilterPrediction($\mathbf{p}_t$)
09: *else if* $\mathbf{p}_t$ is from the UWB localization system *U*
10: $\mathbf{x}_t$ =KalmanFilterCorrection($\mathbf{p}_t$, $\mathbf{x}_{t-1}$)
11: *end if*
12: *end for*

Table 1. Pseudocode of the fusion algorithm based on a Kalman filter.

First of all, the global position measurements registered by both tracking systems have to be represented in the same coordinate system. The static coordinate system *U* of the UWB system has been chosen as reference system and thus the inertial motion capture measurements have to be transformed from their frame G to it. The first two measurements of both systems are used to compute the transformation matrix ${}^{U}\mathbf{T}_{G}$ between their coordinate systems (line 1 of Table 1). If the inertial motion capture system did not have any errors, this initial value of the transformation matrix would always be valid. Nevertheless, as the inertial motion capture measurements accumulate a drift; this transformation matrix has to be updated accordingly. This update process is based on a Kalman filter (Thrun et al., 2005) which aims to reduce the accumulated drift.
The state of the implemented Kalman filter is composed by the 3D position $\mathbf{x}_t = (x_t, y_t, z_t)$ of the human operator. Each time a measurement from one of the tracking systems is received, one of the steps of the Kalman filter (prediction or correction) is executed. The global position measurements $\mathbf{p}_t^G$ of the inertial motion capture system are applied in the prediction step of the Kalman filter (line 8 of Table 1) while the global position measurements $\mathbf{p}_t^U$ of the UWB localization system are applied in the correction step (line 10 of Table 1). An estimate $\hat{\mathbf{x}}_t$ of the position of the human operator is obtained from each of these filter steps. In addition, the transformation matrix ${}^{U}\mathbf{T}_{G}$ is recalculated after each correction step of the filter (line 5 of Table 1). Thereby, the drift accumulated by the inertial motion capture system is corrected because the next measurements of this system are transformed with this new transformation matrix (line 7 of Table 1).

The structure of the developed fusion algorithm takes the most of the complementary features of both tracking systems. On the one hand, the application of the measurements of the motion capture system in the prediction step of the filter maintains their high sampling rate and enables the tracking of quick movements of the human. On the other hand, the application of the measurements of the UWB system in the correction step of the filter removes the drift accumulated by the previous measurements of the motion capture system. Thereby, the resulting tracking system from this fusion algorithm has the high sampling rate of the motion capture system and the precision of the UWB system.

3. Robot control system

Nowadays, the great majority of repetitive assembly or disassembly tasks are commonly developed in the industry by robot manipulators. These tasks are usually defined by a set of poses for the robot manipulator. The robot is isolated and its workspace is constant. The robot is frequently controlled kinematically by different poses that it has to perform in order to complete the task. Nevertheless, when the robot workspace is not constant, the robot needs additional information in order to react to any unexpected situation. This is the case we are dealing with in this chapter. Sight allows the human brain to perceive the shapes, the colours and the movements. Computer vision permits the robot to obtain important information about a changing environment. Visual servoing is a technique that controls the robot movements using the visual information processed by a computer vision system. In Section 3.1 an introduction of visual servoing is presented. Unfortunately, classic visual servoing systems are only adequate for placing the robot in a relative position between it and an object in the environment. In this case, the 3D path that the robot follows to arrive to this goal position is not controlled in any way. When this 3D path followed by the robot must be controlled, classic visual servoing controllers have to be modified in order to precisely track the predefined path. These modifications are described in section 3.2.

3.1 Introduction to visual servoing

Visual servoing is a technique which uses visual information to control the motion of a robot (Hutchinson et al., 1996). It is a technique widely developed in the literature in the last two decades. There are two basic types of visual servoing: the image-based visual servoing and the position-based visual servoing. Image-based visual servoing uses only visual features extracted from the acquired images, $\mathbf{s}$, to control the robot. Therefore, these controllers do not need neither a complete 3D model of the scene nor a perfect camera calibration. The desired visual features, $\mathbf{s}_d$, are obtained from a desired final position of the robot in the scene. Image-based visual servoing controllers minimize the error between any robot position and the goal position by minimizing the error of the visual features computed from the images acquired at each robot position, $\mathbf{e} = (\mathbf{s} - \mathbf{s}_d)$. To minimize exponentially this error $\mathbf{e}$, a proportional control law is used:

$$\dot{\mathbf{e}} = -\lambda \mathbf{e} \tag{1}$$

where λ is a proportional gain.

In a basic image-based visual servoing approach, the velocity of the camera, $\mathbf{v}_c$ is the command input for controlling the robot movements. To obtain the control law, the interaction matrix, $\mathbf{L}_s$, must be firstly presented. The interaction matrix is a matrix that relates the variations of the visual features in the image with the variations of the poses of the camera in the 3D space, i.e. its velocity (Chaumette & Hutchinson, 2006).

$$\dot{\mathbf{s}} = \mathbf{L}_s \mathbf{v}_c \quad (2)$$

From Equation (1) and Equation (2), the velocity of the camera to minimize exponentially the error in the image is obtained:

$$\mathbf{v}_c = -\lambda \hat{\mathbf{L}}_s^+ \left(\mathbf{s} - \mathbf{s}_d\right) \quad (3)$$

where $\hat{\mathbf{L}}_s^+$ is the pseudoinverse of an approximation of the interaction matrix. This camera velocity is then transformed to obtain the velocity to be applied to the end-effector of the robot. To do this, the constant relation between the camera and the end-effector is used in a camera-in-hand configuration (i.e., the camera is mounted at the end-effector of the robot). If the system uses a camera-to-hand configuration, the relation between the robot base and the camera is usually known, and the forward kinematics provides the relation between the robot base and the end-effector coordinate systems.

Image-based visual servoing systems present singularities and/or local minima problems in large displacements tasks (Chaumette & Hutchinson, 2006). To overcome these drawbacks maintaining the good properties of image-based visual servoing robustness with regard to modeling and camera calibration errors, the tracking of a sufficiently sampled path between the two distant poses can be performed.

3.2 Time-independent visual servoing path tracking

Path planning is a commonly studied topic in visual servoing (Fioravanti, 2008). However, the technique used to perform the tracking of the planned image path is not usually presented. The research effort is usually focused on planning the trajectory of the visual features in the image. The main objective of this planning is to avoid the outliers features or the robot joint limits. Only some of the systems proposed up to now to plan trajectories in the image using visual servoing present the technique chosen to perform the tracking (Chesi & Hung, 2007; Malis, 2004; Mezouar & Chaumette, 2002; Pomares & Torres, 2005; Schramm & Morel, 2006). Most of them resolve the problem by sampling the path in the time (i.e., the trajectory is generated from a path and a time law) (Chesi & Hung, 2007; Malis, 2004; Mezouar & Chaumette, 2002). These systems employ a temporal reference, $\mathbf{s}_d(t)$:

$$\mathbf{v}_c = -\lambda \hat{\mathbf{L}}_s^+ \left(\mathbf{s} - \mathbf{s}_d(t)\right) \quad (4)$$

This control action is similar to the one employed by an image-based visual servoing (3). However, in (4) image features of the desired trajectory are obtained from a time-dependent function $\mathbf{s}_d(t)$. These systems do not guarantee the correct tracking of the path when the robot interacts with its environment. If the robot interacts with an object placed in its workspace, the system continues sending visual references to it even though the robot cannot move. Once the obstruction ends, the references that have been sent up to that moment are lost, not allowing the correct tracking of the path. A robot controller with this time-dependent behaviour is not valid for a global human-robot interaction system. When the safety system takes part in the task because the human and the robot are too near, the robot has to move away from the human and the time-dependent visual servoing system loses the references until the distance is safe again.

Only a few tracking systems based on visual servoing have been found in the literature with a time-independent behaviour (Schramm & Morel, 2006; Pomares & Torres, 2005). However, it is not possible to specify the desired velocity at which the robot tracks the trajectory with these previous visual servoing systems. In particular, (Schramm & Morel, 2006) present a visual tracker which reduces the tracking velocity to zero at every intermediate reference of the tracked trajectory. The movement flow-based visual servoing by (Pomares & Torres, 2005) does not stop the robot at each intermediate reference but it does not guarantee a minimum desired tracking velocity. Furthermore, this method suffers from great oscillations when the tracking velocity is increased. The proposed time-independent visual servoing system overcomes this limitation and the tracking velocity can be adjusted to a desired value $|\mathbf{v}_d|$. In this system, the desired visual features do not depend on the time. They are computed depending on the current position of the camera, $\mathbf{s}_d(\mathbf{q})$. This type of visual servoing does not lose any reference and thus, the robot is able to follow the complete path once the human-robot distance is safe again.
In the research exposed in this chapter, the planning path issue is not relevant. There are a lot of works related with this topic. The robot control system must be provided with a desired image path, and this is obtained here in a previous off-line stage. Before the tracking process, the discrete trajectory of the features in the image to be tracked by the robot is sampled $\boldsymbol{T} = \{ {}^k\mathbf{s}/k\in 1...N \}$, with ${}^k\mathbf{s}$ being the set of M points or features observed by the camera at instant k, ${}^k\mathbf{s} = \{ {}^k f_i / i\in 1...M \}$. For instance, Fig. 1 shows the desired image trajectory that the robot has to track to accomplish the task. This path stores the set of M image features (four laser points) which are observed by the camera at each instant k, ${}^k\mathbf{s}$.

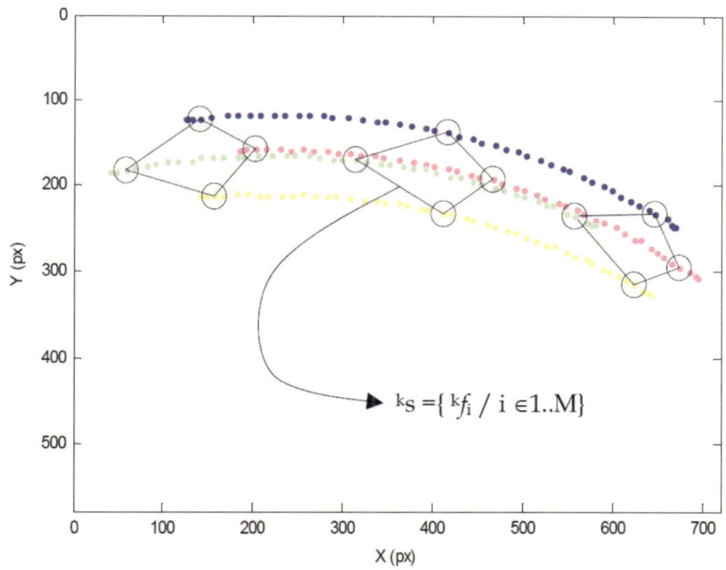

Fig. 1. Image trajectory to be tracked during the disassembly task.

The set of visual features observed at the initial camera position are represented by ${}^1\mathbf{s}$. From this initial set of image features, it is necessary to find an image configuration which provides the robot with the desired velocity $|\mathbf{v}_d|$ by iterating over the set $\boldsymbol{T}$. For each image configuration ${}^k\mathbf{s}$, the corresponding camera velocity is determined considering an image-based visual servoing system (at the first step $\mathbf{s} = {}^1\mathbf{s}$):

$$ {}^{k}\mathbf{v} = -\lambda \hat{\mathbf{L}}_{s}^{+} \left(\mathbf{s} - {}^{k}\mathbf{s} \right) \tag{6} $$

This process continues until the velocity $|{}^{k}\mathbf{v}|$ which is the nearest to $|\mathbf{v}_d|$ is obtained.
At this moment, the set of features ${}^{k}\mathbf{s} = {}^{i}\mathbf{s}$ will be the desired features to be used by an image-based visual servoing system (see Equation (3)) in order to at least maintain the desired velocity. However, the visual features which provide the exact desired velocity are between ${}^{k}\mathbf{s}$ and ${}^{k-1}\mathbf{s}$. To obtain the correct image features the interpolation method described in (Garcia et al., 2009b) is proposed. This interpolation method searches for a valid camera configuration between the poses of the camera in the k and k-1 image path references. To reconstruct the 3D pose of the camera from which the ${}^{k}\mathbf{s}$ and ${}^{k-1}\mathbf{s}$ sets of visual features are observed, virtual visual servoing is employed, taking advantage of all the images acquired until that moment and all the background of this technique to calibrate the camera during the visual servoing task (Marchand & Chaumette, 2001).
Therefore, once the control law represented in Equation (3) is executed, the system searches again for a new image configuration which provides the desired velocity. This process continues until the complete trajectory is tracked.

4. Human-robot integration

4.1 Human-robot interaction behaviour

The human-robot interaction behaviour implemented in this chapter is based on two main components: a hierarchy of bounding volumes and a safety strategy. The hierarchy of bounding volumes is used to model approximately the bodies of the human operator and the robot. The safety strategy computes the minimum distance between these bounding volumes and changes the behaviour of the robot accordingly. In the following sub-sections both elements are described in detail.

4.1.1 Hierarchy of bounding volumes

As it has been stated before, all the limbs of the human operator and all the links of the robotic manipulator have to be localized precisely in order to assure the safety of the human and develop flexible human-robot interaction tasks. The tracking system described in section 2 combines the measurements of an inertial motion capture suit and a UWB localization system (see Fig. 2a) in order to obtain the orientation of all the limbs of the human operator and her/his global position in the environment. All these measurements are applied over a skeleton structure (see Fig. 2b) which represents the kinematic structure (links and joints) of the human's body. The positions of the two ends of each link of this skeleton can be calculated by applying forward kinematics to the measurements of the tracking system.
However, this skeletal representation only considers the length of each link but it does not take into account the 3D dimensions of the link's surface. Therefore, an additional geometric representation is necessary to model the surface of the human's body and localize the human operator completely and precisely. This geometric representation has to be based on bounding volumes because a detailed mesh representation based on polygons would be too expensive for real-time operation. The selected bounding volume should fulfill two requirements: tight fitting to the links' surface and inexpensive distance computation. Previous similar human-robot interaction systems (Balan & Bone, 2006; Martinez-Salvador et al., 2003) implement sphere-based geometric representations to achieve this goal. These

sphere-based representations have an inexpensive distance computation but they do not fit tightly to the links' surface. A high number of spheres are needed to solve this problem and thus the final computational efficiency of this geometric representation is reduced.

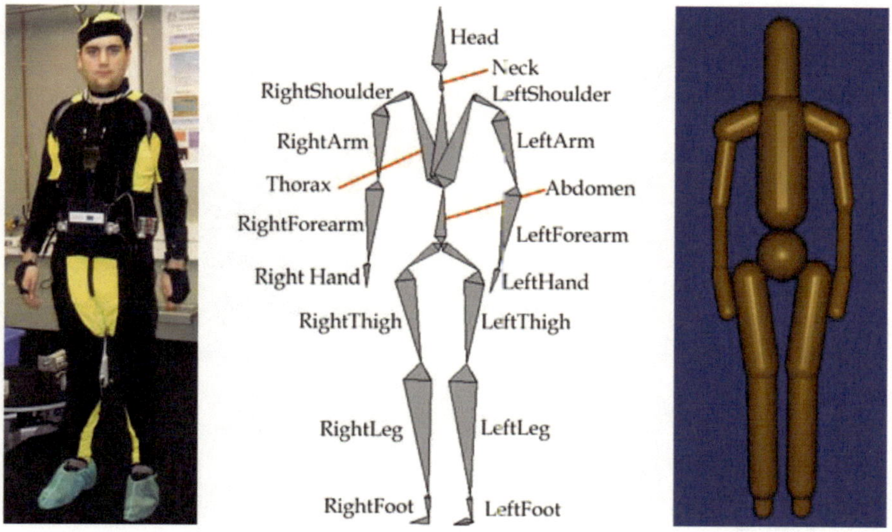

(a) (b) (c)

Fig. 2. Model for the human operator: (a) Human operator with the tracking system; (b) Links of the human; (c) SSL bounding volumes which cover the links of the human.

In this chapter, a new bounding volume representation based on Sphere-Swept Lines (SSLs) is presented to overcome the limitations of previous spherical models. A SSL is a bounding volume obtained from the Minkowski sum of a sphere and a segment (Ericson, 2005). SSLs have a better tight than spheres because they have a similar shape to the limbs. In addition, the distance computation between two SSLs has a low cost because it is calculated from the difference between the SSLs' radii and the distance between the inner segments of the SSLs (Schneider & Eberly, 2003). The geometric representation of the human operator is computed by covering each of the links of the skeleton obtained from the tracking system with a SSL (see Fig. 2c). The location of each SSL is updated on real-time by matching the two ends of the inner segment of the SSL with the ends of the corresponding link.

The robotic manipulator (see Fig. 3a) which collaborates with the human operator is modelled in a similar way. The positions of the links of the robot (see Fig. 3b) are computed by applying forward kinematics to the joint values registered by the robot controller. However, as in the case of the human operator, this procedure only obtains the locations of the ends of each link but it does not consider its surface. A SSL will cover each link in order to model all the dimensions of it.

This geometric representation consists of 18 SSLs for the human operator and 8 SSLs for the robotic manipulator. This fact implies that each time the minimum distance between the human and the robot is computed, 144 pairwise distance tests are required. In order to reduce this number of distance tests and increase the performance of the distance computation, a three-level hierarchy of bounding volumes has been developed. Fig. 4 and Fig. 6 depict the relations between the components of this hierarchy for the human operator and the robotic manipulator, respectively. Fig. 5 and Fig. 7 show the 3D representation of the bounding volumes which compose each level of this hierarchy for the human operator and the robotic manipulator, respectively.

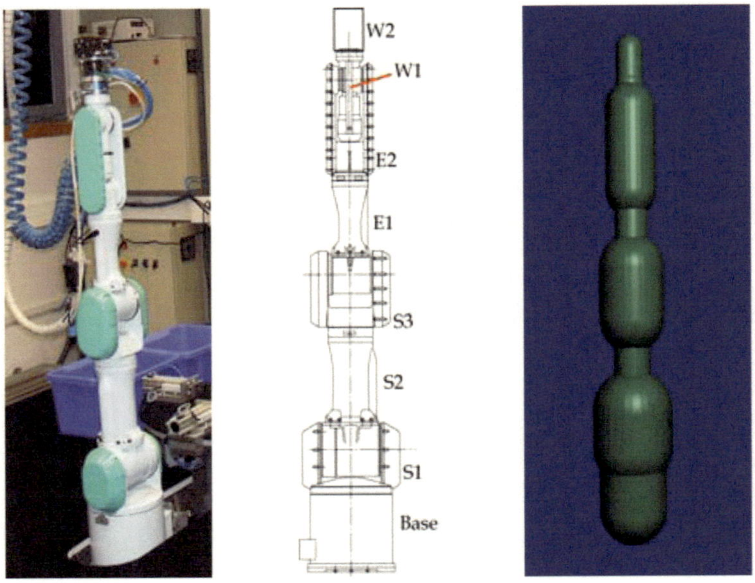

(a) (b) (c)

Fig. 3. Model for the robotic manipulator: (a) PA-10 manipulator; (b) Links of the robot; (c) SSL bounding volumes which cover the links of the robot.

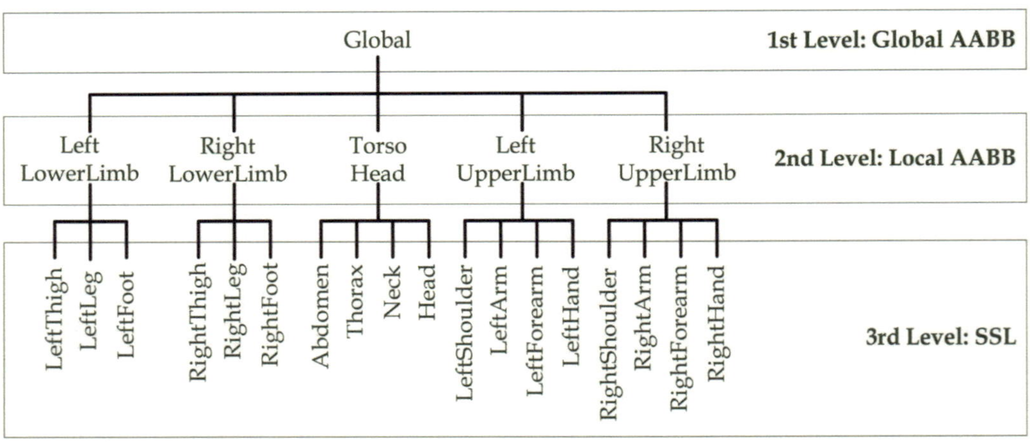

Fig. 4. Hierarchy of bounding volumes which cover the body of the human operator.

The third level of this hierarchy is composed by the SSL bounding volumes described above (see Fig. 2c and Fig. 3c) and it will only be used when the human operator and the manipulator are working quite close to each other. A distance threshold will be established in order to determine whether the third level or the second level of this hierarchy have to be applied to obtain the minimum human-robot distance. The second level of the bounding volume hierarchy is composed by AABBs (Axis-Aligned Bounding Boxes) which cover

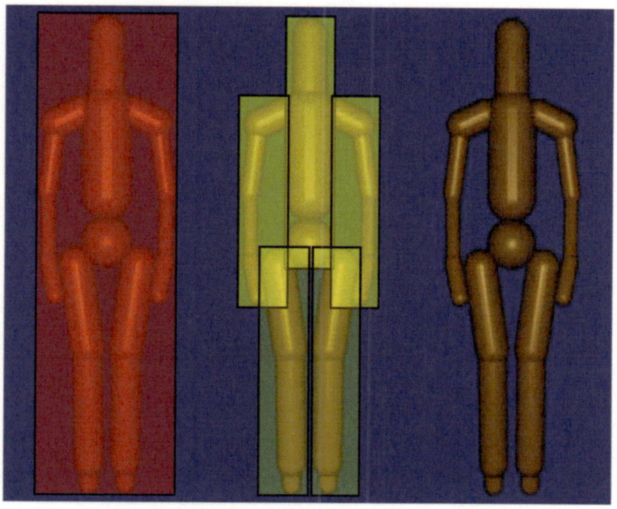

Fig. 5. 3D representation of the hierarchy of bounding volumes for the human operator: (a) Global AABB (level 1); (b) Local AABBs (level 2) and (c) SSLs (level 3).

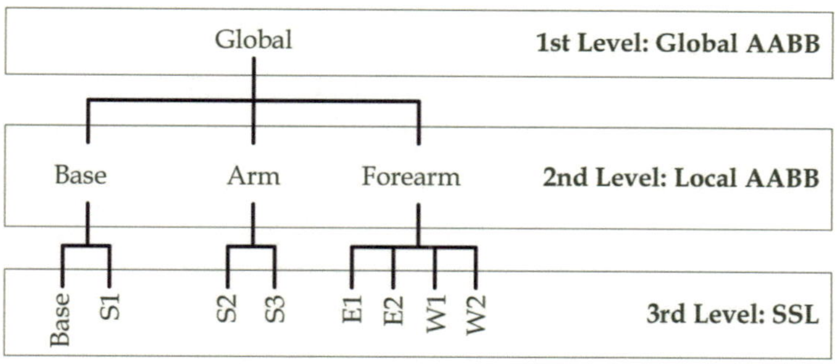

Fig. 6. Diagram of the hierarchy of bounding volumes for the robotic manipulator.

several SSLs of the third level that are related kinematically. In particular, the second level of the bounding volume hierarchy for the human operator is composed by 5 AABBs (left lower limb, right lower limb, torso-head, left upper limb and right upper limb) and by 3 AABBs (base, arm and forearm) for the robot. Therefore, the number of pairwise distance tests is reduced to 15 for the second level of the hierarchy. The AABBs of this level are computed by looking for the maximum and minimum coordinates of the contained links and adding to them the maximum radius of the contained SSLs. The first level of this hierarchy is composed by one global AABB which covers the bodies of the human or the robot and which is computed from the maximum and minimum coordinates of all the links. This global AABB will be used when the human and the robot are far from each other. In a similar way to the second and third levels, a distance threshold will be established in order to determine if the first or the second level of the bounding hierarchy is used to compute the human-robot distance.

The selection of the hierarchy level to be used depends on the distance between the human and the robot. When the distance between the human operator and the robot is high, a geometric representation based on AABBs (levels 1 or 2 of the hierarchy) is used. This

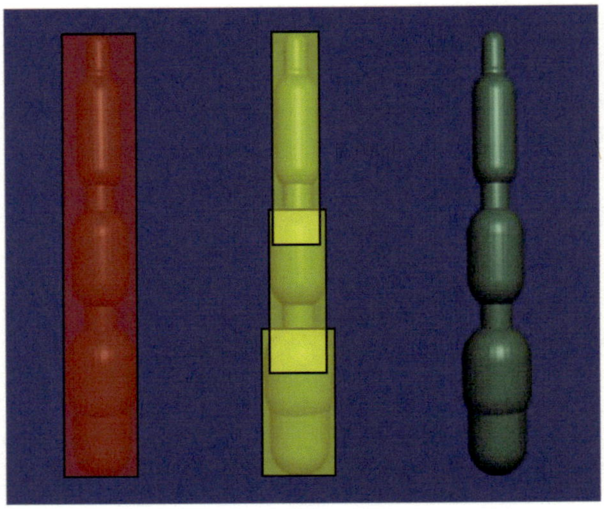

(a) (b) (c)

Fig. 7. 3D representation of the hierarchy of bounding volumes for the robotic manipulator: (a) Global AABB (level 1); (b) Local AABBs (level 2) and (c) SSLs (level 3).

representation increases the efficiency of the distance computation because the number of distance pairwise tests is reduced and AABB-AABB distance tests are less expensive than SSL-SSL distance tests. When the human-robot distance is small, the SSL bounding volumes are necessary to compute a good approximation of the distance because they are a more precise and detailed representation of the human and robot's links. However, as stated before, the SSL-based representation requires more pairwise distance tests. In order to reduce the number of SSL pairwise distance tests, a combination of the levels 2 and 3 of the hierarchy is implemented for the minimum distance computation. This minimum distance algorithm and the processing of the obtained value for modifying the robot behaviour are explained in detail in the following section.

4.1.2 Safety strategy for human-robot interaction

The main goal of the hierarchy of bounding volumes described in the previous section is to represent the bodies of the human operator and the robot in such a way that the human-robot distance computation is precise and efficient for the development of cooperative tasks on real-time. The algorithm which has been implemented to compute the human-robot distance is shown in Table 2.

The first step of this algorithm (line 1 of Table 2) is to initialize the two distance thresholds: $\mathbf{DIST}_{1>2}$ and $\mathbf{DIST}_{2>3}$. The first threshold $\mathbf{DIST}_{1>2}$ determines whether the minimum human-robot distance is obtained from the AABBs of the first level or from the AABBs of the second level. Similarly, the second threshold $\mathbf{DIST}_{2>3}$ determines whether the minimum human-robot distance is obtained from the AABBs of the second level or from the SSLs of the third level. After having initialized the thresholds, this algorithm generates the two global AABBs of the first level (lines 2 and 3 of Table 2) by looking for the minimum and maximum coordinate values of all the links. Next, the minimum distance **mdist1** between both global AABBs is calculated (line 4 of Table 2) and compared with the first threshold $\mathbf{DIST}_{1>2}$. If **mdist1** is bigger than the threshold, this distance value will be used as minimum human-robot distance (line 6 of Table 2) because they are so far from each other that the global AABB representation is sufficient.

However, if the human-robot distance is smaller than the threshold **$DIST_{1>2}$**, the more precise AABB representation of the second level is required. First of all, the local AABBs of the second level are generated (lines 8 and 9 of Table 2) from the minimum and maximum coordinate values of the links contained in each AABB. As there are 5 AABBs for the human and 3 AABBs for the robot, they are stored in two arrays of these lengths: **$AABB2_H$[1..5]** and **$AABB2_R$[1..3]**. All the possible distance values between each pair of AABBs are computed and stored in array **dist2[1..15]** (line 10 of Table 2). This array is sorted in ascending order (line 11 of Table 2) and its minimum component is used as the minimum distance value **mdist2** for the second level of the bounding volume hierarchy (line 12 of Table 2). This value **mdist2** is used as minimum human-robot distance if it is higher than the second threshold **$DIST_{2>3}$** (line 14 of Table 2).

```
01: Initialize distance thresholds: DIST_1>2, DIST_2>3
02: Compute Global AABB (Level 1) for Human: AABB1_H
03: Compute Global AABB (Level 1) for Robot: AABB1_R
04: mdist1= MinimumDistance (AABB1_H, AABB1_R)
05: if (mdist1 > DIST_1>2)
06:   finalDist= mdist1 //Distance in Level 1
07: else
08:   Compute Local AABBs (Level 2) for Human: AABB2_H[1..5]
09:   Compute Local AABBs (Level 2) for Robot: AABB2_R[1..3]
10:   dist2[1..15]= PairwiseDistances(AABB2_H[1..5], AABB2_R[1..3])
11:   dist2[1..15]= SortArrayInAscendingOrder(dist2[1..15])
12:   mdist2= MinimumValue(dist2[1..15])
13:   if (mdist2 > DIST_2>3)
14:     finalDist= mdist2 // Distance in Level 2
15:   else
16:     mdist3= FLOAT_MAX_VALUE
17:     for each element i in dist2[1..15]
18:       if (mdist3 < dist2[i])
19:         break //Finish for loop in line 17
20:       end if
21:       Compute SSLs (Level 3) contained by AABB2_H[i]: SSL3_H[1..nLinksH_i]
22:       Compute SSLs (Level 3) contained by AABB2_R[i]: SSL3_R[1..nLinksR_i]
23:       for each element j in SSL3_H[1..nLinksH_i]
24:         for each element k in SSL3_R[1..nLinksR_i]
25:           distSSL= MinimumDistance(SSL3_H[j], SSL3_R[k])
26:           mdist3= MinimumValue(distSSL, mdist3)
27:         end for
28:       end for
29:     end for
30:     finalDist= mdist3 //Distance in Level 3
31:   end if
32: end if
33: return finalDist
```

Table 2. Pseudocode of the minimum distance algorithm based on the three-level hierarchy of bounding volumes.

When **mdist2** is smaller than **DIST$_{2>3}$**, the SSL representation is applied because AABB bounding volumes do not provide a sufficient level of detail for precise distance computation between very close links. In fact, the SSL bounding volumes have a better tight to the links than the AABB bounding volumes. In order to reduce the number of distance tests between each pair of SSLs, the distance values **dist2[1..15]** between the AABBs of the second level are used as a lower threshold for the SSLs distance values. The SSLs are generated (lines 21 and 22 of Table 2) from the corresponding AABBs of the second level, which have been ordered according to their pairwise distances (line 11 of Table 2). After generating the SSLs contained by each pair of AABBs, the minimum distance **mdist3** between them is computed (lines 25 and 26 of Table 2). If the SSL-SSL distance **mdist3** becomes smaller than the distance **dist2[i]** between the following pair of AABBs (line 18 of Table 2), the minimum distance search process will end (line 19 of Table 2) because the following SSLs will be contained by those AABBs and they will be further away. Thereby, in most cases, not all the 144 SSL-SSL tests are required.
The minimum distance value computed by the algorithm presented in Table 2 is used by the safety strategy in order to guarantee that there are no collisions between the human operator and the robot. In particular, the safety strategy verifies that the human-robot distance is higher than a safety threshold. While this condition is fulfilled, the robot tracks its trajectory normally. Nevertheless, when the human-robot distance is lower than the safety threshold, the robot tracking process is temporarily stopped and a safety behaviour is activated. This safety behaviour moves the robot away from the human operator in order to maintain the human-robot distance above the safety threshold.

4.2 Software architecture

A multi-threaded software architecture has been developed in order to implement the human-robot interaction behaviour described in the previous sections. It has been programmed as a C++ program which is executed in the controller PC. The robot controller, the motion capture system, the UWB localization system and the vision system are connected to this PC in order to avoid any synchronisation problem between their measurements.
Three threads compose this software architecture (see Fig. 8): the distance computation thread, the path tracking thread and the safety strategy thread. These three threads are executed simultaneously and share a common memory space where they interchange information. The distance computation thread obtains the links orientation measurements from the human tracking system (see Section 2) and the joints angles of the robot from the robot controller. The positions of the links of the human and the robot are computed from these measurements through a forward kinematics algorithm and they are stored in the common memory space. Finally, these links positions are used by the algorithm described in Table 2 in order to generate the hierarchy of bounding volumes and calculate the minimum distance between the human operator and the robot manipulator. The distance computation thread will update this human-robot distance value each time new measurements from the human tracking system and the robot controller are registered by the controller PC.
The minimum human-robot distance calculated by the distance computation thread is stored in the common memory space where it is checked by the other two threads. On the one hand, when this human-robot distance is greater than the safety threshold, the path tracking thread performs the visual servoing path tracking process (see Section 3.2). The

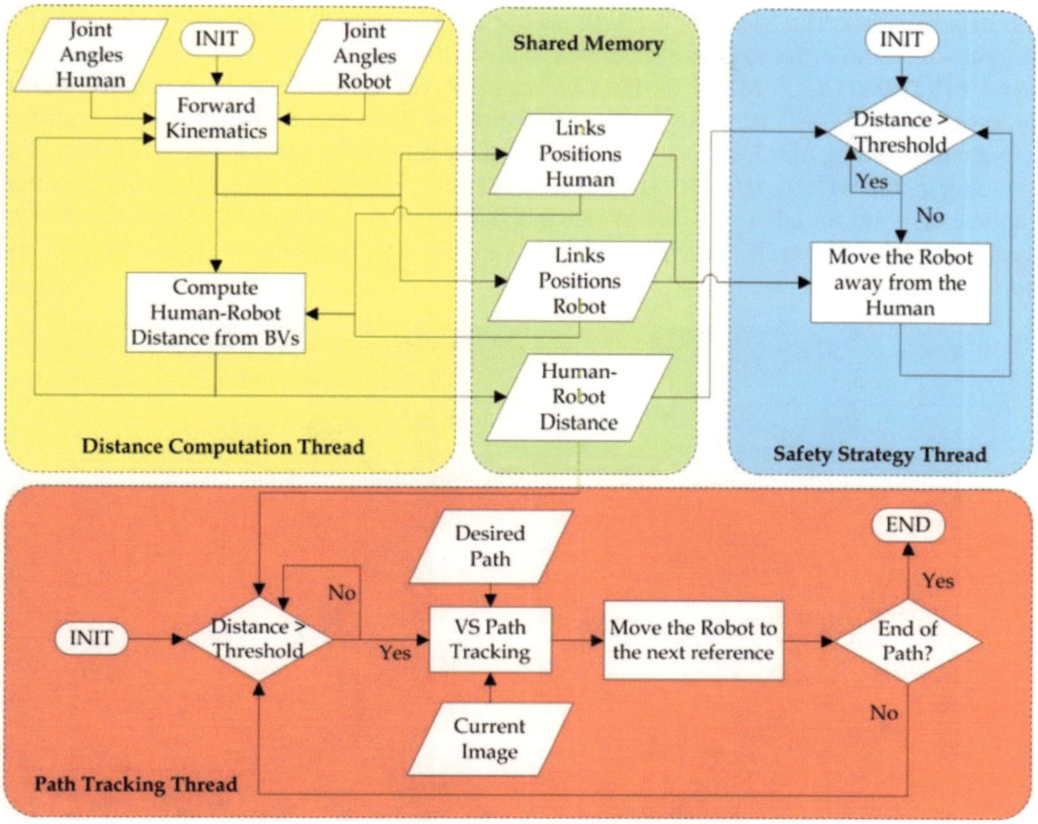

Fig. 8. Components of the software architecture which implements the human-robot interaction behaviour.

time-independent path tracker compares the current image from the eye-in-hand camera with the corresponding image of the desired path and calculates the robot velocity required to make them coincide. This tracking process ends when the desired path is completed and is paused when the safety behaviour is activated. On the other hand, when the human-robot distance is smaller than the safety threshold, the safety strategy thread executes the safety behaviour. This safety behaviour moves the robot away from the human operator in order to keep the human-robot distance above the safety threshold. This escape trajectory is calculated from the line which links the two closest links. While the safety behaviour is executed, the time-independent path tracking process is paused, and vice versa.

5. Experimental results

5.1 Task description

The human-robot interaction system proposed here can be applied in any collaborative task where the human enters in the robot workspace. In particular, the described system has been applied to a disassembly task. The object to be disassembled is a small fridge. The main elements which take part in this task are depicted in Fig. 9: the fridge, the Mitsubishi PA-10 robotic manipulator which unscrews the fridge lid, the human operator who extracts the internal tray and the storage box where the different parts of the fridge are stored after they are disassembled. The Mitsubishi PA-10 robotic manipulator has three devices installed at its end-effector in order to perform the task: a screwdriver, a JR3 force sensor and an eye-in-hand PHOTONFOCUS MV-D752-160-CL-8 camera. The camera is able to acquire and to process up to 100fps using an image resolution of 320x240px. The image trajectory is generated by using four laser points projected on the floor as extrated features for the visual servoing path tracking system. The human operator wears an *Animazoo* inertial motion capture suit and an *Ubisense* Ultra-WideBand (UWB) radio localization system in order to track precisely all her/his movements and compute the human-robot distance with the bounding volume hierarchy.

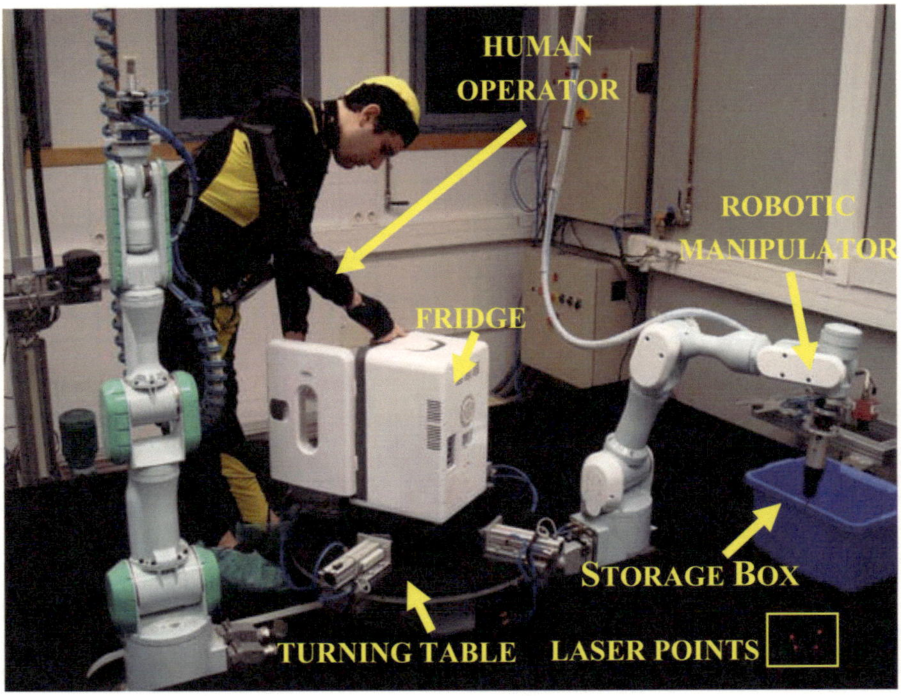

Fig. 9. Experimental setup for the fridge disassembly task.

The disassembly task can be split in the following subtasks to be performed by the human operator and the robot:

- Robot: The robotic manipulator has to remove the screws from the rear lid of the fridge. Firstly, the robotic manipulator goes from the storage box to the unscrew position by tracking a predefined path using the image path tracker based on visual servoing described in Section 3.2. Secondly, the robot unscrews the corresponding screw and then the robot is again guided to the storage box where the screw is left. This task is repeated until it removes all the screws.
- Human operator: Meanwhile, the human operator has to empty the fridge's contents.

The tasks described can be performed simultaneously. The global safety system presented in this chapter is employed during all the disassembly task. In Fig. 10, the sequence of the disassembly task is depicted. The image path tracker guides the robot from the storage box to the next screw (Frame 1 of Fig. 10). During the tracking, the human approaches the fridge in order to empty its content (Frame 2 of Fig. 10). The distance between the human operator

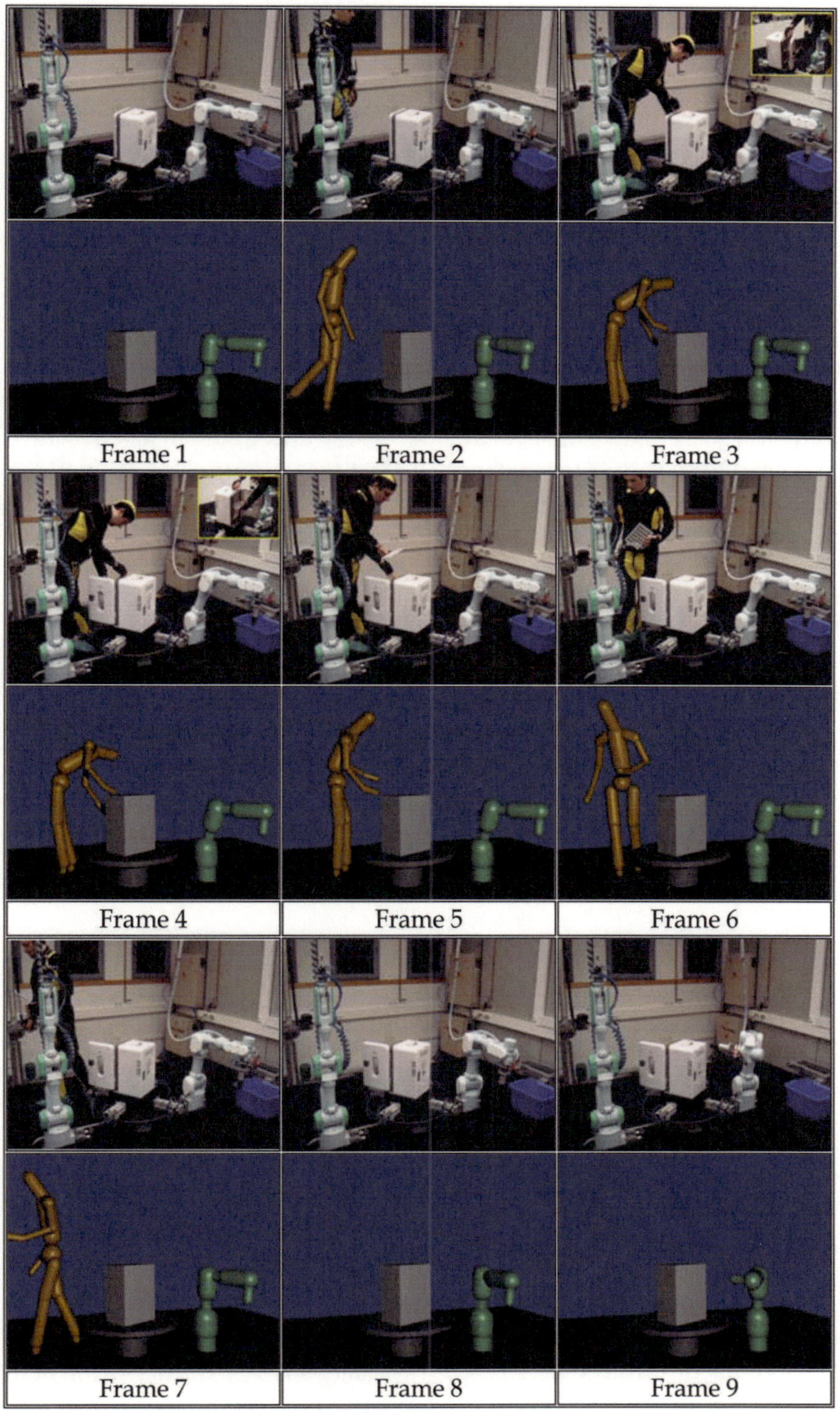

Fig. 10. Disassembly sequence with human-robot interaction. Each frame of the sequence is shown with a photograph of the workspace and the corresponding SSLs bounding volumes.

and the robot falls under the safety threshold activating the safety strategy. The robot goes away from the human operator (Frames 3 and 4 of Fig. 10) and stops until the distance exceeds the safety threshold again. Meanwhile, the human operator opens the fridge's door (Frame 4 of Fig. 10) and takes the internal tray out of the fridge (Frames 5 and 6 of Fig. 10) in order to carry it to a storage box which is out of the workspace. While the human operator is going away from the robot, the human-robot distance is again greater than the safety threshold and the visual servoing path tracking is re-activated (Frames 7 and 8 of Fig. 10). Thanks to the time-independent behaviour, the path is then tracked correctly and the robot can arrive at the unscrew position by following the predefined path (Frame 9 of Fig. 10).

5.2 Robot controller results

To show the correctness of the proposed time-independent image path tracker described in Section 3.2, a comparative between this time-independent and a time-dependent systems is next presented. Fig. 11 shows the evolution of the features in the image obtained with the two different methods. Both of the methods perform the tracking correctly until the moment when the obstruction begins. Nevertheless, from the moment when the robot is released, the time-dependent system is not able to return to the exact point in the trajectory where the obstruction began. This is due to the loss of temporal references. Therefore, the time-independent method described in this chapter is adequate for the tracking of image paths in tasks where the robot interacts with a human or with any object in its workspace.

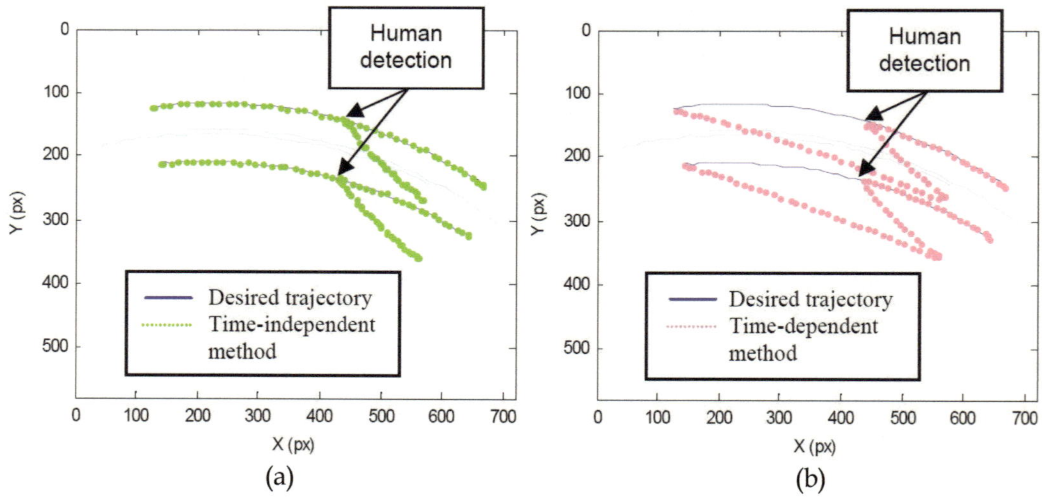

Fig. 11. (a) Evolution (from right to left)of the features in the image space using the time-independent tracking; (b) Evolution (from right to left) of the features in the image using a time-dependent method.

5.3 Human-robot integration results

The main goal of the safety strategy is to maintain the human-robot distance above a safety threshold (0.5m for this disassembly task). This safety strategy calculates the human-robot distance on real-time by executing the algorithm in Table 2 based on the three-level hierarchy of bounding volumes described in Section 4.1.1. The distance threshold $\mathbf{DIST}_{1>2}$ of this algorithm is set to 2m while the distance threshold $\mathbf{DIST}_{2>3}$ is set to 1m. Therefore, the global AABBs (level 1) are used for distances greater than 2m; the local AABBs (level 2) are used for the distances between 2m and 1m and the SSLs (level 3) are used for distances smaller than 1m.

Fig. 12 depicts the evolution of the minimum human-robot distance obtained by the distance algorithm for the disassembly task. This plot shows how the human operator approaches the robot while the robot performs the time-independent visual servoing path tracking from iteration 1 to iteration 289. In iteration 290, the safety strategy starts and the robot controller pauses the path tracking. The safety strategy is executed from iteration 290 to iteration 449 and it tries to keep the human-robot distance above the safety threshold (0.5m). In iteration 450, the robot controller re-activates path tracking because the human-robot distance is again greater than the threshold when the human is going away from the workspace.

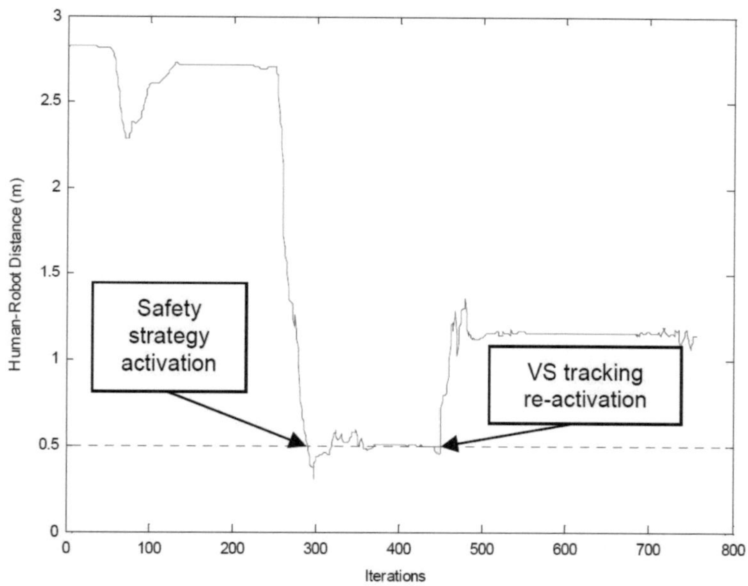

Fig. 12. Evolution of the minimum human-robot distance during the disassembly task.

Fig. 13.a. depicts the error evolution of the distance obtained by the algorithm in Table 2 with regard to the distance values obtained from the SSL bounding volumes, which are used as ground-truth. This figure shows an assumable mean error of 4.6cm for distances greater than 1m. For distances smaller than 1m (between iterations 280 and 459), the error is null because the SSLs are used for the distance computation.

The proposed distance algorithm obtains more precise distance values than previous research. In particular, Fig. 13.b. shows the difference between the distance values obtained by the algorithm in Table 2 and the distance values computed by the algorithm in (Garcia et al., 2009a), where no bounding volumes are generated and only the end-effector of the robot is taken into account for the distance computation instead of all its links.

Fig. 14 shows the histogram of distance tests which are performed for the distance computation during the disassembly task. In 64% of the executions of the distance algorithm, a reduced number of pairwise distance tests is required (between 1 and 16 tests) because the bounding volumes of the first and/or second level of the hierarchy (AABBs) are used. In the remaining 36%, between 30 and 90 distance tests are executed for the third level of the hierarchy (SSLs). This fact demonstrates that the hierarchy of bounding volumes involves a significant reduction of the computational cost of the distance computation with regard to a pairwise strategy where 144 distance tests would always be executed.

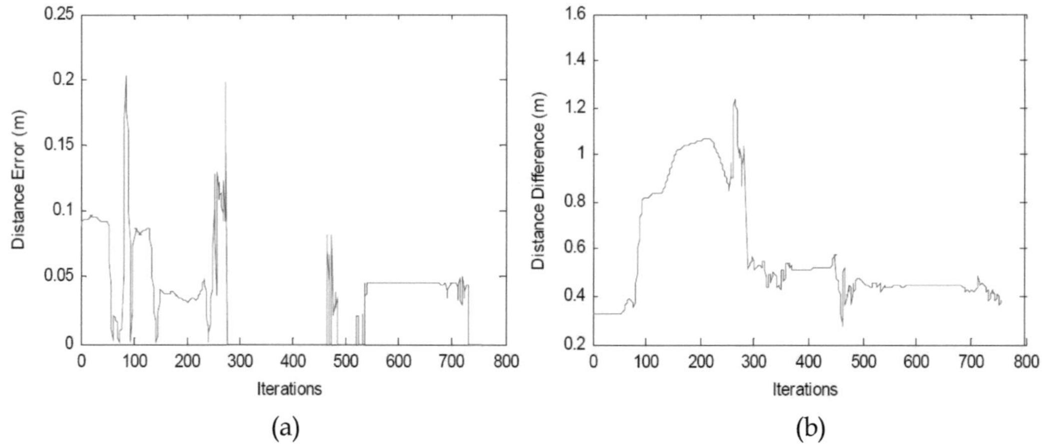

Fig. 13. (a) Evolution of the distance error from the BV hierarchy; (b) Evolution of the distance difference between the BV hierarchy algorithm and (Garcia et al., 2009a).

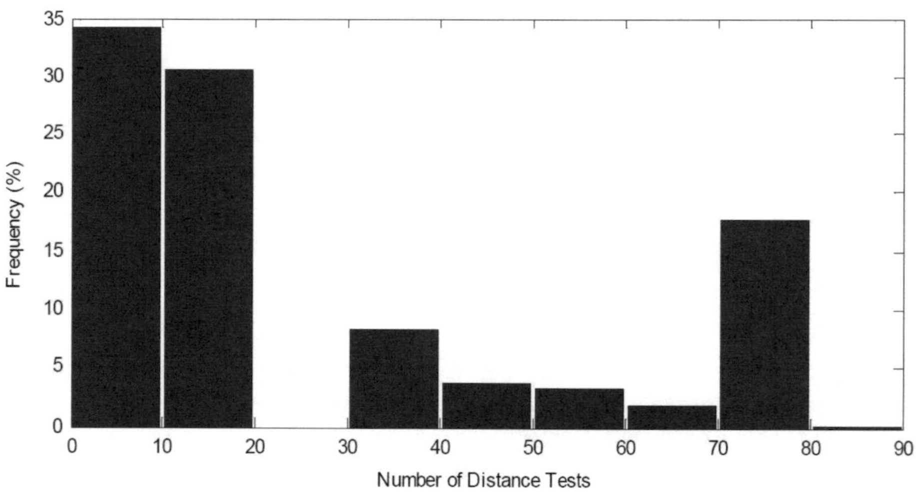

Fig. 14. Histogram of the number of distance tests required for the minimum human-robot distance computation.

6. Conclusions

This chapter presents a new human-robot interaction system which is composed by two main sub-systems: the robot control system and the human tracking system. The robot control system uses a time-independent visual servoing path tracker in order to guide the movements of the robot. This method guarantees that the robot tracks the desired path completely even when unexpected events happen. The human tracking system combines the measurements from two localization systems (an inertial motion capture suit and a UWB

localization system) by a Kalman filter. Thereby, this tracking system calculates a precise estimation of the position of all the limbs of the human operator who collaborates with the robot in the task.
In addition, both sub-systems have been related by a safety behaviour which guarantees that no collisions between the human and the robot will take place. This safety behaviour computes precisely the human-robot distance by a new distance algorithm based on a three-level hierarchy of bounding volumes. If the computed distance is below a safety threshold, the robot's path tracking process is paused and a safety strategy which tries to maintain this separation distance is executed. When the human-robot distance is again safe, the path tracking is re-activated at the same point where it was stopped because of its time-independent behaviour. The authors are currently working at improving different aspects of the system. In particular, they are considering the use of dynamic SSL bounding volumes and the development of more flexible tasks where the human's movements are interpreted.

7. Acknowledgements

The authors want to express their gratitude to the Spanish Ministry of Science and Innovation and the Spanish Ministry of Education for their financial support through the projects DPI2005-06222 and DPI2008-02647 and the grant AP2005-1458.

8. References

Balan, L. & Bone, G. M. (2006). Real-time 3D collision avoidance method for safe human and robot coexistence, *Proceedings of IEEE/RSJ International Conference on Intelligent Robots and Systems*, pp. 276-282, Beijing, China, Oct. 2006.

Chaumette, F. & Hutchinson, S. (2006). Visual Servo Control, Part I: Basic Approaches. *IEEE Robotics and Automation Magazine*, Vol. 13, No. 4, 82-90, ISSN: 1070-9932.

Chesi, G. & Hung, Y. S. (2007). Global path-planning for constrained and optimal visual servoing. *IEEE Transactions on Robotics*, Vol. 23, 1050-1060, ISSN: 1552-3098.

Corrales, J. A., Candelas, F. A. & Torres, F. (2008). Hybrid tracking of human operators using IMU/UWB data fusion by a Kalman filter, *Proceedings of 3rd. ACM/IEEE International Conference on Human-Robot Interaction*, pp. 193-200, Amsterdam, March 2008.

Ericson, C. (2005). *Real-time collision detection*, Elsevier, ISBN: 1-55860-732-3, San Francisco, USA.

Fioravanti, D. (2008). Path planning for image based visual servoing. Thesis.

Foxlin, E. (1996). Inertial head-tracker sensor fusion by a complementary separate-bias Kalman filter, *Proceedings of IEEE Virtual Reality Annual International Symposium*, pp. 185-194, Santa Clara, California, 1996.

Garcia, G.J., Corrales, J.A., Pomares, J., Candelas, F.A. & Torres, F. (2009). Visual servoing path tracking for safe human-robot interaction, *Proceedings of IEEE International Conference on Mechatronics*, pp. 1-6, Malaga, Spain, April 2009.

Garcia, G.J., Pomares, J. & Torres, F. (2009). Automatic robotic tasks in unstructured environments using an image path tracker. *Control Engineering Practice*, Vol. 17, No. 5, May 2009, 597-608, ISSN: 0967-0661.

Hutchinson, S., Hager, G. D. & Corke, P. I. (1996). A tutorial on visual servo control. *IEEE Transactions on Robotics and Automation*, Vol. 12, No. 5, 651-670, ISSN: 1042-296X.

Malis, E. (2004). Visual servoing invariant to changes in camera-intrinsic parameters. *IEEE Transactions on Robotics and Automation*, Vol. 20, No. 1, February 2004, 72-81, ISSN: 1042-296X.

Marchand, E. & Chaumette, F. (2001). A new formulation for non-linear camera calibration using VVS. Publication Interne 1366, IRISA, Rennes, France.

Martinez-Salvador, B., Perez-Francisco, M. & Del Pobil, A. P. (2003). Collision detection between robot arms and people. *Journal of Intelligent and Robotic Systems*, Vol. 38, No. 1, Sept. 2003, 105-119, ISSN: 0921-0296.

Mezouar, Y. & Chaumette, F. (2002). Path planning for robust image-based control. *IEEE Transactions on Robotics and Automation*, Vol. 18, No. 4, 534-549, ISSN : 1042-296X.

Pomares, J. & Torres, F. (2005). Movement-flow based visual servoing and force control fusion for manipulation tasks in unstructured environments. *IEEE Transactions on Systems, Man, and Cybernetics – Part C*, Vol. 35, No. 1, 4-15, ISSN: 1094-6977.

Schneider, P. J. & Eberly, D. H. (2003). *Geometric tools for computer graphics*, Elsevier, ISBN: 1-55860-594-0, San Francisco, USA.

Schramm, F. & Morel, G. (2006). Ensuring visibility in calibration-free path planning for image-based visual servoing. *IEEE Transactions on Robotics*, Vol. 22 No. 4, 848-854, ISSN : 1552-3098.

Thrun, S., Burgard, W. & Fox, D. (2005). *Probabilistic Robotics*, MIT Press, ISBN: 978-0-262-20162-9, Cambridge, USA.

Welch, G., & Foxlin, E. (2002). Motion tracking: no silver bullet but a respectable arsenal. *IEEE Computer Graphics and Applications*, Vol. 22, No. 6, Nov. 2002, 24-38, ISSN: 0272-1716.

16

Capturing and Training Motor Skills

Otniel Portillo-Rodriguez[1,2], Oscar O. Sandoval-Gonzalez[1], Carlo Avizzano[1], Emanuele Ruffaldi[1] and Massimo Bergamasco[1]
[1]*Perceptual Robotics Laboratory, Scuola Superiore Sant'Anna, Pisa,*
[2]*Facultad de Ingeniería, Universidad Autonóma del Estado de México, Toluca,*
[1]*Italy*
[2]*México*

1. Introduction

Skill has many meanings, as there are many talents: its origin comes from the late Old English scele, meaning knowledge, and from Old Norse skil (discernment, knowledge), even if a general definition of skill can be given as "the learned ability to do a process well" (McCullough, 1999) or as the acquired ability to successfully perform a specific task.
Task is the elementary unit of goal directed behaviour (Gopher, 2004) and is also a fundamental concept -strictly connected to "skill"- in the study of human behaviour, so that psychology may be defined as the science of people performing tasks. Moreover skill is not associated only to knowledge, but also to technology, since technology is -literally in the Greek- the study of skill.
Skill-based behaviour represents sensory-motor performance during activities following a statement of an intention and taking place without conscious control as smooth, automated and highly integrated patterns of behaviour. As it is shown in Figure 1, a schematic representation of the cognitive-sensory-motor integration required by a skill performance, complex skills can involve both gesture and sensory-motor abilities, but also high level cognitive functions, such as procedural (e.g. how to do something) and decision and judgement (e.g. when to do what) abilities. In most skilled sensory-motor tasks, the body acts as a multivariable continuous control system synchronizing movements with the behavioural of the environment (Annelise Mark Pejtersen, 1997). This way of acting is also named also as, action-centred, enactive, reflection-in-action or simply know-how.
Skills differ from talent since talent seems native, and concepts come from schooling, while skill is learned by doing (McCullough, 1999). It is acquired by demonstration and sharpened by practice. Skill is moreover participatory, and this basis makes it durable: any teacher knows that active participation is the way to retainable knowledge.
The knowledge achieved by an artisan throughout his/her lifelong activity of work is a good example of a skill that is difficult to transfer to another person. At present the knowledge of a specific craftsmanship is lost when the skilled worker ends his/her working activity or when other physical impairments force him/her to give up. The above considerations are valid not only in the framework of craftsmanship but also for more general application domains, such as the industrial field, e.g. for maintenance of complex mechanical parts, surgery training and so on.

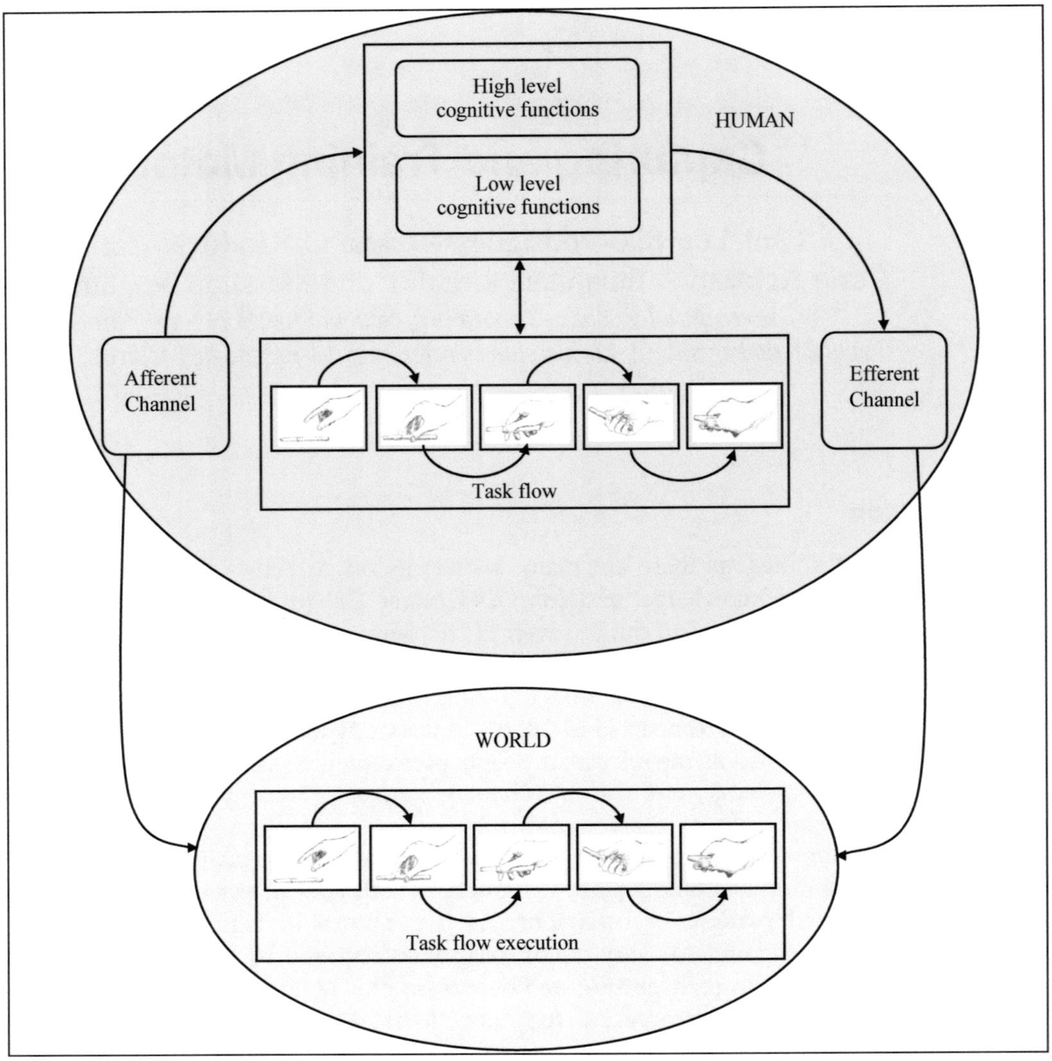

Fig. 1. A schematic representation of the cognitive-sensory-motor integration required by a skill performance

The research done stems out from the recognition that technology is a dominant ecology in our world and that nowadays a great deal of human behaviour is augmented by technology. Multimodal Human-Computer Interfaces aim at coordinating several intuitive input modalities (e.g. the user's speech and gestures) and several intuitive output modalities.

The existing level of technology in the HCI field is very high and mature, so that technological constrains can be removed from the design process to shift the focus on the real user's needs, as it is demonstrated by the fact that nowadays the user-centered design has became fundamental for devising successful everyday new products and interfaces. (Norman, 1986; Norman, 1988), fitting people and that really conforming their needs.

However, until now most interaction technologies have emphasized more input channel (afferent channel in Figure 1 The role of HCI in the performance of a skill), rather than output (efferent channel); foreground tasks rather than background contexts.

Advances in HCI technology allows now to have better gestures, more sensing combinations and improve 3D frameworks, and so it is possible now to put also more emphasis on the output channel, e.g. recent developments of haptic interfaces and tactile effector technologies. This is sufficient to bring in the actual context new and better instruments and interfaces for doing better what you can do, and to teach you how to do something well: so interfaces supporting and augmenting your skills. In fact user interfaces to advanced augmenting technologies are the successors to simpler interfaces that have existed between people and their artefacts for thousands of years (M. Chignell & Takeshit, 1999).

The objectives is to develop new HCI technologies and devise new usages of existing ones to support people during the execution of complex tasks, help them to do things well or better, and make them more skilful in the execution of activities, overall augmenting the capability of human action and performance.

We aimed to investigate the transfer of skills defined as the use of knowledge or skill acquired in one situation in the performance of a new, novel task, and its reproducibility by means of VEs and HCI technologies, using actual and new technology with a complete innovative approach, in order to develop and evaluating interfaces for doing better in the context of a specific task.

Figure 2 draws on the scheme of Figure 1, and shows the important role that new interfaces will play and their features. They should possess the following functionalities:

- Capability of interfacing with the world, in order to get a comprehension of the status of the world;
- Capability of getting input from the humans through his efferent channel, in a way not disturbing the human from the execution of the main task (transparency);
- Local intelligence, that is the capability of having an internal and efficient representation of the task flow, correlating the task flow with the status of the environment during the human-world interaction process, understanding and predicting the current human status and behaviour, formulating precise indications on next steps of the task flow or corrective actions to be implemented;
- Capability of sending both information and action consequences in output towards the human, through his/her afferent channel, in a way that is not disturbing the human from the execution of the main task.

We desire improving both input and output modalities of interfaces, and on the interplay between the two, with interfaces in the loop of decision and action (Flach, 1994) in strictly connection with human, as it is shown clearly in Figure 2. The interfaces will boost the capabilities of the afferent-efferent channel of humans, the exchange of information with the world, and the performance of undertaken actions, acting in synergy with the sensory-motor loop.

Interfaces will be technologically invisible at their best -not to decrease the human performance-, and capable of understanding the user intentions, current behaviour and purpose, contextualized in the task.

In this chapter a multimodal interface capable to understand and correct in real-time hand/arm movements through the integration of audio-visual-tactile technologies is presented. Two applications were developed for this interface. In the first one, the interface acts like a translator of the meaning of the Indian Dance movements, in the second one the interface acts like a virtual teacher that transfers the knowledge of five Tai-Chi movements

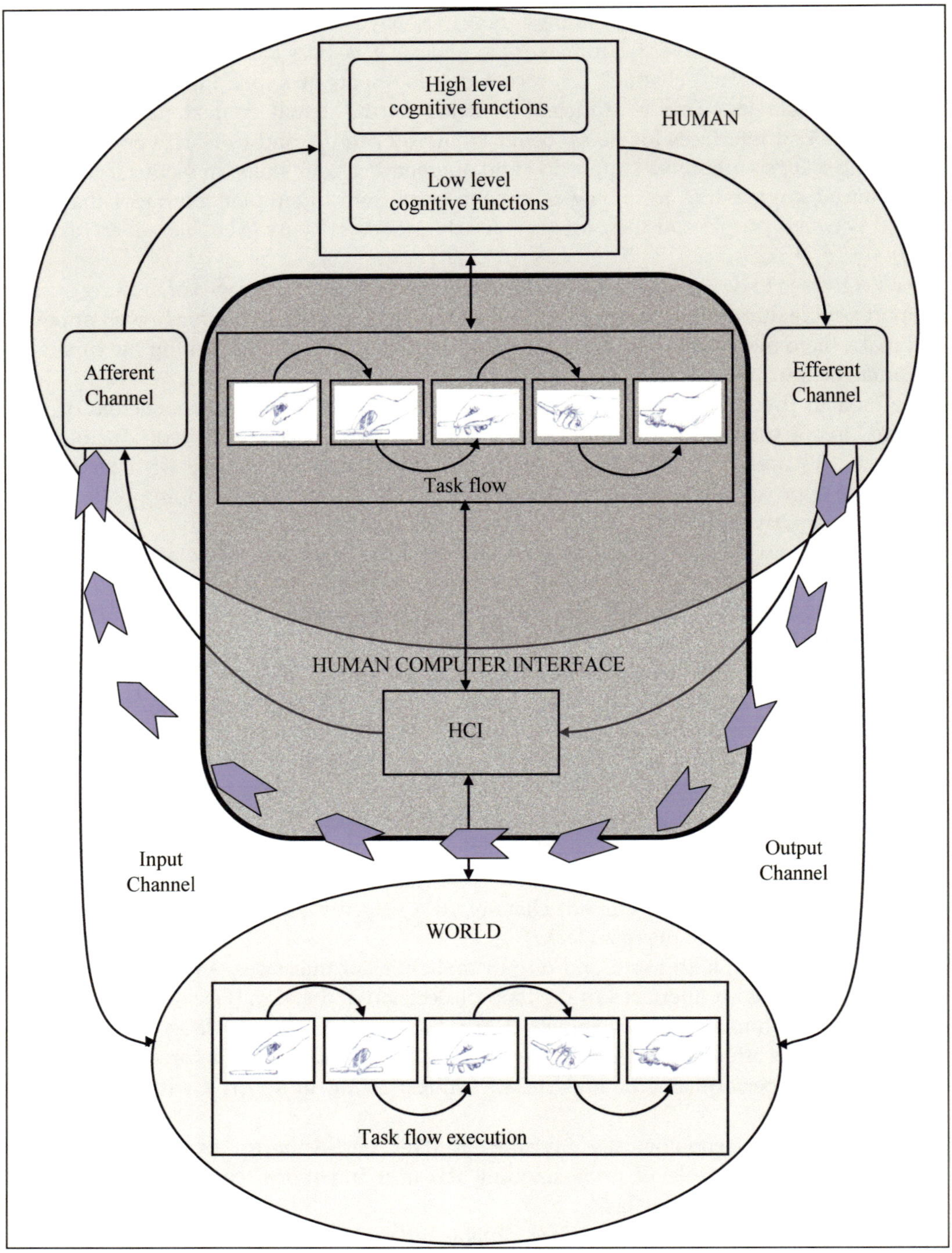

Fig. 2. The role of HCI in the performance of a skill

using feed-back stimuli to compensate the errors committed by a user during the performance of the gesture (Tai-Chi was chose due its movements must be performed precise and slow).

In both applications, a gesture recognition system is its fundamental component, it was developed using different techniques such as: k-means clustering, Probabilistic Neural Networks (PNN) and Finite State Machines (FSM). In order to obtain the errors and qualify the actual movements performed by the student respect to the movements performed by the master, a real-time descriptor of motion was developed. Also, the descriptor generate the appropriate audio-visual-tactile feedbacks stimuli to compensate the users' movements in real-time. The experiments of this multimodal platform have confirmed that the quality of the movements performed by the students is improved significantly.

2. Methodology to recognize 3D gestures using the state based approach

For human activity or recognition of dynamic gestures, most efforts have been concentrated on using state-space approaches (Bobick & Wilson, 1995) to understand the human motion sequences. Each posture state (static gesture) is defined as a state. These states are connected by certain probabilities. Any motion sequence as a composition of these static poses is considered a walking path going through various states. Cumulative probabilities are associated to each path, and the maximum value is selected as the criterion for classification of activities. Under such a scenario, duration of motion is no longer an issue because each state can repeatedly visit itself. However, approaches using these methods usually need intrinsic nonlinear models and do not have closed-form solutions. Nonlinear modeling also requires searching for a global optimum in the training process and a relative complex computing. Meanwhile, selecting the proper number of states and dimension of the feature vector to avoid "underfitting" or "overfitting" remains an issue.

State space models have been widely used to predict, estimate, and detect signals over a large variety of applications. One representative model is perhaps the HMM, which is a probabilistic technique for the study of discrete time series. HMMs have been very popular in speech recognition, but only recently they have been adopted for recognition of human motion sequences in computer vision (Yamato et al., 1992). HMMs are trained on data that are temporally aligned. Given a new gesture, HMM use dynamic programming to recognize the observation sequence (Bellman, 2003).

The advantage of a state approach is that it doesn't need a large set of data in order to train the model. Bobick (Bobick, 1997) proposed an approach that models a gesture as a sequence of states in a configuration space. The training gesture data is first manually segmented and temporally aligned. A prototype curve is used to represent the data, and is parameterized according to a manually chosen arc length. Each segment of the prototype is used to define a fuzzy state, representing transversal through that phase of the gesture. Recognition is done by using dynamic programming technique to compute the average combined membership for a gesture.

Learning and recognizing 3D gestures is difficult since the position of data sampled from the trajectory of any given gesture varies from instance to instance. There are many reasons for this, such as sampling frequency, tracking errors or noise, and, most notably, human variation in performing the gesture, both temporally and spatially. Many conventional gesture-modeling techniques require labor-intensive data segmentation and alignment work.

The attempt of our methodology is develop a useful technique to segment and align data automatically, without involving exhaustive manual labor, at the same time, the representation used by our methodology captures the variance of gestures in spatial-temporal space, encapsulating only the key aspect of the gesture and discarding the intrinsic variability to each person's movements. Recognition and generalization is spanned from very small dataset, we have asked to the expert to reproduce just five examples of each gesture to be recognized.

As mentioned before, the principal problem to model a gesture using the state based approach is the characterization of the optimal number of states and the establishment of their boundaries. For each gesture, the training data is obtained concatening the data of its five demonstrations. To define the number of states and their coarse spatial parameters we have used dynamic k-means clustering on the training data of the gesture without temporal information (Jain et al., 1999). The temporal information from the segmented data is added to the states and finally the spatial information is updated. This produces the state sequence that represents the gesture. The analysis and recognition of this sequence is performed using a simple Finite State Machine (FSM), instead of use complex transitions conditions as in (Hong et al., 2000), the transitions depend only of the correct sequence of states for the gesture to be recognized and eventually of time restrictions i.e., minimum and maximum time permitted in a given state.

For each gesture to be recognized, one PNN is create to evaluate which is the nearest state (centroid in the configuration state) to the current input vector that represents the user's body position. The input layer has the same number of neurons as the feature vector (Section 3) and the second layer has the same quantity of hidden neurons as states have the gesture. The main idea is to use the states' centroids obtained from the dynamic k-means as weights in its correspondent hidden neuron, in a parallel way where all the hidden neurons computes the similarities of the current student position and its corresponding state. In our architecture, each class node is connected just to one hidden neuron and the number of states in which the gesture is described defines the quantity of class nodes. Finally, the last layer, a decision network computes the class (state) with the highest summed activation. A general diagram of this architecture is presented in the Figure 3.

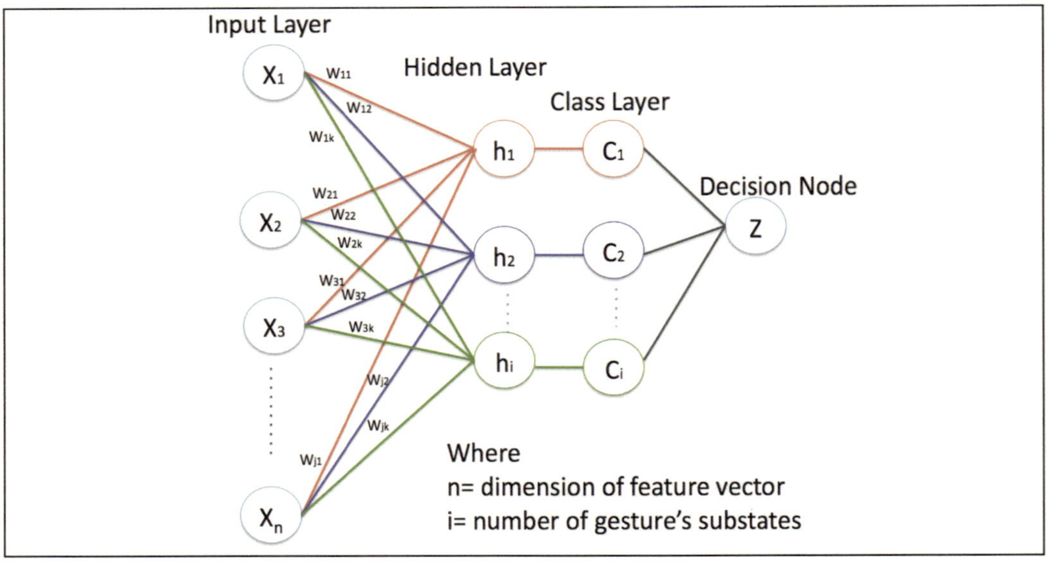

Fig. 3. PNN architecture used to estimate the most similar gesture's state from the current user's body position.

This approach allows real-time recognition while avoiding the classical disadvantages of this network: big computational resources (storage and time) during execution than many other models. An alternative for such computation is the use of RNN. In (Inamura et al., 2002) the authors tested the use of RNNs for motion recognition, according to their results, more than 500 nodes and more than 200,000 weight parameters between each node are needed in order to integrate the memorization process in the RNN. The RNN consist of motion elements neurons, symbol representation neurons and buffer neurons for treating time-series data. The required number of weights increases in proportion to the square of the number of all nodes. On the contrary in our methodology the number of parameters is proportional to the product of the number of hidden nodes (states in the gesture) and the dimension of the feature vector. To give a concrete example, a gesture typically has 10 states and the dimension of the feature vector is 13 (Section 3), resulting that with only 130 parameters a gesture is modeled, given as result a high information compression ratio.
The creation of the finite state machine is fast and simple; the transitions depend only of the correct sequence of states for the gesture to be recognized and eventually of time restrictions. If the FSM reaches its final state then the algorithm concludes that a gesture is recognized.

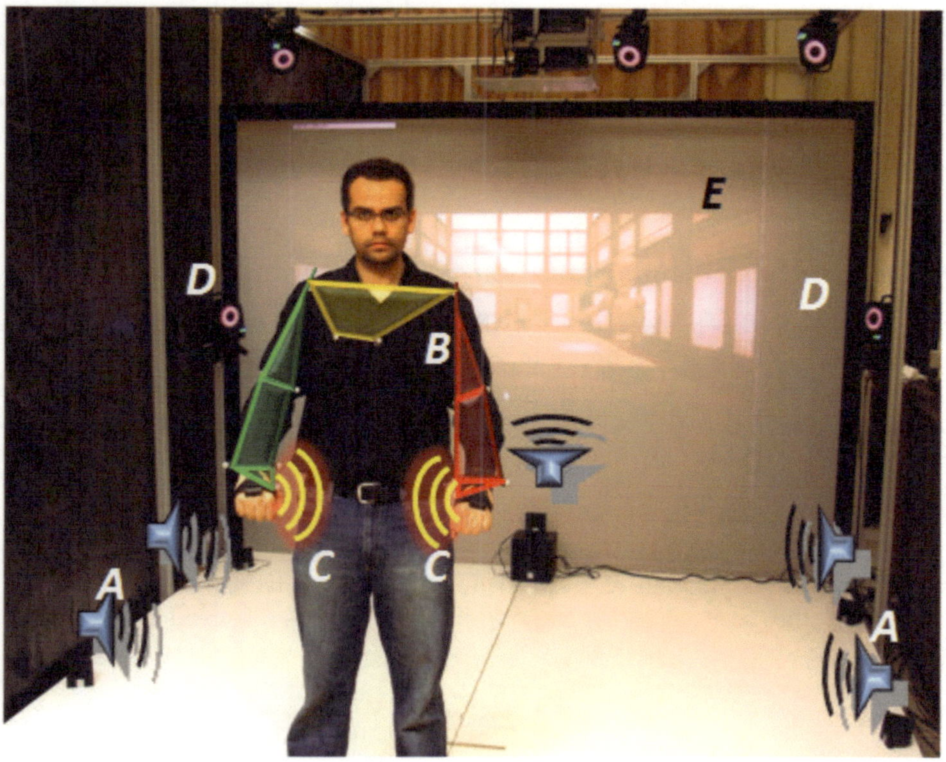

Fig. 4. Multimodal Platform set up, A) 3D sound, B) Kinematics Body C) Vibrotactile device (SHAKE) D) Vicon System E) Virtual Environment

3. Architecture of the multimodal interface

The hardware architecture of our Interface is composed of five components: VICON capture system (VCS), a host PC (with processor of 3 Ghz Intel Core Duo and 2 gigabytes of RAM

memory), one video projector, a pair of wireless vibrotactile stimulators and a 5.1 sound system. In the host PC, four applications run in parallel: Vicon Nexus (VICON, 2008), Matlab Simulink® and XVR (VRMedia, 2008). Depending of the application some of the components are not used.
The user has to perform a movement inside the capture system's workspace, depending of the application it could be an Indian Dance movement or a Tai Chi movement. Both applications where built using almost the same components, and uses the same methodology to create the recognizer blocks (pair of PNN-FSM) described in previous section. The Indian Dance application is an interactive demo in which five basic movements of the Indian dance can be recognized and their meaning can be represented using images and sound, its scope is more artistic than interactive, its development was done to test our methodology to recognize complex gestures.
In the Tai Chi application the objective was create a multimodal transfer skill system of Tai Chi movements, in this case, the gesture to be performed is known, the key idea is understand what far the current position of the user limbs is respect with the ideal position inside the reference movement. Once the distance (error) is calculated, it is possible modulate feedback stimuli (vision, audio and tactile) to indicate to the user the correct configuration of the upper limbs. It is important mention that the feature variables for gesture recognition and for gesture error are not the same.
It is possible observed from the Figures 8 and 11 that the acquisition data and recognition system (enclosed by a blue frame) in both applications are composed for the same blocks. For this reason in the next sections all the common elements in both applications there will be described, then each application it will be presented.

3. Modeling the upper limbs of the body

In order to track the 3D position of the markers using the VICON it is necessary create a kinematics model of the user's upper limbs. The model used is shown in Figure 5; it is composed of 13 markers united for hinge and balls joints. When the user is inside of the workspace of tracking system, it searches the correspondence between the model and the tracked markers, if a match exists; VICON sends the 3D position of all markers to Simulink via UDP.
The tracked positions by themselves are not useful in information for recognition due they are dependent of the position of the user in the capture system's workspace. We have chosen another representation of these elements that are invariant to the position and orientation of the user, allowing having enough degrees of freedom to model the movements of the user without over-fitting.
The 13 elements chosen are:

- Right & left elbows angles
- Right & left wrists pitch angles
- Right & left Shoulders angles
- The magnitude of the distance between palms
- Three spatial vector components from the back to the right palm
- Three spatial vector components from the back to the left palm

In order to recognized movements from different persons, it is necessary normalize the data to the "pattern user". The normalization relies in the fact that in the gestures to be recognized, the ensemble of angles of the feature vector have approximately the same temporal behavior and their range of values are similar for different users independently of their body sizes due their arms can be seen as kinematics chains with the same joint variables with different lengths. The key idea is normalize just the components of the feature vector that involves distances. The normalize factor is obtained each time that a new user interact with the system measuring the length of his/her right arm.

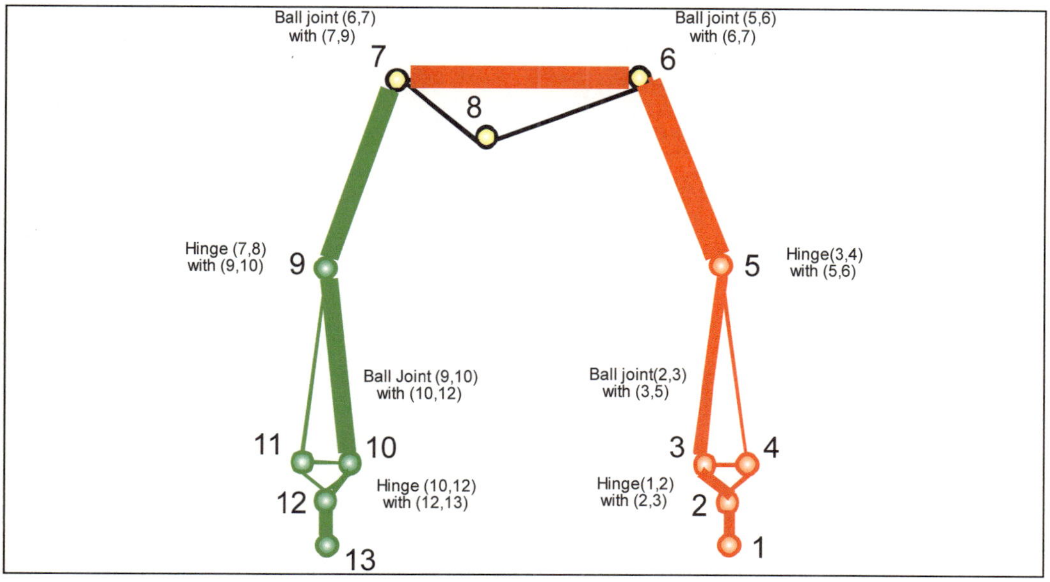

Fig. 5. Kinematics for the upper limbs based on the marker placement on the arms and hands

4. Cleaning and autocalibration algorithms

The motion of the user is tracked with the eight cameras and the electronics acquisition unit of the VICON system at 300 Hz. Sometimes, due to the markers obstructions in the human motion, the data information is lost. For this reason, the "cleaning algorithm" described in (Qian 2004), was implemented.

A calibration process was implemented in order to identify the actual position of the markers and adjust the kinematics model to the new values. Therefore, a fast (1ms) auto calibration process was designed in order to obtain the initial position of the markers of a person placed in a military position called "stand at attention". The algorithm checks the dimension of his/her arms and the position of the markers. The angles are computed and finally this information is compared with to the ideal values in order to compensate and normalize the whole system.

5. Capturing and recognizing Indian Dance movements in real time

In order to test our methodology to recognize gestures, we have implemented a system that recognizes seven basic movements (temporal gestures) of the Indian Dance. Each one has a meaning; thus, the scope of our system is to discover if the user/dancer has performed a valid known movement in order to translate their meaning in an artistic representation using graphics and sounds.

The gestures to be recognized are: earth, fish, fire, sky, king, river and female. In Figure 6, different phases of each gesture are presented. The gestures were chosen from the Indian Dance and their spatial-temporal complexity was useful to test our gesture recognition methodology.

Fig. 6. The seven Indian Dance movements that the system recognizes

The interaction with the user is simple and easy to learn (Figure 7). At the beginning the user remains in front of the principal screen in a motionless state. Once the system has sensed the user's inactivity, it sends a message indicating that it is ready to capture the movement of the user. If the system recognizes the movement, it renders a sound and image corresponding to its meaning. After that, the system is again ready to capture a new movement.

It is possible observe how the cameras track the user movements using the reflective markers that are mounted on the suit dressed by the user. The user's avatar and the image that represents the meaning (a King) of the recognized gesture are displayed in the screen.

The operation of the recognition system is as follow (it is useful see the Figure 8 to check the flow of the information): The eight cameras and their electronics acquisition that compose the VCS acquire the 3D positions of reflective markers attached to a suit that the user dresses on its upper limbs and send the positions to Simulink through UDP protocol.

In Matlab Simulink's Real-Time Windows Target, we have developed a real-time recognition system with a sample rate of 50 Hz. Its operation at each frame rate it's as follows: the current 3D position of the markers is read from Nexus, then the data are filtered to avoid false inputs, then, the data are send simultaneously to XVR (to render a virtual avatar) and other block that converts the 3D positions of the 13 markers (vector of 39 elements) in a normalized feature vector (values from 0 to 1). The elements of feature vector were described in the section 3.

Fig. 7. The recognition system in action.

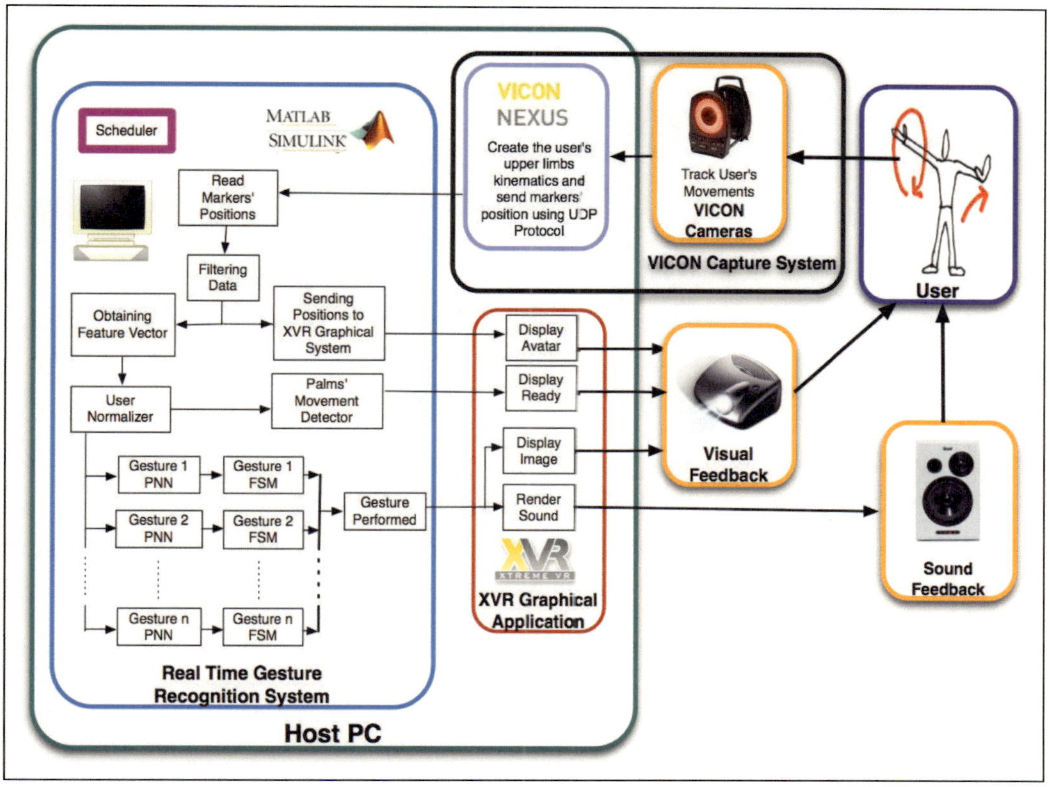

Fig. 8. Architecture of the Indian Dance Recognition System

Then, the normalized feature vector is sent simultaneously to a group of PNN-FSM couples that work in parallel. Each couple is used to recognize one gesture; the PNN is used to determinate which is the nearest state respect to the current body position for one particular gesture. Recognition is performed using a FSM, where, its state transitions depend only of the output of the its PNN. If the FSM arrives to its final state the gesture is recognized.
The movements of both palms are analyzed in order to determine if the user is or is not in motion, this information is useful in order to interact correctly with the user. All the FSM are initialized when no motion of the palms is detected.
XVR, a virtual environment development platform is used to display visual and sound information to the user. An avatar shows the users' movements, a silhouette is used to render the sensation of performing movements along the time. The graphical application also indicates when is ready to recognized a new movement. If the recognition system has detected a valid gesture, the meaning of the gesture recognized is displayed to the user using an appropriate image and sound (Figure 9).

Fig. 9 .The gesture that represents "Sky" has been recognized. The user can see the avatar, hear the sound of a storm and see the picture of the firmament.

6. Development of a real-time gesture recognition, evaluation and feed-forward correction of a multimodal Tai-Chi platform

The learning process is one of the most important qualities of the human being. This quality gives us the capacity to memorize different kind of information and behaviors that help us to analyze and survive in our environment. Approaches to model learning have interested researches since long time, resulting in such a way in a considerable number of underlying representative theories.
One possible classification of learning distinguishes two major areas: Non-associative learning like habituation and sensitization, and the associative learning like the operant conditioning (reinforcement, punish and extinction), classical conditioning (Pavlov Experiment), the observational learning or imitation (based on the repetition of a observed

process) (Byrne & Russon, 1998), play (the perfect way where a human being can practice and improve different situations and actions in a secure environment) (Spitzer, 1988), and the multimodal learning (dual coding theory) (Viatt & Kuhn, 1997).

Undoubtedly, the imitation process has demonstrated a natural instinct action for the acquisition of knowledge that follows the learning process mentioned before. One example of multimodal interfaces using learning by imitation in Tai-chi has been applied by the Carnegie Mellon University in a Tai-Chi trainer platform (Tan, 2003), demonstrating how through the use of technology and imitation the learning process is accelerated.

The human being has a natural parallel multimodal communication and interaction perceived by our senses like vision, hearing, touch, smell and taste. For this reason, the concept of Human-Machine Interaction HMI is important because the capabilities of the human users can be extended and the process of learning through the integration of different senses is accelerated (Cole & Mariani, 1995; Sharma et al., 1998; Akay et al., 1998)). Normally, any system that pretends to have a normal interaction must be as natural as possible (Hauptmann & McAvinney, 1993). However, one of the biggest problems in the HMI is to reach the transparency during the Human-Machine technology integration.

In such a way, the multimodal interface should present information that answers to the "why, when and how" expectations of the user. For natural reasons exists a remarkable preference for the human to interact multimodally rather than unimodally. This preference is acquired depending of the degree of flexibility, expressiveness and control that the user feels when these multimodal platforms are performed (Oviatt, 1993). Normally, like in real life, a user can obtain diverse information observing the environment. Therefore, the Virtual Reality environment (VR) concept should be applied in order to carry out a good Human-Machine Interaction. Moreover, the motor learning skills of a person is improved when diverse visual feedback information and correction is applied (Bizzi et al., 1996).

For instance the tactile sensation, produced on the skin, is sensitive to many qualities of touch. (Lieberman & Breazeal, 2007) carried out, for first time, an experiment in real time with a vibrotactile feedback to compensate the movements and accelerate the human motion learning. The results demonstrate how the tactile feedback induces a very significant change in the performance of the user. In the same line of research Boolmfield performed a Virtual Training via Vibrotactile Arrays (Boolmfield & Badler, 2008).

Another important perception variable is the sound because this variable can extend the human perception in Virtual Environments. The modification of parameters like shape, tone and volume in the sound perceived by the human ear (Hollander & Furness, 1994), is a good approach in the generation of the description and feedback information in the human motion.

Although a great grade of transparency and perception capabilities are transmitted in a multimodal platform, the intelligence of the system is, unquestionably, one of the key parts in the Human-Machine interaction and the transfer of a skill. Because of the integration, recognition and classification in real-time of diverse technologies are not easy tasks, a robust gesture recognition system is necessary in order to obtain a system capable to understand and classify what a user is doing and pretending to do.

6.1 Tai-Chi system implementation

This application teaches to novel students, five basic Tai Chi movements. Tai-Chi movements were chosen because they have to be performed slowly with high precision.

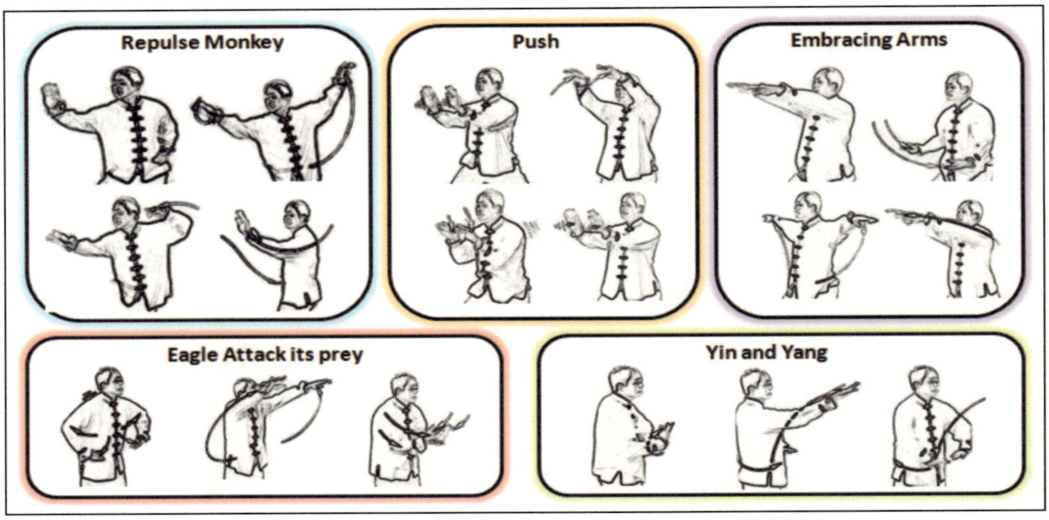

Fig. 10. The 5 Tai-Chi Movements

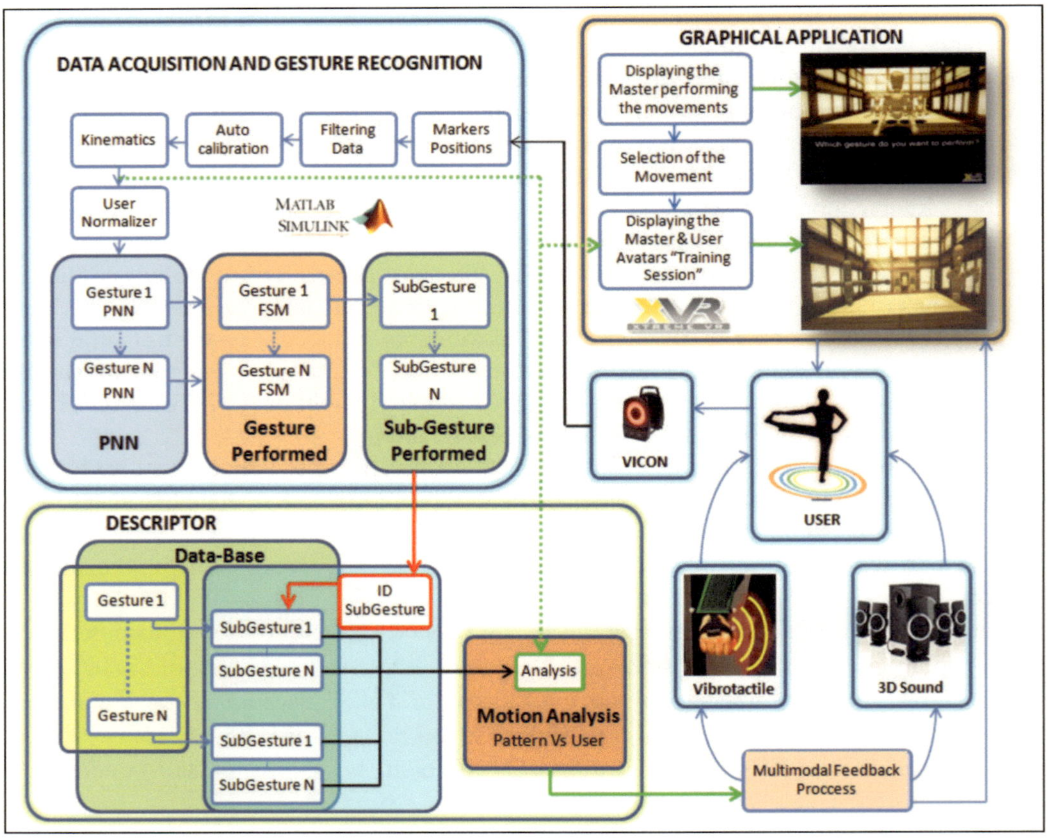

Fig. 11. Architecture of the Multimodal Tai Chi Platform System

Each movement is identified and analyzed in real time by the gesture recognition system. The gestures performed by the users are subdivided in n-spatial states (time-independent)

and evaluated step-by-step in real time by the descriptor system. Finally, the descriptor executes audio-visual-tactile feedback stimuli in order to try to correct the user's movements. The general architecture of the multimodal platform is shown in Figure 11.
The data acquisition and gesture recognition blocks of this application practically are identical to the Indian Dance Application. The PNNs and FSMs used to recognize the gestures were changed to recognize the new movements. A new virtual environment was developed and in this application a vibrotactile feedback in addition to the visual and auditive was employed.

6.1.1 Real-time descriptor process

The comparison and qualification in real-time of the movements performed by the user is computed by the descriptor system. In other words, the descriptor analyzes the differences between the movements executed by the expert and the movement executed by the student, obtaining the error values and generating the feedback stimuli to correct the movement of the user.
The Real-Time Descriptor is formed by a database where for each gesture *n* instances of 26 variables (where *n* it is number of states of the gesture) with the following information are saved:

- Right & left elbows angles
- Right & left pitch and yaw wrists angles
- Right & left pitch, yaw and roll shoulders angles
- The magnitude of the distance between palms
- The magnitude of the distance between elbows
- Three spatial vector components from the back to the right palm
- Three spatial vector components from the back to the left elbow
- Three spatial vector components from the back to the right elbow

The descriptor's database is created through an offline process as follow:

1. For each gesture it is necessary capture from a pattern movement performed by the expert samples of the 26 mentioned variables
2. Each sample must be classified in its corresponding state normalizing the 13 variables described in the section 3 and feeding them to their corresponding PNN. Once that all the samples were classified, the result it is a sequence of estates to which each sample belongs
3. The sequence's indexes of the *n*-1 transitions between states plus the last index that conform the sequence are detected
4. With the *n* indexes founded, their corresponding data in the original samples are extracted to conform the data used to describe the gesture
5. Change of gesture and repeat the steps 1-5 until finish all the gestures
6. Create the descriptor database of all gestures

When the application is running, each state or subgesture is recognized in real time by the gesture recognition system during the performance of the movement. Using the classic feedback control loop during the experiments was observed that the user feels a delay in the corrections. For that reason, a feed-forward strategy was selected to compensate this perception. In this methodology when a user arrives at one state of the gesture, the descriptor system carries out an interpolation process to compare the actual values with respect to the values in the descriptor for the following state, creating a feed-forward loop which estimates in advance the next correction values of the movement. The error is computed by:

$$\theta_{error} = \left[P_{n+1} - U_n \right] * F_n \tag{1}$$

Where θ_{error} is the difference between the pattern and the user, P is the pattern value obtained from the descriptor, U is the user value, F_u is the normalize factor and n is the actual state.

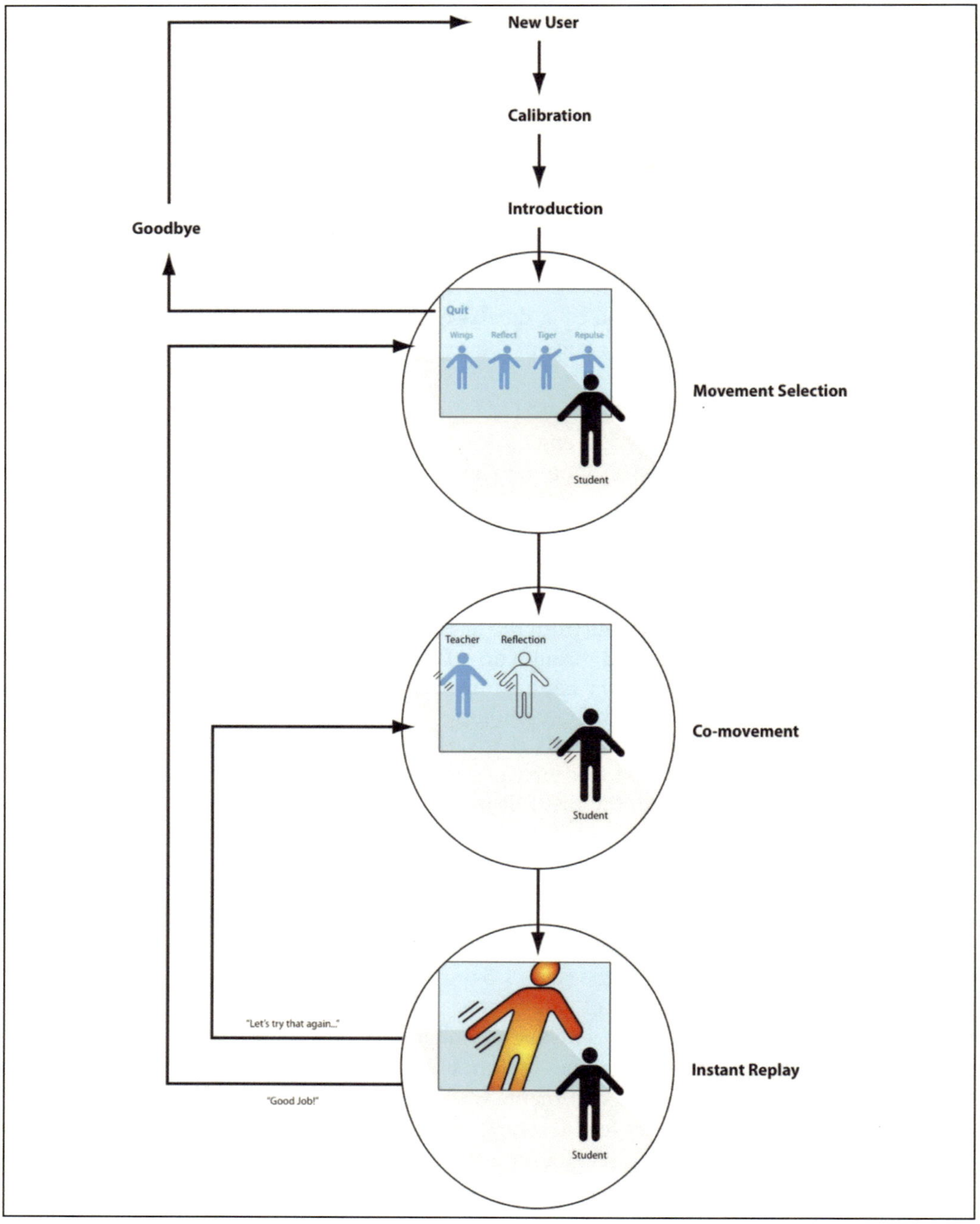

Fig. 12. Storyboard for the Multimodal Interface for Tai Chi Training

6.1.2 Virtual reality platform

The virtual environment platform provides the visual information to the user was programmed in XVR. There are 3 different sequences involved in this scenery. The first one is the initial screen that shows 5 avatars executing different Tai Chi movements. When a user tries to imitate one movement, the system recognizes the movement through the gesture recognition algorithm and passes the control to the second stage called "training session". In this part, the system visualizes 2 avatars, one represents the master and the other one is the user. Because learning strategy is based on the imitation process, the master performs the movement one step forward to the user. The teacher avatar remains in the state n+1 until the user has reached or performed the actual state n.

With this strategy the master gives the future movement to the user and the user tries to reach him. Moreover, the graphics displays a virtual energy line between the hands of the user. The intensity of this line is changing proportionally depending on the error produced by the distance between the hands of the student. When a certain number of repetitions have been performed, the system finishes the training stage and displays a replay session that shows all the movements performed by the student and the statistical information of the movement's performance. Figure 12 shows the storyboard for the interaction with the user and Figure 13 (A)(B) shows the virtual Tai-Chi environment.

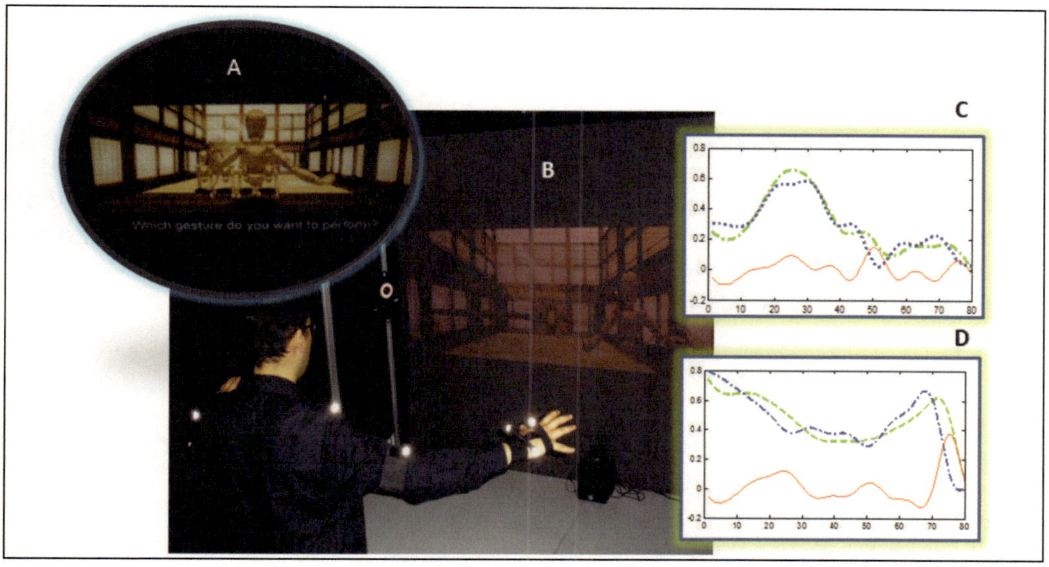

Fig. 13. VR environment, A) Initial Screen, 5 avatars performing Tai-Chi movements, B) Training session, two avatars, one is the master and second is the user. C) Distance of the Hands, D) Right Hand Position.

6.1.3 Vibrotactile feedback system

The SHAKE device was used to obtain wireless feedback vibrotactile stimulation. This device contains a small motor that produces vibrations at different frequencies. In this process, the descriptor obtains the information of the distance between the hands, after this, the data is compared with the pattern and finally sends a proportional value of the error. The SHAKE varies proportionally the intensity of the vibration according to error value produced by the descriptor (1 Hz - 500 Hz). This constraint feedback is easy to understand for the users when the arms have reached a bad position and need to be corrected. Figure 13 (C) shows the ideal distance between the hands (green), the distance between the hands performed by the user (blue) and the feed-back correction (red).

6.1.4 Audio feedback system

The position of the arms in the X-Y plane is analyzed by the descriptor and the difference in position between the pattern and the actual movement in each state of the movement is computed. A commercial Creative SBS 5.1 audio system was used to render the sound through 5 speakers (2 Left, 2 Right, 1 Frontal) and 1 Subwoofer. In this platform was selected a background soft-repetitive sound with a certain level of volume. The sound strategy performs two major actions (volume and pitch) when the position of the hands exceeds the position of the pattern in one or both axes. The first one increases, proportionally to the error, the volume of the speakers in the corresponding axis-side (Left-Center-Right) where is found the deviation and decreases the volume proportionally in the rest of the speakers. The second strategy varies proportionally the pitch of the sound (100-10KHz) in the corresponding axis-side where was found the deviation. Finally, the user through the pitch and the volume can obtain information which indicates where is located the error and its intensity in the space.

7. Experimental results

The experiments were performed capturing the movements of 5 Tai-Chi gestures (Figure 10) from 5 different subjects. The tests were dived in 5 sections where the users performed 10 repetitions of the each one of the 5 movements performed. In the first section was avoid the use of technology and the users performs the movement in a traditional way, only observing a video of a professor performing one simple tai-chi movement. The total average error TAVG is calculated in the following way:

$$TAVG = \frac{1}{Ns}\sum_{s=0}^{Ns}\frac{1}{n}\sum_{i=0}^{n}(\theta_{Teacher} - \theta_{Student}) \qquad (2)$$

Where N_s is the total number of subjects, n is the total number of states in the gesture and θ is the error between the teacher movement and the student.

Figure 14 (A) shows the ideal movements (Master Movements) of the gesture number 1 and (B) represents the TAVG of the gesture 1 executed by the 5 subjects without feedback. The TAVG value the 5 subjects without feedback was around 34.79% respect to the ideal movement. In the second stage of the experiments, the Virtual Reality Environment was activated. The TAVG value for the average of the 5 subjects in the visual feedback system presented in Figure 14(C) was around 25.31%. In the third section the Visual-Tactile system was activated and the TAVG value was around 15.49% respect to the ideal gesture. In the next stage of the experiments, the visual- 3D audio system was performed and the TAVG value for the 5 subjects in the audio-visual feedback system was around 18.42% respect to the ideal gesture. The final stage consists in the integration of the audio, vibrotactile and visual systems. The total mean error value for the average of the 5 subjects in the audio-visual-tactile feedback system was around 13.89% respect to the ideal gesture. Figure 14 (D) shows the results using the whole integration of the technologies.

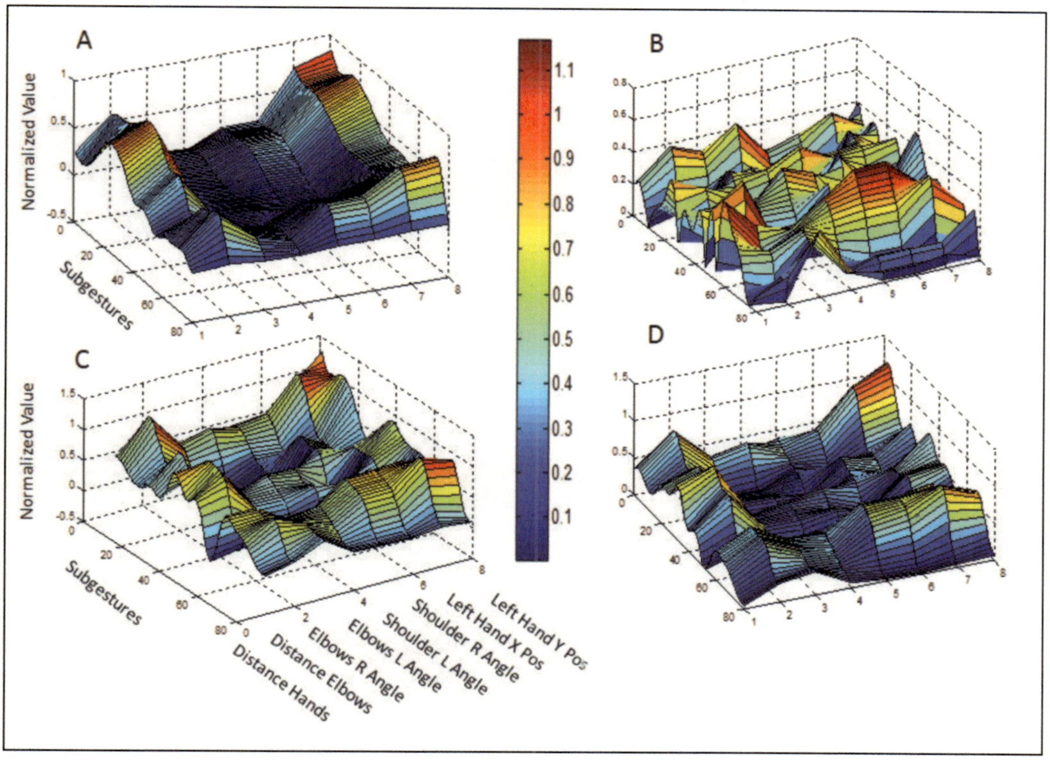

Fig. 14. Variables of Gesture 1, A) Pattern Movement, B) Movement without feedback, C) Movement with Visual feedback and D) Signals with Audio-Visual-Tactile feedback.

Figure 15 presents an interesting graph where the results of the four experiments are indicated. In one hand, as it was expected, the visual feedback presented the major error. In the other hand the integration of audio-visual-vibrotactile feedback has produced a significant reduction of the error of the users. The results of the experiments show that although the process of learning by imitation is really important, there is a remarkable improvement when the users perform the movements using the combination of diverse multimodal feedbacks systems.

8. Conclusion

We have built an intelligent multimodal interface to capture, understand and correct in real time a complex hand/arm gestures performed inside its workspace. The interface is formed by a commercial vision tracking system, a commercial PC and feedback devices: 3D sound system, a cave like VE and a pair of wireless vibrotactile devices. The interface can capture the upper part limbs kinematics of the user independently of the user's size and high. The interface recognizes complex gestures due a novel recognition methodology based on several machine-learning techniques such as: dynamic k-means, probabilistic neural networks and finite state machines. This methodology is the main contribution of this research Human Hand Computer Interaction research area, its working principle is simple: a gesture is split in several states (a state is an ensemble of variables that define an static position or configuration), the key is obtain the optimal number of states that define

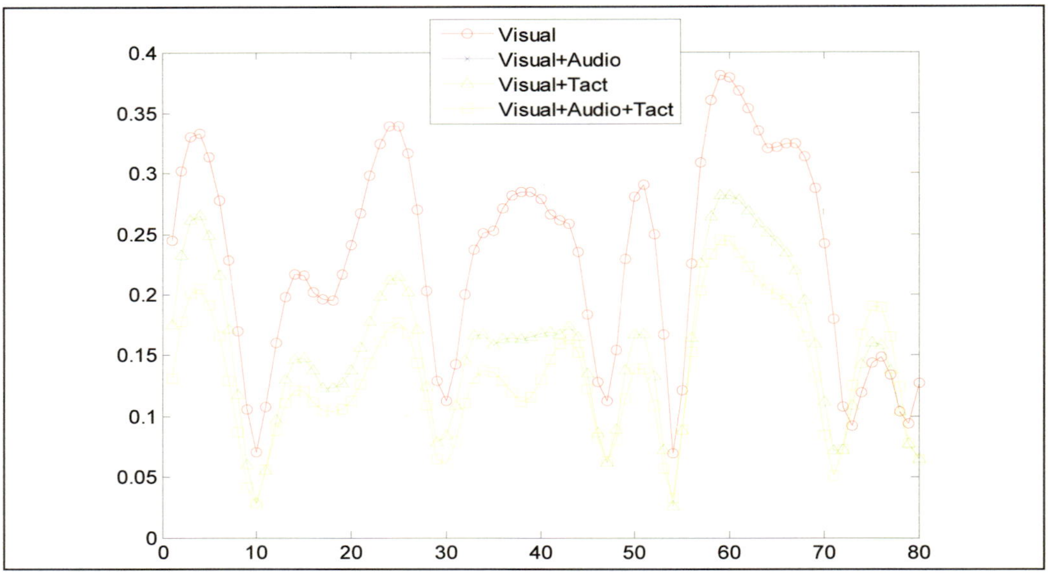

Fig. 15. Average Errors

correctly a gesture and develop an algorithm that recognize which is the most similar state to the current position of the user limbs; then the gesture recognition is simple due that just it is necessary check the sequence of states that the user generated with his/her movement, if the sequence is correct and arrives to the gesture's last state without error, the gesture is recognized.

The methodology proposed showed the effectiveness of dynamic k-means to obtain the optimal number and spatial position of each state. To calculate the boundaries of each state instead to use complex sequential algorithms such as Hidden Markov Models or recurrent neural networks, we have employed Probabilistic Neural Networks. For each gesture a PNN was created using as a hidden neurons the states founded by the dynamic k-means algorithm, this way a gesture can be modeled with few parameters enabling compress the information used to describe the gesture.

Furthermore the PNN is used not only to model the gesture but also to recognize it, avoiding use two algorithms. For example when a recognizer is developed with HMM its necessary at least executed two algorithms, the first one defines the parameters of the HMM given a dataset of sequences using the Baum-Welch algorithm and then, online the forwad-backward algorithm computes the probability of a particular output sequence and the probabilities of the hidden state values given that output sequence. This approach it is neither intuitive nor easy to implement when the sequence of data is multidimensional, to solve this problem, researchers that desire recognize complex gestures use dimension reductions algorithms (such as principal components analysis, independent components analysis or linear discriminant analysis) or transform the time dependent information to its frequential representation destroying their natural representation (positions, angles, distances, etc). Our methodology shown its effectiveness to recognize complex gesture using PNN with a feature vector of 16 dimensions without reduce its dimensionality.

The comparison and qualification in real-time of the movements performed by the user is computed by the descriptor system. In other words, the descriptor analyzes the differences between the movements executed by the expert and the movements executed by the student, obtaining the error values and generating the feedback stimuli to correct the movements of the student. The descriptor can analyze step by step the movement of the user and creates a comparison between the movements by the master and user. This descriptor can compute the comparison up to 26 variables (angles, positions, distances, etc). For the Tai-Chi skill transfer system, only four variables were used which represents X-Y deviation of each hand with respect to the center of the body, these variables were used to generate spatial sound, vibrotactile and visual feedback. The study shown that with the use of this interface, the Tai Chi students improve to its capability to imitate their movements.

A lot of work must be done, first is still not clear the contribution of each feedback stimuli to correct the movements, seems that the visual stimuli (Master avatar) dominate to the auditive and vibrotactile feedbacks. A separate studies in which auditive and vibrotactile feeback will be the only stimulus must be done in order to understand their contributions to create the multimodal feeback. For the auditive study, a 3D spatial sound system must be developed putting emphasis in the Z position. For the vibrotactile study, a upper limbs suit with tactors distributed along the arm/hands must be developed, the position of the tactors must be studied through a psychophysical tests.

Once the multimodal platform has demonstrated the feasibility to perform the experiments related to the transfer of a skill in real-time, the next step will be focused in the implementation of a skill methodology which consists, in a brief description, into acquire the data from different experts, analyze their styles and the descriptions of the most relevant data performed in the movement and, through this information, select a certain lessons and exercises which can help the user to improve his/her movements. Finally it will be monitored these strategies in order to measure the progress of the user and evaluate the training. These information and strategies will help us to understand in detail the final effects and repercussions that produce each multimodal variable in the process of learning.

9. References

Akay, M., Marsic, I., & Medl, A. (1998). A System for Medical Consultation and Education Using Multimodal Human/Machine Communication. *IEEE Transactions on information technology in Biomedicine*, 2.

Annelise Mark Pejtersen, J. R. (1997). Ecological Information Systems and Support of Learning: Coupling Work Domain Information to user Characteristics. *Handbook of Human-Computer Interaction.* North/Holland.

Bellman, R. (2003). *Dynamic Programming.* Princeton University Press.

Bizzi, E., Mussa-Ivaldi, F., & Shadmehr, R. (1996). *Patent n. 5,554,033.* United States of America.

Bloomfield, A., & Badler, N. (2008). Virtual Training via vibrotactile arrays. *Teleoperator and Virtual Environments*, *17.*

Bobick, A. F., & Wilson, A. D. (1995). A state-based technique for the summarization and recognition of gesture. *5th International Conference on Computer Vision*, (p. 382-388).

Bobick, A., & Davis, J. (1996). Real-Time Recognition of Activity Using Temporal Templates. *Proc. Int'l Conf. Automatic Face and Gesture Recognition.* Killington, Vt.

Byrne, R., & Russon, A. (1998). Learning by imitation: a Hierarchical Approach. *Behavioral and Brain Sciences*, *21*, 667-721.

Cole, E., & Mariani, J. (1996). Multimodality. Survey of the State of the Art of Human Language Technolgy.

Flach, J. M. (1994). Beyond the servomechanism: Implications of closed-loop, adaptive couplings for modeling humna-machine systems. *Symposium on Human Interaction with Complex Systems.* North Carolina A&T State University.

Gopher, D. (2004). Control processes in the formation of task units. *28th International Congress of Psychology.* Beijing, China.

Hauptmann, A., & McAvinney, P. (1993). Gesture with Speech for Graphics Manipulation. *Man-Machines Studies , 38.*

Hollander, A., & Furness, T. A. (1994). Perception of Virtual Auditory Shapes. *Proceedings of the International Conference on Auditory Displays.*

Hong, P., Turk, M., & Huang, T. S. (2000). Gesture Modeling and Recognitition Using Finite State Machines. *Proceedings of the Fourth IEEE International Conference on Automatic Face and Gesture Recognition.*

Inamura, T., Nakamura, Y., & Shimozaki, M. (2002). Associative Computational Model of Mirror Neurons that connects mising link betwwen behavior and symbols. *Proceedings of the IEEE/RSJ International Conference on Intelligent Robots and Systems. .* Lausanne, Switzerland.

Jain, A., Murty, M. N., & Flynn, P. (1999). Data Clustering: A Review. *ACM Computing Surveys , 31* (3).

Lieberman, J., & Breazeal, C. (2007). Development of a wearable Vibrotactile FeedBack Suit for Accelerated Human Motor Learning. *IEEE International Conference on Robotics and Automation.*

McCullough, M. (1999). *Abstracting Craft.* MIT Press.

Norman, D. (1986). *User-centered systems design.* Hillsdale.

Norman, D. (1988). *The design of everyday things.* New York: Basic Books.

M. Chignell, P., & Takeshit, H. (1999). Human-Computer Interaction: The psychology of augmented human behavior. In P. Hancock (A cura di), *Human performance and Ergonomics.* Academic Press.

Oviatt, S. (1993). User Centered Modeling and Evaluation of Multimodal Interfaces. *Proceedings of the IEEE , 91.*

Qian, G. (2004). A gesture-Driven Multimodal Interactive Dance System. *IEEE International Conference on Multimedia and Expo.*

Sharma, R., Huang, T., & Pavlovic, V. (1998). A Multimodal Framework for Interacting With Virtual Environments. In C. Ntuen, & E. Park (A cura di), *Human Interaction With Complex Systems* (p. 53-71). Kluwer Academic Publishers.

Spitzer, M. (1998). *The mind within the net: models of learning, thinking and acting.* The MIT press.

Tan Chau, P. (2003). Training for physical Tasks in Virtual Environments: Tai Chi. *Proceedings of the IEEE Virtual Reality.*

Viatt, S., & Kuhn, K. (1997). Integration and synchronization of input modes during multimodal human-computer interaction. *Proc. Conf. Human Factors in Computing Systems CHI.*

VICON. (2008). Seen at December 29, 2008 from http://www.vicon.com

VRMedia. (2008). Seen at December 29, 2008 from EXtremeVR: virtual reality on the web: http://www.vrmedia.com

Yamato, J., Ohya, J., & Ishii, K. (1992). Recognizing human action in time-sequential images using Hidden Markov Model. *IEEE Conference CPVPR,* (p. 379-385). Champaign, IL.

17

Robot-Aided Learning and r-Learning Services

Jeonghye Han
Department of Computer Education
Cheongju National University of Education
Republic of Korea

1. Introduction

To date, there have been many studies that have deployed robots as learning and teaching assistants in educational settings to investigate their pedagogical effects on learning and teaching. Hendler (2000) categorized the robots with which learners may interact in the future into five categories, i.e., toy robotics, pet robotics, interactive displays, service robotics including assistive ones, and educational robotics. Goodrich and Schultz (2007) classified the educational service robots into assistive and educational robotics. The robots that can serve for educational purposes can be divided into two categories: educational robotics (also referred to as hands-on robotics), and educational service robotics. The difference between these two types of robotics stems from the primary user groups. Educational robotics has been used by prosumers, a blend of producers and consumers, while educational service robots show a clear boundary between the producers and consumers. In general, the latter takes anthropomorphized forms to substitute or support teachers. It can also add more than what computers have offered to aid language learning because their anthropomorphic figures lower the affective filter and provide Total Physical Response (TPR) in terms of actions, which may lead to form social interactions. This chapter focuses on educational service robots.

Taylor (1980) emphasized that computers have played important roles as educational tutors, tools and tutees. It seems that educational service robots can act as emotional tutors, tutoring assistants (teaching assistants), and peer tutors. The tutor or teaching assistant robots can also be a kind of assistant for innovative educational technologies for blended learning in order to obtain the knowledge and skills under the supervision and support of the teacher inside and outside the classroom. Examples of this include computers, mobile phones, Sky TV or IP TV channels and other electronics. The studies of Mishra and Koehler (2006) probed into teachers' knowledge, building on the idea of Pedagogical Content Knowledge (PCK) suggested by Shulman (1987). They extended PCK to consider the necessary relationship between technology and teachers' subject knowledge and pedagogy, and called this Technological Pedagogical Content Knowledge (TPCK), as shown in Fig. 3. An educational service robot as a teaching and learning assistant for blended learning is divided into three categories: the tele-operated (or tele-conference, tele-presence) type, autonomous type, and transforming type, according to the location of TPCK, as displayed in Table 1.

Types of Robots	The location of TPCK	Applications	Tele-operator
tele-operated (tele-presence, tele-conferenc)	tele-operator's brain	PEBBLES SAKURA Giraffe Some Korean robots	a child children and teacher parents native speakers
Autonomous	Robot's intelligence	Irobi, Papero, RUBI	
Transforming (Convertible)	tele-operator's brain or robot's intelligence	iRobiQ	

Table 1. Educational Service Robots for Blended Learning

Tele-operated robots in educational environments have substituted teachers in remote places, and have provided the tele-presence of educational services through instructors' remote control. The PEBBLES (Providing Education By Bringing Learning Environments to Students) of Telebotics Inc., which are remote-controlled mobile video conferencing platforms, enable a child due to illness or for other reasons, who is far away, to enjoy all the benefits of real-school life face to face (Williams et al., 1997). The Giraffe of HeadThere Inc. provides the service of babysitter supervision, and it can be used like PEBBLES. The physical version of the speech-driven embodied group-entrained communication system SAKURA with InterRobots and InterActor (Watanabe et al., 2003; 2007) is one of this kind of robot. Since 2008, some tele-operated robots have been commercialized to teach foreign languages to Korean children by English-speakers in the USA or Australia. Since the robots' anthropomorphic forms resemble the English-speakers, it may reduce the language learners' affective filter and strengthens the argument for a robot-based education that is remotely controlled by a native speaker. Furthermore, tele-operated robots, because of their anthropomorphic bodies, might fairly overcome the two outstanding issues of videoconferencing, eye contact and appearance consciousness. These issues are preventing videoconferencing from becoming the standard form of communication, according to Meggelen (2005).

With respect to the autonomous robots, the TPCK acts as the robots' intelligence. Hence, it can function as an instructor, instructor assistant, and peer tutor. Because robots have technological limitations in artificial intelligence, robot-based education should prefer focusing more on children's education. Although current autonomous robots narrowly have TCK, and not TPCK, many previous studies (Kanda et al., 2004; Han et al., 2005; Hyun et al., 2008; Movellan et al., 2009) have displayed positive results in using iRobiQ, Papero, RUBI in teaching children. This will be discussed further in the next chapter. Convertible robot can provide both tele-operation and autonomous control, and converts between the two depending on the surroundings or the command. These robots speak in TTS when they are in the autonomous mode, but in the voice of a remote instructor when it is in the tele-operated mode. The conversion between machine and natural voices might confuse children about the robot's identity. Therefore, the mode of transformation should be explicitly recognizable to children.

Robotic learning (r-Learning) is defined as learning by educational service robots, and has been identified as robot-aided learning (RAL), or robot assisted learning, in this study. The collection of educational interaction offered by educational service robots can be referred to as r-Learning Services (Han et al., 2009a; Han & Kim, 2009; Han & Kim, 2006). The purpose of this chapter was to describe the service framework for r-Learning, or RAL. This study begins by a review of literature on educational service robots to classify the r-Learning taxonomy. Then, this study demonstrates case studies for the adoption of r-Learning services in an elementary school. Also, this study discusses the results, focusing on how r-Learning services teachers and students feel, and the possibility of commercialization of this technology. Finally, this study discusses future work in this field.

2. Related works

A growing body of work investigates the impact on RAL through educational service robots. In Table 2, the mains of existing studies are categorized into groups by the type of robot, the role of the robot, the target group, subjects taught, use of visual instruction material (such as Computer Aided Instruction, or CAI, and Web-based Instruction, or WBI), the type of educational service provided, and the number and duration of each field experiment.

		Fels & Weiss	Kanda et al.	Han & Kim	Watanabe et al.	Osada	Hyun et al.	You et al.	Movellan et al.	YuJin
Type	Autonomous		•	•		•	•		•	
	Tele-operated	•			•			•		
	Transforming									•
Role	Tutor				Avatar					
	Tutoring Assistant			•				•		•
	Peer Tutor	Not tutor (Peer)	•		Avatar	•	•		•	•
Target Group	Toddler				•	•	•		•	•
	Children	•	•	•	•			•		•
	Silver									
Subject	English		•	•				•		•
	Domestic Language			•	Any Subject		•		•	•
	Etc	•		•		Nursing				•
Instruct-ion	Visual			•	•		•		•	•
	Not Visual	•	•			•		•		
Services	Conversation	•	•	•	•	•	•	•	•	•
	Edutainment		•		•	•	•		•	•
	Showing Instruction			•	•		•		•	•
	Calling User		•		•	•			•	•
	VR or AR			AV	VR, AR					AR
Experi-ment	Term	6 weeks	2 weeks	40mins X3	185 days	185 days	1 month	40mins X2	2 weeks	N/A
	Effect	•	Motive	•	N/A	N/A	•		•	

N/A: we did not obtain related information in detail

Table 2. Some Reviews of Literature on RAL

Robot Types

Most of the recent studies about the types of robots (e.g., Kanda et al., 2004; Han et al., 2005; Han & Kim, 2006; You et al., 2006; Hyun et al., 2008; Movellan et al., 2009; Han et al. 2009a) have concentrated on the autonomous types of educational service robots. Tele-operated robots for educational purposes were shown in Williams et al. (1997), Fels and Weiss (2001), Watanabe et al. (2003), and You et al. (2006). The tele-operators of these studies were students or parents, not teachers or teaching assistants, except in You et al. (2006). iRobiQ, made by Yujin Robot Inc., has commercialized a transforming type that can act as both an autonomous and a tele-operated unit. In the study by Fels and Weiss (2001), the perception of the remote sick students' attitude toward the PEBBLES interactive videoconferencing system became more positive over time, although there appeared to be an increasing trend that is not significant for their health, individuality, and vitality. Watanabe (2001, 2007) and Watanabe et al. (2003) developed a speech-driven embodied communication system that consisted of a virtual system with InterActor and a physical system with InterRobot. The system was operated by speech of tele-operators that might be teachers or students.

Robot Roles and Target Group

With respect to the role of a robot, peer-tutor took the dominant form (e.g., Kanda et al., 2004; Han et al., 2005; Hyun et al., 2008; Movellan et al., 2009) followed by teaching assistant robots (e.g., Han & Kim, 2006, 2009; You et al., 2006; Yujin, 2008) as shown in Fig 1. Study targets comprised pre-school children (e.g., Hyun et al., 2008; Movellan et al., 2009; Yujin, 2008), and elementary school children (Kanda et al., 2004; Han et al., 2005; Han & Kim, 2006, 2008; You et al., 2006; Han et al., 2009a). Some robots, such as Papero, embraced a wide range of user targets, including pre-school children, adults, and even elders (Osada, 2005) taking the role of a younger partner, an assistant, an instructor, and an elder partner, respectively.

Fig. 1. Roles: Teaching Assistant Robot in English and Peer Tutoring Robot

Subject Suitability

Han and Kim (2006) performed a Focus Group Interview (FGI) study with 50 elementary school teachers who were relatively familiar with robots and information technology. The survey results showed that the classes that teach foreign language, native language, and music are suitable for r-Learning services. Most teachers used educational service robots for language courses, such as English class (Kanda et al., 2004; Han et al., 2005; Han & Kim, 2006), native language class to acquire vocabulary (Hyun et al., 2008; Movellan, 2009), Finnish vocabulary (Tiffany Fox, 2008), and Chinese class (Yujin, 2008). However, robots also assisted other classes, including ethnic instrument lessons (Han and Kim, 2006), and music class (Han et al., 2009a). In addition, Yujin (2008) expanded the range of robot learning in various areas, such as teaching science, learning how to cook, and supervising homework.

Visual Instruction

The teaching interaction provided by these robots may or may not include teaching materials from a screen-based robot. A teaching assistant robot that uses screen-based material can share much of its educational frame with e-Learning (electronic-Learning). Indeed, the teaching interactions of robots often request screen-based teaching materials such as CAI or WBI. r-Learning services may often need screen-based materials. Educational service robots have provided instruction materials in the following literature: Han et al. (2005), Han and Kim (2006, 2008), Hyun et al. (2008), Han et al. (2009a, 2009b), Movellan et al. (2009), and Yujin (2008). Most of the instruction materials were created via Flash, which has often been used as an authorized tool for WBI. Thus, instructors can interact by displaying instruction materials based on e-Learning for robots that have a touch screen. The screen- based interaction has the advantage of being able to also be used as the replacement for failed voice or vision recognition, which makes this technology appropriate for r-Learning services.

r-Learning Services

Robovie in Kanda et al. (2004) offered voice-based casual English conversation, and non-verbal communication such as playing rock-paper-scissors. Also, Osada (2005) made Papero perform the following activities: conversation, different reactions for touching different points, roll-call of attendees, quizzes, communication over mobile phone, and making stories. Kanda et al. (2007) revealed that robots might need to use children's native language in order to establish relationships as well as to teach foreign languages. Robovie serves hundreds of interactive behaviours such as shaking hands, hugging, playing rock-paper-scissors, exercising, greeting, kissing, singing, briefly conversing, and pointing to an object in the surroundings. Moreover, Robovie could call children's names by RFID tags and confide personal matters to the children who had interacted with it for an extended period of time. Those services made children interact with Robovie over the long-term.

The teaching assistant robots in Han and Kim (2009), and Han et al. (2009a) provided class management services, such as checking students' attendance, getting attention, being a time clock for activities, selecting random presenters, and instructing classes, such as giving quizzes. In these studies, RUBI, a fun-looking, and bandana-wearing robot tutor interacted with children to teach them numbers, colours, vocabulary and other basic concepts. RUBI sang popular songs and danced, while presenting a related video clip on the screen, and provided Flash-based educational games with its physical activities for improving toddler's vocabulary. You et al. (2006) from Taiwan utilized tablet PC and blue-tooth technology in order to deploy Robosapien as a teaching assistant. Robosapien had five models: storytelling model, Q&A model, cheerleader model, let's act model, and pronunciation leading model.

Robots with touch screens can provide augmented reality (AR), the overlaying of computer graphics onto the real worldview, as well as augmented virtuality (AV), the overlaying of real objects into the virtual space view, as presented in Milgram's virtual reality continuum (Milgram & Kishino, 1994). Han and Kim (2006) identified that the service using robots' touch screens can positively stimulate children in class. In their study, they made the robot display children's photos to use them for checking attendance and selecting a presenter.

Han et al. (2009b) suggested that there is a high potential for the commercialization of robots in educational settings, and expected that the AV service of robots to be a positive influence in opening the robot markets. Similarly, Movellan et al. (2009) offered an AV service that allowed clicking on children's faces. Hence, the relationship between robots and children may become more intimate and solid, one that will last for the long-term when in the future such AR service based screens become technically affordable. Yet, not all robot services utilize touch screens. Watanabe (2007) introduced the AR version of SAKURA, which activated group communication between a virtual teacher, InterActor, students, and InterRobot in the same classroom. This study forecasts that AR can enhance human-robot collaboration, particularly in learning and teaching because AR technology has many benefits that may help create a more ideal environment and communication channels such as an ego-centric view and ex-centric view mentioned by Green et al. (2008).

Field Experiment

Field experiments of educational service robots require the robots' stable operation, parental and school's consent, and a means to protect privacy. If the robot acts as an assistant, the study may need a lot of resources in terms of time and money, and the interaction capability of the instructor remains as a factor of non-sampling error. Also, a field study that involves a robot as an assistant may be largely influenced by the degree of technology acceptance of the main teacher. A theoretical basis for new technology spread, using the Innovation Diffusion Theory of Rogers (1995), can be found in the Technology Acceptance Model (TAM) proposed by Davis (1989). Davis (1989) attempted to explain an individual's actual behavior or behavioral intentions, based upon the user's perception of the usefulness and ease of use of a particular piece of Information Technology (IT). Furthermore, a study that involves children should be carefully devised to control the factor of non-sampling error in children's responses. Also, an existing theory on the adoption and distribution of a new technology can be applied to educational robots since their software design and manufacture are based upon IT. Therefore, in experiments where robots act as peer tutors, they interact with children for two weeks straight as in Kanda et al. (2004) or Movellan et al. (2009). In experiments using a robot as an assisant teacher, robots are either exposed for a longer period with four repetitive sessions per month, such as in Hyun et al. (2008), or robots interact with a larger group of participants in a single session with some pre-study activities to miminze novelty effect, as in Han et al. (2005).

Whether it lasts for a single session or for a month, all these studies have suggested meaningful results in movitation to learn language and in academic achievement. In general, studies have primarily assessed how effectively robots convey verbal communication, non-verbal communcation, such as gestures, and TPR and so forth. Han et al. (2005) reported that the RAL group showed the biggest achievement among the non-computer-based learning group, web-based learning group, and RAL group. Also, Hyun et al. (2008) reiterated the improved linguistic ability of pre-teen children through a series of story making tests, vocabulary tests, story understanding tests, and word recognition tests. Movellan et al. (2009) investigated the effects on knowledge of target words taught by RUBI, which is fully autonomous and has a touch screen, to toddlers over a period of two weeks. They reported that there was a significant improvement in the learning of 10 words to the 9 toddlers (aged 15 through 23 months). Finally, Tomio (2007) and Yujin (2008) began commercializing educational assistant robots for toddlers.

3. r-Learning services

3.1 r-Learning services

In general, r-Learning services can be currently defined as pedagogical and interactive activities which can be reciprocally conducted and interacted between learners and anthropomorphic educators, i.e. robots, in both the real and virtual worlds. There exist some significant differences between totally internet-based e-Learning and robot-based r-Learning. First, there is reciprocal authority to start learning. E-learning is passive and can only start when the learner logs in of his own accord. However, robots have the ability to suggest that the learner start, making r-Learning somewhat active. Second, there is more responsiveness of teaching and learning activities. General computers do not take action but instead let the learner's action happen to them, but robots based on autonomous recognition can be responsive to the learner. Third, there is greater frequency of physical and virtual space. E-Learning can only occur in the virtual world, but r-Learning can be conducted not only physically in the real world, such as in classroom, but also in the virtual world, such as through the touch screen of robots or on TV. Furthermore, unlike computers, robots can interact with learners by direct contact, i.e. warmly hugging learners. Fourth, the anthropomorphism of media for learning is better in robots. Robots can function as both learning media and anthropomorphic educators that can make relationships and interactions between learners and robots. On the other hand, computers used for e-learning are simply learning media. Moreover, robots are avatars themselves both in the real world and in the virtual world, whereas there is nothing for computers to do but to insert avatars into the virtual world. Fifth, robots can provide physical activities to reinforce learning. Learners can directly and physically contact robots by seeing and touching them, imitating them and moving with them. Robots are very suitable for TPR learning. So the learners' physical interaction with robots is much better than in e-Learning, where learners have to stay in front of a computer screen. Sixth, robots serve as a convenient means of communication between teachers and parents, one that may incidentally reinforce the relationship between children and their parents. Robots can take photos of the children engaged in classroom activities for the children's class portfolios, and then send the photos to parents via e-mail or mobile phone. Seventh, robots simplify providing fantasy for immersion learning. Robots can more automatically serve augmented virtuality via their camera, mobility and search functions as shown in Figure 7. Computers cannot compete at the same level. The service of augmented virtuality can help learners enhance both their motivation and immersion.

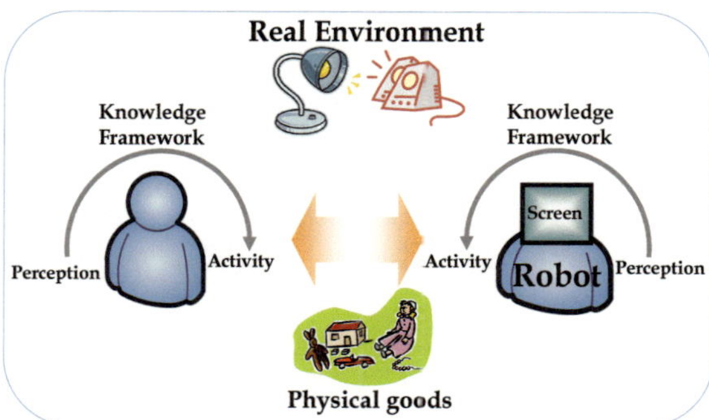

Fig. 2. r-Learning services: Interaction between Learner and Teaching Robot

R-Learning services are based on the interactions between a teaching robot and learners. The interaction occurs through the knowledge framework of a teaching (or teaching assistant) robot after perceiving data from external sensors, as shown in Figure 2. The sensing data may be influenced by real environmental factors, such as light, sound, and obstacles. Educational service robots should possess some level of artificial intelligence, so that they are equipped to perceive physical events and take appropriate action based on the TPCK framework.

However, the knowledge framework of a teaching robot is still at an early stage due to the limitations of today's technology. For a tele-operated TA robot, the TPCK framework may be supported as shown in (a) of Figure 3 (the TPCK framework image is from the site http://www.tpck.org/). Moreover, the autonomous robots are a long way from the teacher's framework of TPCK. That is, the r-Learning services of autonomous robots would be in the teaching assistant framework of TCK along with pedagogical knowledge from the teacher. Autonomous robots have TCK that autonomously shares content via touch screen and TV with children but present robots have technological limitations by not being equipped with TPCK, as shown in (b) of Figure 3.

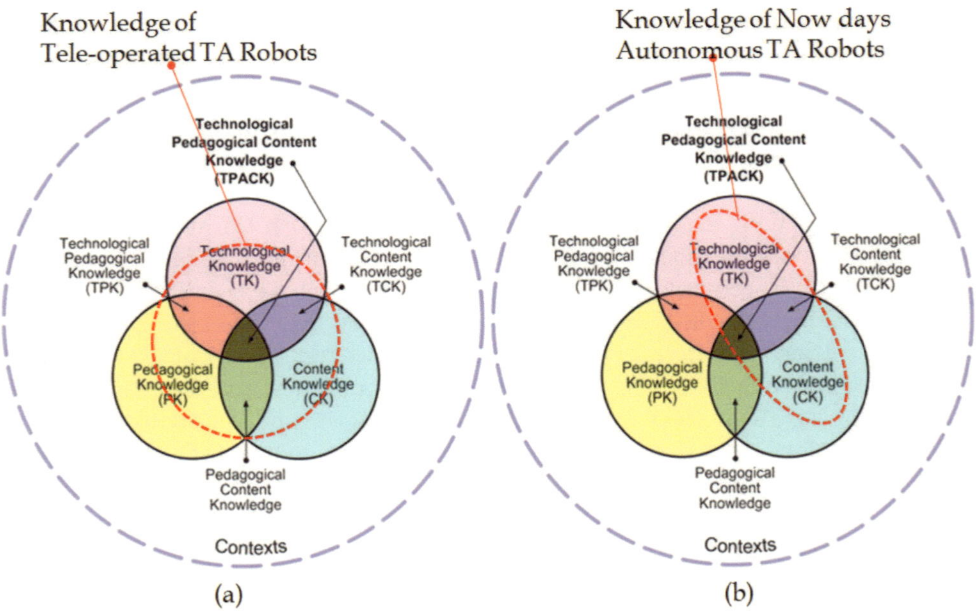

Fig. 3. Teaching Assistant Robot in TPCK framework

R-Learning services include learning activities that utilize a direct physical experience, such as chanting and dancing (Kanda et al., 2004; Han et al., 2009a; Yujin, 2008), and learning that uses teaching props such as toys (Yujin, 2008; Movellan et al., 2009), and that delivers multi-media contents through a touch screen (Han et al., 2005; Han and Kim, 2006; Hyun et al., 2008). This final type of activity for delivery of multi-media contents can sub-divided into two categories: class management and class instruction (Han et al., 2009a). Class instruction can further be sub-divided into contents delivery type (Han et al. 2005; Han and Kim, 2006; Hyun et al., 2008) and participatory type through augmented virtuality (Han et al., 2009b), depending on the participation of the learners.

3.2 Services framework

A service is a non-material product that is well consumed and utilized by the requesting consumer to support his need. Webservice is a software system designed to support interoperable machine-to-machine interaction over a network defined by the W3C. Robot service consists of a collection of interactions between human and robot in authentic environments. Robot service is similar to a pure and pure commodity goods service. Although some utilities actually deliver physical goods, utilities are usually treated as services. For example, robots bring a physical good (a cup of water and medicine to elders when in need), but also provide services (calling their attention to their health). Similarly, r-Learning services refer to the interactions that robots provide for educational purposes. To date, most of the reviews of literature from previous studies in this field have not mentioned the framework of r-Learning services. In this study I have constructed an r-Learning services framework for autonomous robots based on web-based services because r-Learning services have stemmed from e-Learning services. The framework of tele-operated robot services is similar to that of videoconferencing, and consists of tele-operated robots and the supporting system to control their multi-modal interface via the Internet.

Web services refer to those that provide a variety of services to users by inter-connecting devices and applications through networks. Web service is required for users to access the system through any device. A web-based e-Learning system interacts with an operational environment, called the Learning Management System (LMS), through SCORM (Sharable Content Object Reference Model) standards, which are a collection of standards and specifications for web-based e-Learning. These define communications between the client side content and a host system, called the run-time environment (commonly a function of the learning management system). LMS is a system that plans, communicates and manages educational materials in on-line and virtual classrooms both. LCMS, however, is a multi-user system that can produce, store, recycle, and manage digital learning materials, and transfer them to users. The following figure suggests a framework for e-Learning and r-Learning services.

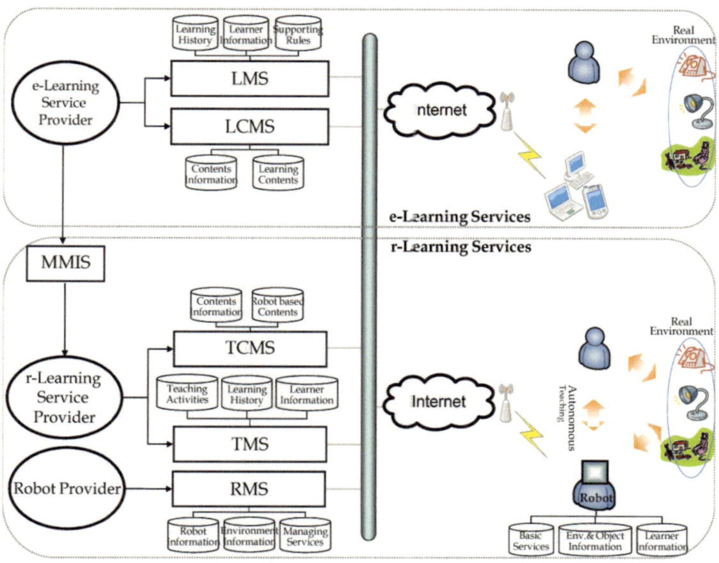

Fig. 4. Service Framework for Teaching Robots

In the e-Learning service framework, users learn educational materials provided by LCMS and LMS in their personal devices, such as a computer, and a PDA. In this case, learners only initiate the interaction because the devices do not have sensors. In the r-Learning service framework, the Teaching Content Management System (TCMS) is necessary for managing robot-based contents. The contents in TCMS differ from those in LCMS because robot-based learning requires emotional interaction that has expression and action. Therefore, in order for LCMS contents to transfer to robot-based multi-modal contents, it needs MMI (Multi Modal Interface). In the same way, e-Learning service providers should convert e-Learning contents into robot-based contents through MMIS. Teaching robots should download appropriate robot-based contents and utilize them autonomously. To this end, the robot service provider should support TMS (Teaching Management System), which allows robots to interact with TCSM on the network and be able to instruct. TMS not only manages teaching based on learner information and learning history, but also contains various teaching activities. The r-Learning service framework should have an RMS (Robot Management System) that can manage robots on the network. Namely, a robot provider can even manage information (e.g. model number and software version) on schools that bought robots, on the physical environment in which it operates, and on the service record. Also, it may have information on its hardware capability, information on the user (school or teacher), information on some objects (e.g. text books, teaching props) and environment (e.g. number of classrooms, position of robots among classrooms) and essential knowledge to support the basic service when it is off the network.

3.3 Design of r-Learning services

This section discusses the factors that need to be considered when designing r-Learning services. There is an example of dynamic services of a teaching assistant robot in Figure 5.

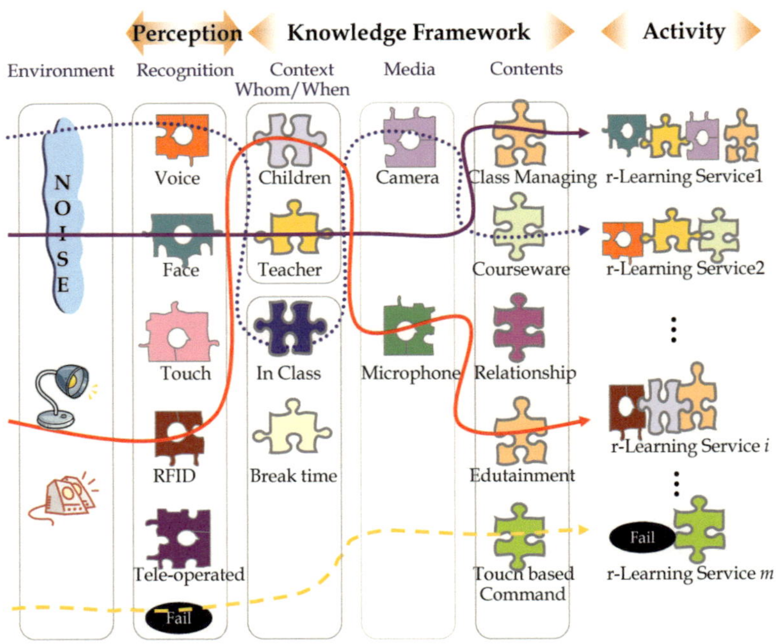

Fig. 5. Examples of Constructing Dynamic r-Learning Services of Teaching Assistant Robots

The first r-Learning service operates the command of class management from an instructor through a facial recognition system. The second r-Learning service provides an augmented virtuality service using a camera that operates on voice commands given in the class. The ith r-Leaning service is an educational game, or edutainment, that a child plays after he or she has been recognized through a RFID tag. The mth r-Learning service uses the touch screen for command because of failed-voice recognition.

This study explains the design requirements of the three types of services (physical activities, teaching props, multimedia contents based touch screen) discussed in Section 3.1. For r-Learning services that utilize physical activities, such as TPR, it is important to make children practice with their own bodies. For r-Learning services that utilize teaching props, such as books or balls, the design is required to include tactile sensing and object recognition. For the r-Leaning services that use multi-media contents, the design needs to support tools, such as Flash, and should utilize both vision and voice recognition to understand a learner's progress. Consequently, this type of service requires a multi-modal script design consisting of TTS and expressions modelled based on pedagogical knowledge, such as educational psychology. Furthermore, it requires a high quality augmented virtuality service, which will enhance the motivation of children in class.

The process of design for r-Learning services is as follows: (1) analysis of the technical capacity of vision, voice, emotion, non-verbal, and object recognition to maximize the autonomy of the robot hardware and software; (2) planning of a knowledge framework for the robot within given technical specifications; (3) developing the teaching scenario, which consists of a series of robot actions and Flash-based (in most of the cases) visual materials for the touch screen, created based on the knowledge framework; (4) GUI design for the visual material in the scenario occurs; (5) final assessment to confirm whether the teaching scenario has maximized the autonomy of the robot hardware, including anthropomorphism, and to consider the collaborative efforts with the teachers; (6) design reiteration begins after this evaluation.

4. Case study on r-Learning services of teaching assistant robots

This chapter discusses case studies relating to r-Learning services using an assistive robot. Two field studies were conducted: one in an elementary school and the other in a kindergarten. The r-Leaning services in each arena aimed to provide class management and learning materials, respectively. Also, the study explains the insights from children and instructors with more detail in the following sections.

4.1 In elementary schools

The teaching assistant robot, Tiro can help teachers as an educational media in class, and a classmate of the children for English learning. This study tried to reflect the concept of a *human-friendly Internet-connected robot with e-Learning technology*. The study developed the educational materials of 'How many cows?' for a 40 minute English class for 3rd grade students. Robot-based contents were uploaded to the TCMS server for public sharing by robot service providers. The instructor taught with a blackboard and a TV connected to a computer in the classroom. The classroom layout in the field work is displayed in Figure 6. Tiro could be downloaded from the TCMS server by the teacher's voice command through wireless internet, and then transferred to projection TV. A classroom is a noisy environment that can influence the recognition rate for voice and vision. If recognition failed, teachers converted to the touch screen-based interaction. Teachers could also use a remote-control to interact with the robot if they were across the classroom. Tiro displayed and explained the learning material while the teacher roamed around the classroom and attended to any individuals who needed extra help.

Fig. 6. Scene of Role Playing and Experiment in Classroom

Tiro's services contained multimedia objects such as the Flash files (SWF format), children's photos and names. Tiro's services were divided into two categories: (1) class management, such as checking students' attendance, getting attention, acting as a timer for activities, and selecting presenters; (2) learning materials transmitted to TV, such as providing lesson objectives, conversation scripts, English chants and dancing, storytelling and role playing, praising and cheering up, providing reviews and quiz games, and more. This study matched these types of services with the models suggested by You et al. (2006) as illustrated in Table 3.

Services in classroom	Han et al. (2006, 2009a, 2009b), Han and Kim (2009), Yujin (2008)	You et al. (2006)
Class Management	Calling the roll Concentration Timer for activity Selection of presenters Checking homework Recording the history of rewards	
Language Learning	Storytelling and role play English chant and dance Augmented reality role playing Quiz for checking pronunciation Quiz Cheering up or praising after the quiz	Storytelling/let's act model Let's act model Pronunciation leading model Q&A model Cheerleader model

Table 3. Comparison with You et al. (2006)'s model for a teaching assistant robot

The following Figure 7 illustrates Tiro's teaching services. In the first column, *calling the roll* shows the situation where the robot calls children's names to check their attendance. The next column, *augmented reality role playing* shows how the robot provides a role playing service by graphically treating children's pictures to become animation characters in order to increase motivation for learning, as suggested by Han et al. (2009b). The third column, *conversation and role play,* describes how the robot and children play a game together as a team.

Calling the Roll | *Getting Attention* | *Conversation and Role Play*

Selecting Children to Present | *Augmented Reality Role Playing* | *English Chant and Dance*

Fig. 7. Some of Tiro Services as Teaching Assistant

Experiments were executed in three English classes in a Korean elementary school. Each class lasted for 40 minutes. In total, 117 children participated in the classes while 20 teachers monitored the classes. To investigate children's responses, the study randomly selected 27 out of the 117 children, and then surveyed them. The study asked questions such as 'Which do you prefer among Tiro's services?', and the children could choose plural preferences. Figure 8 displays the four favourite services that came out of the study, and categorizes them according to who used the service and how much they cooperated. The services preferred by children are illustrated in a yellow circle, those preferred by teachers are represented in blue hexagons and those preferred by both groups are represented in a green rectangular. In general, children liked services that were centred around them, while teachers preferred those that focused on them. Attendance checking was one exception. It was guessed that children preferred it because of the novelty effect of robots recognizing each individual. Teachers spoke of it less desirably because robots had difficulties in recognizing children's voices and faces in the noisy and busy classroom environment, which made this activity more time-consuming than before.

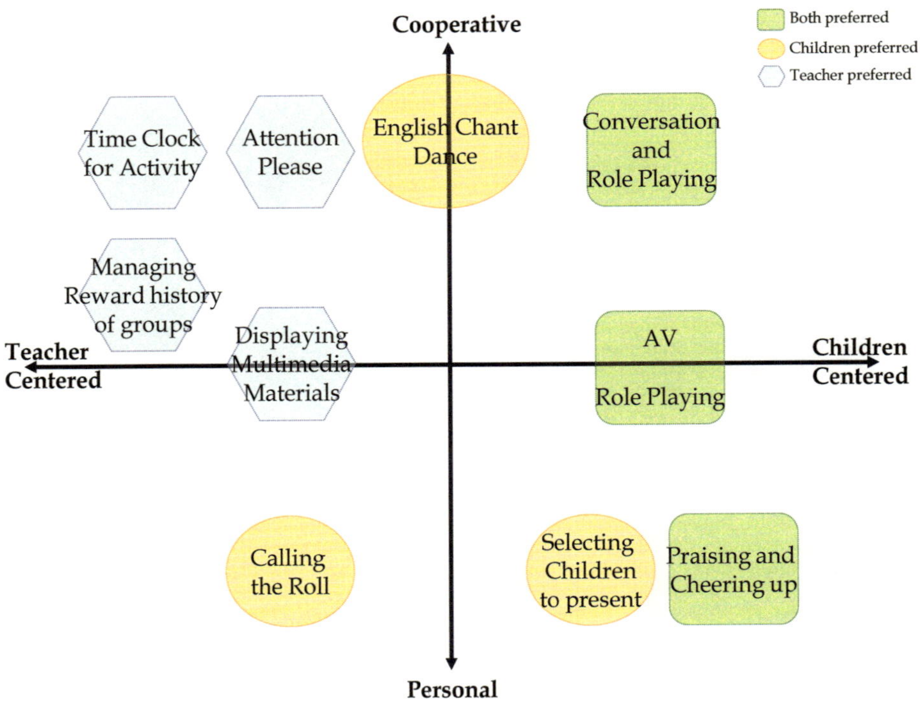

Fig. 8. Popularity of Tiro's Services in English Class

The most popular service was Tiro's praising and cheering up, which enhanced the relationship between children and Tiro in English class. The second one was face-to-face conversation and role play with Tiro. Particularly, the skits using identification based augmented virtuality appear to have had effects on children's motivation for learning. Next in rank was English chant and dance. Calling the roll and selecting a child to present ranked similarly, but the former carried a bigger impact because all children could interact with Tiro. For the teacher group, teachers preferred Tiro's accommodating services rather than those that enhanced the relationship between children and Tiro. The results also showed that teachers preferred face-to-face conversation and role play with Tiro because the robot could pronounce English words better than them. Tiro's praising and cheering up service and display of multimedia presentations were ranked next. Through Tiro's assistance, teachers expected to have more time to interact with and help students individually by reducing the amount of time they spent at the front of the classroom to change slides on the computer. The teachers' expectation on the learning effect with Tiro was positive, with an average value of 3.6 (p-value=0.0001) on a five-point-scale. In follow-up interviews, the teachers also recommended that the encyclopaedia services of Tiro be used during break time and that Tiro could take photos of classroom activities and automatically upload these photos to the school web server for viewing by parents.

4.2 In kindergartens

Up to now, Yujin robotics has begun r-Learning services using iRobiQ in about a hundred kindergartens (Yujin, 2008). The following figure (Figure 9) shows an example of a TMS server. This system comprises a set of menus, such as an introduction for this r-Learning service and iRobiQ, uploading and downloading of r-Learning contents, classroom management for teachers, and general information for parents.

Fig. 9. An Example of TMS Server (http://www.edurobot.net/)

Under teachers' supervision, iRobiQ performs such activities as checking attendance, supporting English learning, reading books, playing music, guiding and arranging daily activities (e.g., making beds, eating, and cleaning), compiling academic portfolios, and then transferring them to parents. If iRobiQ fails to recognize children's faces when checking attendance, it provides a photo menu for children themselves to enter the information. The following Figure 10 illustrates a situation in which iRobiQ dances with children in a TPR dance class. To instruct daily activities, teachers let iRobiQ inspect the cleaning, sing a lullaby during naptime, and teach general eating etiquette, such as standing a line in a cafeteria, washing hands before eating, and more. Also, iRobiQ takes photos of the children engaged in classroom activities for the children's class portfolios, and then send the photos to parents via e-mail or mobile phone. The compiled portfolios are historically valuable to both children and parents, and allow the parents to understand and follow the kindergarten life of their children more closely.

A teacher said that "Although iRobiQ does not walk on two feet like a humanoid and delivers a cup of water, it supports linguistic development by interacting with children as an assistant teacher and instructs everyday knowledge by how to behave in daily life.

Furthermore, it is very important that iRobiQ can share emotional experiences with children as their friend. Most importantly, iRobiQ frees up more time for the teachers to give extra attention to students because it shares our workload." Another teacher commented that children who are the only child tend to think of iRobiQ as a younger sibling and try to become a role model for it. A growing number of studies with similar topics have been conducted, and the results will soon be available through publications. Also, the teacher added that vicarious reinforcement occurred at the beginning as children mimicked the robot and sang and danced in the rigid form that iRobiQ demonstrated. However, the teacher assured that it happened out of curiosity, and faded away soon after.

Chant and Dance with iRobiQ *iRobiQ Compiling Children's Portfolios*

Fig. 10. iRobiQ's Services as Teaching Assistant

5. Conclusion and discussion

Very recently, many researchers have shown much more interest in the pedagogical effects of educational service robots. Depending on the location of the knowledge framework of educational service robots, the robots are categorized into three types: autonomous, tele-operated, and convertible. Most of the current educational service robots inter-connect their knowledge framework with a web service. These types of robotic services are referred to as r-Learning, or robot-aided learning.

In this study, r-Learning services are defined as the interaction between a learner and a robot that occurs for educational purposes. To date, the knowledge framework of educational service robots primarily consists of technology and subject contents with almost no pedagogical knowledge, making the teacher's pedagogical knowledge still important. Therefore, referring to it as r-Learning instead of R-Learning may be more appropriate until the day when there is unity between artificial intelligence and human intelligence, as forecast by Kurzweil (2005).

Also, this study reviewed previous studies relating to r-Learning, and categorized them into r-Learning services according to the types of robots, their role, and so on. A literature review revealed that most of the existing r-Learning services utilize web-based contents as the information that robots provide. Many of them confirm that the use of robots can positively contribute to improving learners' motivation for learning, which has led to the commercialization of a teaching assistant robot.

This study concludes r-Learning has the seven advantages: reciprocal authority to start learning, responsiveness of teaching and learning activities, greater frequency of physical and virtual space, the anthropomorphism of media for learning, providing physical activities, convenient communication for teachers and parents, providing fantasy for immersion learning. It was proposed r-Learning service frameworks based on the frameworks of web-based services and teacher's knowledge. Also, this study defined r-Learning services as a set of activities in the knowledge frameworks built around the perceived sensor data, and divided the activities of r-Learning services into three types: physical experience type, using teaching prop type, and multimedia content based on screen type.

The design of r-Learning services are made up of five steps: the design of vision, voice, emotion, non-verbal, and object recognition; the construction of robots' knowledge framework within a given technical circumstance; the creation of a robot education scenario within the boundary of the knowledge framework of robots, in which the scenario normally includes robot actions and visual materials (normally Flash-based) for the touch screen; the design of GUI for the visual material in the scenario; and confirming whether the teaching scenario maximized the autonomy of the robot hardware, included anthropomorphism, and considering the collaborative efforts with the teachers. Design reiteration begins after this evaluation.

Case studies conducted on r-Learning services development in an elementary school and a kindergarten were introduced. By observing how students and teachers interacted with r-Learning services, the study found an r-Learning paradigm based on its educational impact and emotional communication in the upcoming future.

However, challenges remain. The challenges for tele-presence robots include ethical violations that may come from the field. These robots may invade privacy by intruding into personal school lives of students. Other challenges include protecting the system from misuse outside of a class led by a tele-presence system with a remote instructor, such as information leaks on the classroom itself, unapproved visual and audio recordings, and distribution of such recordings. Next, in the case of an autonomous robot, the recognition technology and the knowledge framework of a teaching robot are still limited. Robot expressions are minimized to meet the minimal hardware specifications required for commercialization. Recognition often fails in a real environment. The cost benefit and uniqueness have been controversial in comparison with computer based content services that also utilize camera and recognition techniques. A high level of TPCK is required for teachers to constantly interact with robots.

Finally, among TPCK, the PK that can elicit a long-term interaction beyond the novelty effect needs to be studied in depth. Several possibilities exist to overcome these challenges including the improvement of a recognition technology, such as using RFID, the development of a new interaction service between the physical activity type and teaching prop type, the development a means to increase the relationship with a robot, continuous studies on an acceptance model of teachers to use a teaching assistant robot.

6. Acknowledgment

This work is supported by Korea Evaluation Institute of Industrial Technology Grant # KEIT-2009-S-032-01.

7. References

Goodrich, M.A. and Schultz, A.C. (2007). Human-robot interaction: a survey. Foundations and Trends in Human-Computer Interaction, 1(3), pp. 203-275.

Davis, F. D. (1989). Perceived usefulness, perceived ease of use, and user acceptance of information technology, MIS Quarterly, 13(3), pp. 319-340.

Deborah I. Fels, Patrice Weiss. (2001). Video-mediated communication in the classroom to support sick children: a case study, *International Journal of Industrial Ergonomics*, 28, pp.251-263

Eunja Hyun, Soyeon Kim, Siekyung Jang, Sungju Park. (2008). Comparative study of effects of language education program using intelligence robot and multimedia on linguistic ability of young children, *Proceedings of the 14th IEEE International Workshop on Robot and Human Interactive Communication (RO-MAN 2008)*, pp.187-912, Munich, Germany

HeadThere Inc. (2009). Retreived from the web site, http://www.headthere.com/

James Hendler. (2000). Robots for the rest of us: designing systems "out of the box", *Robots for Kids: Exploring new technologies for learning*, ISBN:1-55860-597-5, pp. 2-7, The Morgan Kaufmann Publishers

Javier R. Movellan, Micah Eckhardt, Marjo Virnes, Angelica Rodriguez. (2009). Sociable robot improves toddler vocabulary skills, *Proceedings of the 4th ACM/IEEE Human Robot Interaction*, ISBN:978-1-60558-404-1, pp. 307-308, March 11-13, La Jolla, California, USA

Jeonghye Han, Dongho Kim. (2006). Field trial on robots as teaching assistants and peer tutors for children, *Proceedings of the Asia Pacific International Symposium on Information Technology*, pp. 497-501, January 9-10, Hanzhou, China

Jeonghye Han, Dongho Kim. (2009). r-Learning services for elementary school students with a teaching assistant robot, *Proceedings of the 4th ACM/IEEE Human Robot Interaction*, ISBN:978-1-60558-404-1, pp. 255-256, March 11-13, La Jolla, California, USA

Jeonghye Han, Dongho Kim, Jongwon Kim. (2009a). Physical learning activities with a teaching assistant robot in elementary school music class, *Proceedings of the 5th IEEE International Joint Conference on Network Computing, Advanced Information Management and Service, Digital Contents and Multimedia Technology and its Application (NCM 2009)*, pp. , August 25-27, Seoul, Korea

Jeonghye Han, Eunja Hyun, Miryang Kim, Hyekyung Cho, Takayuki Kanda, Tatsuya Nomura. (2009b). The cross-cultural acceptance of tutoring robots with augmented reality services, *International Journal of Digital Content Technology and its Applications*, IBSN:1975-9339, pp.95-102

Jeonghye Han, Miheon Jo, Sungju Park, Sungho Kim. (2005). The educational use of home robots for children. *Proceedings of the 14th IEEE International Workshop on Robot and Human Interactive Communication (RO-MAN 2005)*, pp. 378-383, August 13-15, Nashville, TN, USA

Jim Van Meggelen. (2005). The problem with video conferencing, Retrieved from the web site http://www.oreillynet.com/etel/blog/2005/04/the_problem_with_video_confere.html

Junichi Osada. (2005). Scenario design for children care robot, Papero. *Proceedings of Robot Design Forum*, pp.29-36, November 4, Daejeon, Korea

Laurel A. Williams, Deborah I. Fels, Graham Smith, Jutta Treviranus, Roy Eagleson. (1997). Using PEBBLES to facilitate remote communication and learning, *Proceedings of the 41st Annual Meeting of Human Factors and Ergonomics Society*, Communications, 5, pp. 320-324

Milgram, P. and F. Kishino. (1994). Taxonomy of mixed reality visual displays, *IEICE Transactions on Information and Systems*, E77-D(12), pp.1321-1329.

Punya Mishra, Matthew J. Koehler. (2006). Technological pedagogical content knowledge: a framework for teacher knowledge, Teachers College Record, 108(6), pp. 1017-1054.

Ray Kurzweil. (2005). The singularity is near: when humans transcend biology, ISBN:0670033847, Viking adult press.

Rogers, E. M. (1995). Diffusion of Innovation, *New York*, The Free Press.

Scott A. Green, Mark Billinghurts, XiaoQi Chen, J. Geoffrey Chase. (2008). Human-robot collaboration: a literature review and augmented reality approach in design, *International Journal of Advanced Robotics Systems*, ISSN: 1729-8806, 5(1), pp.1-18.

Shulman, L.S. (1987). Knowledge and teaching: foundations of the new reform, *Havard Educational Review*, 57(1), pp.1-22.

Takayuki Kanda, Rumi Sato, Naoki Saiwaki and Hiroshi Ishiguro (2007). A two-month field trial in an elementary school for long-term human-robot interaction, *IEEE Transactions on Robotics (Special Issue on Human-Robot Interaction)*, 23(5), pp. 962-971.

Takayuki Kanda, Takayuki Hirano, Daniel Eaton, Hiroshi Ishiguro. (2004). Interactive robots as social partners and peer tutors for children: a field trial, *Human-Computer Interaction*, 19(1&2), pp. 61-84

Taylor, R. P. (Ed.). (1980). *The computer in the school: Tutor, tool, tutee*, New York: Teacher's College Press.

Telebotics Inc. http://www.telebotics.com/

Tiffany Fox. (2008). Meet RUBI the robot tutor, July 30, University of California News.

Tomio Watanabe. (2001). E-COSMIC: Embodied Communication System for Mind Connection, *Usability Evaluation and Interface Design*, 1, pp. 253~257.

Tomio Waranabe. (2007). Human-entrained embodied interaction and communication technology for advanced media society, *Proceedings of the 16th IEEE International Conference on Robot & Human Interactive Communication*, pp.31~p36, Aug, 2007.

Tomio Watanabe, Masahi Okubo, Ryusei Danbara. (2003). InterActor for human interaction and communication support, *Human Computer Interaction*, pp.113~120, M. Rauterberg et al.(Eds.), IOS Press.

Yujin Robotics Inc. (2008). Ubiquitous home robot IROBI: Teacher Guide, white paper retrieved in 2008.

Zhenjia You, Chiyuh Shen, Chihwei Chang, Bawjhiune Liu, Gwodong Chen. (2006). A robot as a teaching assistant in an English class, *Proceedings of the 6th IEEE International Conference on Advanced Learning Technologies (ICALT'06)*, pp.87-91, July 5-7, Kerkrade, The Netherlands

18

Design of a Neural Controller for Walking of a 5-Link Planar Biped Robot via Optimization

Nasser Sadati[1,2], Guy A. Dumont[1], and Kaveh Akbari Hamed[2]
[1]*Electrical and Computer Engineering Department, The University of British Columbia, Vancouver,*
[2]*Electrical Engineering Department, Sharif University of Technology, Tehran,*
[1]*BC Canada*
[2]*Iran*

1. Introduction

Underactuation, impulsive nature of the impact with the environment, the existence of feet structure and the large number of degrees of freedom are the basic problems in control of the biped robots. Underactuation is naturally associated with dexterity [1]. For example, headstands are considered dexterous. In this case, the contact point between the body and the ground is acting as a pivot without actuation. The nature of the impact between the lower limbs of the biped walker and the environment makes the dynamic of the system to be impulsive. The foot-ground impact is one of the main difficulties one has to face in design of robust control laws for biped walkers [2]. Unlike robotic manipulators, biped robots are always free to detach from the walking surface and this leads to various types of motions [2]. Finally, the existence of many degrees of freedom in the mechanism of biped robots makes the coordination of the links difficult. According to these facts, designing practical controller for biped robots remains to be a challenging problem [3]. Also, these features make applying traditional stability margins difficult.

In fully actuated biped walkers where the stance foot remains flat on the ground during single support phase, well known algorithms such as the Zero Moment Point (ZMP) principle guarantees the stability of the biped robot [4]. The ZMP is defined as the point on the ground where the net moment generated from ground reaction forces has zero moment about two axes that lie in the plane of ground. Takanishi [5], Shin [6], Hirai [7] and Dasgupta [8] have proposed methods of walking patterns synthesis based on ZMP. In this kind of stability, as long as the ZMP lies strictly inside the support polygon of the foot, then the desired trajectories are dynamically feasible. If the ZMP lies on the edge of the support polygon, then the trajectories may not be dynamically feasible. The Foot Rotation Indicator (FRI) [9] is a more general form of the ZMP. FRI is the point on the ground where the net ground reaction force would have to act to keep the foot stationary. In this kind of stability, if FRI is within the convex hull of the stance foot, the robot is possible to walk and it does not roll over the toe or the heel. This kind of walking is named as fully actuated walking. If FRI is out of the foot projection on the ground, the stance foot rotates about the toe or the heel. This is also named as underactuated walking. For bipeds with point feet [10] and

Passive Dynamic walkers (PDW) [11] with curved feet in single support phase, the ZMP heuristic is not applicable. Westervelt in [12] has used the Hybrid Zero Dynamics (HZD) [13], [14] and Poincaré mapping method [15]-[18] for stability of RABBIT using underactuated phase. The controller proposed in this approach is organized around the hybrid zero dynamics so that the stability analysis of the closed loop system may be reduced to a one dimensional Poincaré mapping problem. HZD involves the judicious choice of a set of holonomic constraints that were imposed on the robot via feedback control [19]. Extracting the eigenvalues of Poincaré return map is commonly used for analyzing PDW robots. But using of eigenvalues of Poincaré return maps assumes periodicity and is valid only for small deviation from limit cycle [20].
The ZMP criterion has become a very powerful tool for trajectory generation in walking of biped robots. However, it needs a stiff joint control of the prerecorded trajectories and this leads to poor robustness in unknown rough terrain [20] while humans and animals show marvelous robustness in walking on irregular terrains. It is well known in biology that there are Central Pattern Generators (CPG) in spinal cord coupling with musculoskeletal system [21]-[23]. The CPG and the feedback networks can coordinate the body links of the vertebrates during locomotion. There are several mathematical models which have been proposed for a CPG. Among them, Matsuoka's model [24]-[26] has been studied more. In this model, a CPG is modeled by a Neural Oscillator (NO) consisting of two mutually inhibiting neurons. Each neuron in this model is represented by a nonlinear differential equation. This model has been used by Taga [22], [23] and Miyakoshi [27] in biped robots. Kimura [28], [29] has used this model at the hip joints of quadruped robots.
The robot studied in this chapter is a 5-link planar biped walker in the sagittal plane with point feet. The model for such robot is hybrid [30] and it consists of single support phase and a discrete map to model the frictionless impact and the instantaneous double support phase. In this chapter, the goal is to coordinate and control the body links of the robot by CPG and feedback network. The outputs of CPG are the target angles in the joint space, where P controllers at joints have been used as servo controllers. For tuning the parameters of the CPG network, the control problem of the biped walker has been defined as an optimization problem. It has been shown that such a control system can produce a stable limit cycle (i.e. stride). The structure of this chapter is as follows. Section 2 models the walking motion consisting of single support phase and impact model. Section 3 describes the CPG model and tuning of its parameters. In Section 4, a new feedback network is proposed. In Section 5, for tuning the weights of the CPG network, the problem of walking control of the biped robot is defined as an optimization problem. Also the structure of the Genetic algorithm for solving this problem is described. Section 6 includes simulation results in MATLAB environment. Finally, Section 7 contains some concluding remarks.

2. Robot model

The overall motion of the biped involves continuous phases separated by abrupt changes resulting from impact of the lower limbs with the ground. In single support phase and double support phase, the biped is a mechanical system that is subject to unilateral constraints [31]-[33]. In this section, the biped robot has been assumed as a planar robot consisting of n rigid links with revolute and parallel actuated joints to form a tree structure. In the single support phase, the mechanical system consists of $n+2$ DOF, where $n-1$

DOF associated with joint coordinates which are actuated, two DOF associated with horizontal and vertical displacements of the robot in the sagittal plane which are unactuated, and one DOF associated with orientation of the robot in sagittal plane which is also unactuated. With these assumptions, the generalized position vector of the system (q_e) can be split in two subsets q and r. It can be expressed as

$$q_e := (q^T, r^T)^T, \tag{1}$$

where $q := (q_0, q_1, \dots, q_{n-1})^T$ encapsulates the joint coordinates and q_0 which is the unactuated DOF between the stance leg and the ground. Also $r := (x, y)^T \in \mathbb{R}^2$ is the Cartesian coordinates of the stance leg end.

A. Single support phase

Figure 1 depicts the single support phase and configuration variables of a 5-link biped robot ($n = 5$). In the single support phase, second order dynamical model immediately follows from Lagrange's equation and the principle of virtual work [34]-[36]

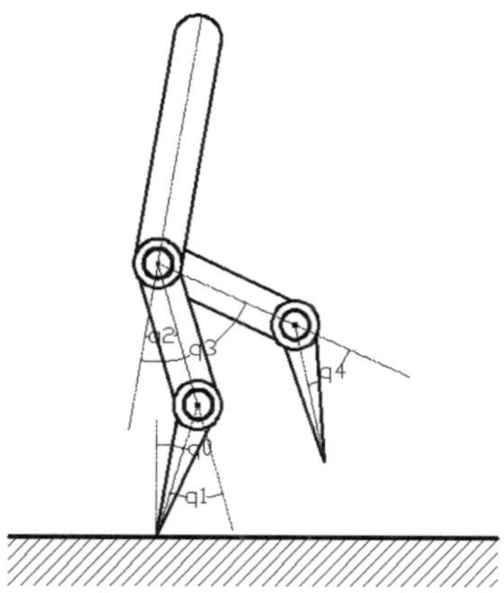

Fig. 1. Single support phase and the configuration variables.

$$M_e(q_e)\ddot{q}_e + H_e(q_e, \dot{q}_e) + G_e(q_e) = B_e u - B_e F_e(q_e, \dot{q}_e) + J_e^{st}(q_e)^T F^{ext,st}, \tag{2}$$

where $M_e(q_e) \in \mathbb{R}^{(n+2)\times(n+2)}$ is the symmetric and positive definite inertia matrix, $H_e(q_e, \dot{q}_e) \in \mathbb{R}^{n+2}$ includes centrifugal and Coriolis terms and $G_e(q_e) \in \mathbb{R}^{n+2}$ is the vector containing gravity terms. Also $u := (u_1, u_2, \dots, u_{n-1})^T \in \mathbb{R}^{n-1}$ includes the joint torques applied at the joints of the robot, $B_e \in \mathbb{R}^{(n+2)\times(n-1)}$ is the input matrix, $F_e(q_e, \dot{q}_e) \in \mathbb{R}^{n-1}$ includes the joint frictions modeled by viscous and static friction terms, $J_e^{st}(q_e) := \partial r^{st} / \partial q_e \in \mathbb{R}^{2\times(n+2)}$ is the Jacobian at the stance leg end. Also $F^{ext,st} := (F_x^{ext,st}, F_y^{ext,st})^T \in \mathbb{R}^2$ is the ground reaction force at the stance leg end. With setting $q_e := (q^T, r^T)^T$ in (2), the dynamic equation of the mechanical system can be rewritten as the following form

$$\left[\begin{array}{c|c} M_{11}(q) & M_{12}(q) \\ \hline M_{12}(q)^T & m\,I_2 \end{array}\right]\begin{bmatrix}\ddot{q}\\ \ddot{r}\end{bmatrix}+\begin{bmatrix}H_{e1}(q_e,\dot{q}_e)\\ H_{e2}(q_e,\dot{q}_e)\end{bmatrix}+\begin{bmatrix}G_{e1}(q_e)\\ G_{e2}(q_e)\end{bmatrix}=\begin{bmatrix}0\\ u\\ \hline 0_{2\times1}\end{bmatrix}-\begin{bmatrix}0\\ F(q,\dot{q})\\ \hline 0_{2\times1}\end{bmatrix}+J_e^{st}(q_e)^T\begin{bmatrix}F_x^{ext,st}\\ F_y^{ext,st}\end{bmatrix}, \tag{3}$$

where m is the total mass of the robot. If we assume that the Cartesian coordinates have been attached to the stance leg end and the stance leg end is stationary (i.e. in contact with the ground and not slipping), these assumptions (i.e. $r=0,\ \dot{r}=0,\ \ddot{r}=0$) will allow one to solve for the ground reaction force as explicit functions of $(q,\dot{q},u)$ [37], [38]. Also, the dynamic equation in (3) will be reduced with this assumptions and this will lead to a lower dimensional mechanical model which describes the single support phase if the stance leg end is stationary as follows

$$\begin{aligned} M(q)\ddot{q}+H(q,\dot{q})+G(q) &= \begin{bmatrix}0\\ u\end{bmatrix}-\begin{bmatrix}0\\ F(q,\dot{q})\end{bmatrix} \\ F^{ext,st} &= \Psi(q,\dot{q},u), \end{aligned} \tag{4}$$

where $M(q)=M_{11}(q)$ and $\Psi(.):TQ\times\mathbb{R}^{n-1}\to\mathbb{R}^2$ is a nonlinear mapping of $(q,\dot{q},u)$. Also $TQ:=\{x:=(q^T,\dot{q}^T)^T \mid q\in Q,\ \dot{q}\in\mathbb{R}^n\}$ is the state space of the reduced model where Q is a simply connected, open subset of $[-\pi,\pi)^n$. Note that q_0 is an unactuated DOF in (4) (i.e. without actuation) and hence $\dim u<\dim q$. It can be shown that

$$\Psi(q,\dot{q},u)=\begin{bmatrix}\sum_{i=1}^{n} m_i\ddot{x}_i\\ \sum_{i=1}^{n} m_i(\ddot{y}_i+g)\end{bmatrix}=\begin{bmatrix}m\ddot{x}_{cm}\\ m(\ddot{y}_{cm}+g)\end{bmatrix}, \tag{5}$$

where $r_i:=(x_i,y_i)^T$ and $r_{cm}:=(x_{cm},y_{cm})^T$ are the coordinate of the mass center of link i and the mass center of the robot, respectively, m_i is the mass of the link i and g is the gravitational acceleration. With assumption $x_{cm}=f_1(q)$ and $y_{cm}=f_2(q)$, we have

$$\ddot{r}_{cm}(q,\dot{q},\ddot{q})=\begin{bmatrix}\partial f_1/\partial q\\ \partial f_2/\partial q\end{bmatrix}\ddot{q}+\begin{bmatrix}\dot{q}^T(\partial^2 f_1/\partial q^2)\dot{q}\\ \dot{q}^T(\partial^2 f_2/\partial q^2)\dot{q}\end{bmatrix}. \tag{6}$$

With setting $\ddot{q}=M(q)^{-1}(0,\overline{u}^T)^T-M(q)^{-1}(H(q,\dot{q})+G(q))$ where $\overline{u}:=u-F(q,\dot{q})$ from equation (4) in equation (6) and using equation (5), we have

$$\begin{aligned} \Psi(q,\dot{q},u) &= mJ^c(q)M^{-1}(q)\begin{bmatrix}0\\ u\end{bmatrix}-mJ^c(q)M^{-1}(q)\begin{bmatrix}0\\ F(q,\dot{q})\end{bmatrix}-mJ^c(q)M^{-1}(q)H(q,\dot{q}) \\ &\quad -mJ^c(q)M^{-1}(q)G(q)+m\begin{bmatrix}\dot{q}^TH_1(q)\dot{q}\\ \dot{q}^TH_2(q)\dot{q}\end{bmatrix}+\begin{bmatrix}0\\ mg\end{bmatrix}, \end{aligned} \tag{7}$$

where $J^c(q) := \partial r_{cm} / \partial q = \begin{bmatrix} \partial f_1 / \partial q \\ \partial f_2 / \partial q \end{bmatrix} \in \mathbb{R}^{2\times n}$ is the Jacobian matrix at the center of mass, also $H_1(q) := \partial^2 f_1 / \partial q^2 \in \mathbb{R}^{n\times n}$ and $H_2(q) := \partial^2 f_2 / \partial q^2 \in \mathbb{R}^{n\times n}$. The validity of the reduced model in (4) is dependent on two following conditions

$$\begin{aligned} &i) \;\; \ddot{y}_{cm} + g > 0 \\ &ii) \;\; \left|\ddot{x}_{cm}\right| < \mu \left|\ddot{y}_{cm} + g\right|, \end{aligned} \tag{8}$$

where μ is the static friction coefficient between the stance leg end and the ground. The first condition in (8) is to ensure that the stance leg end is contact with the walking surface and the second condition is to ensure that the slipping dos not occur at the stance leg end [39]. The dynamic equation of (4) in the state-variable is expressed as $\dot{x} = f(x) + g(x)u$ where $x := (q^T, \dot{q}^T)^T \in TQ$ is the state vector. If we assume that $x_1 := q$ and $x_2 := \dot{q}$, we get $x = (x_1{}^T, x_2{}^T)^T$ and

$$\begin{aligned} f(x) &= \begin{bmatrix} x_2 \\ -M^{-1}(x_1)\left(H(x_1,x_2) + G(x_1) + \begin{pmatrix} 0 \\ F(x_1,x_2) \end{pmatrix}\right) \end{bmatrix} \\ g(x) &= \begin{bmatrix} 0_{n\times n-1} \\ M^{-1}(x_1)\begin{pmatrix} 0_{1\times n-1} \\ \hline I_{n-1} \end{pmatrix} \end{bmatrix}. \end{aligned} \tag{9}$$

B. Frictionless impact model

In this section, following assumptions are done for modeling the impact [40]:

A1. the impact is frictionless (i.e. $F(q,\dot{q}) = 0$). The main reason for this assumption is the problem arising of the introducing of dry friction [2];
A2. the impact is instantaneous;
A3. the reaction forces due to the impact at impact point can be modeled as impulses;
A4. the actuators at joints are not impulsive;
A5. the impulsive forces due to the impact may result in instantaneous change in the velocities, but there is no instantaneous change in the positions;
A6. impact results in no slipping and no rebound of the swing leg; and
A7. stance foot lifts from the ground without interaction.

With these assumptions, impact equation can be expressed by the following equation

$$M_e(q_e)\dot{q}_e(t^+) - M_e(q_e)\dot{q}_e(t^-) = J_e^{sw}(q_e)^T \delta F^{ext,sw}, \tag{10}$$

where $\delta F^{ext,sw} := \int_{t^-}^{t^+} F^{ext,sw}(\tau)\, d\tau$ is the impulsive force at impact point and $J_e^{sw}(q_e) := \partial r^{sw} / \partial q_e \in \mathbb{R}^{2\times(n+2)}$ is the Jacobian matrix at the swing leg end. The assumption A6 implies that impact is plastic. Hence, impact equation becomes

$$\begin{aligned} M_e(q_e)\dot{q}_e(t^+) - J_e^{sw}(q_e)^T \delta F^{ext,sw} &= M_e(q_e)\dot{q}_e(t^-) \\ J_e^{sw}(q_e)\dot{q}_e(t^+) &= 0. \end{aligned} \tag{11}$$

This equation is solvable if the coefficient matrix has full rank. The determinant of the coefficient matrix is equal to $\det M_e(q_e) \times \det(J_e^{sw}(q_e)M_e(q_e)^{-1}J_e^{sw}(q_e)^T)$ and it can be shown that the coefficient matrix has full rank iff the robot is not in singular position. The solution of the equation in (11) can be given by the following equation

$$\begin{bmatrix} \dot{q}_e(t^+) \\ \delta F^{ext,sw} \end{bmatrix} = \Lambda(q_e) \begin{bmatrix} M_e(q_e)\dot{q}_e(t^-) \\ 0 \end{bmatrix}, \tag{12}$$

where

$$\Lambda(q_e) := \begin{bmatrix} M_e(q_e) & -J_e^{sw}(q_e)^T \\ J_e^{sw}(q_e) & 0 \end{bmatrix}^{-1}, \tag{13}$$

and also $q_e(t^-) := \left(q(t^-)^T, 0, 0\right)^T$ and $q_e(t^+) := \left(q(t^+)^T, 0, 0\right)^T$. The map from $\dot{q}_e(t^-)$ to $\dot{q}_e(t^+)$ without relabeling is

$$\begin{aligned} \dot{q}_e(t^+) &= \Lambda_{11}(q_e)M_e(q_e)\dot{q}_e(t^-) \\ \delta F^{ext,sw} &= \Lambda_{21}(q_e)M_e(q_e)\dot{q}_e(t^-). \end{aligned} \tag{14}$$

After solving these equations, it is necessary to change the coordinates since the former swing leg must now become the stance leg. Switching due to the transfer of pivot to the point of contact is done by relabeling matrix [39], [40] $R \in \mathbb{R}^{n \times n}$. Hence, we have

$$\begin{aligned} q(t^+) &= Rq(t^-) \\ \dot{q}(t^+) &= R\begin{bmatrix} I_n & 0_{n\times 2} \end{bmatrix}\Lambda_{11}(q_e)M_e(q_e)\dot{q}_e(t^-). \end{aligned} \tag{15}$$

The final result is an expression for x^+ in terms of x^-, which is written as [39]-[41]

$$x^+ = \Delta(x^-). \tag{16}$$

In equation (16), $\Delta(.) : S \to TQ$ is the impact mapping where $S := \{(q, \dot{q}) \in TQ \,|\, y^{sw}(q) = 0\}$ is the set of points of the state-space where the swing leg touches the ground. $x^+ := (q(t^+)^T, \dot{q}(t^+)^T)^T$ and $x^- := (q(t^-)^T, \dot{q}(t^-)^T)^T$ are the state vector of the system after impact and the state vector of the system before impact, respectively. Also, we have

$$\Delta(x^-) := \begin{bmatrix} Rx_1(t^-) \\ \Sigma(x_1(t^-))x_2(t^-) \end{bmatrix}, \tag{17}$$

where $\Sigma(.) : Q \to \mathbb{R}^{n\times n}$ by $\Sigma(x_1(t^-)) := R\begin{bmatrix} I_n & 0_{n\times 2} \end{bmatrix}\Lambda_{11}(q_e)M_e(q_e)\begin{bmatrix} I_n \\ 0_{2\times n} \end{bmatrix}$. The ground reaction force due to the impact can be shown as the following form

$$\delta F^{ext,sw} = \Gamma(x_1(t^-))x_2(t^-), \tag{18}$$

where $\Gamma(.) : Q \to \mathbb{R}^{2\times n}$ by $\Gamma(x_1(t^-)) := \Lambda_{21}(q_e)M_e(q_e)\begin{bmatrix} I_n \\ 0_{2\times n} \end{bmatrix}$. The validity of the results of equation (17) depends on two following conditions

$$\begin{aligned} &i) \quad \Theta(x_1(t^-))x_2(t^-) > 0 \\ &ii) \quad \mu\left|\Gamma_2(x_1(t^-))x_2(t^-)\right| - \left|\Gamma_1(x_1(t^-))x_2(t^-)\right| > 0, \end{aligned} \tag{19}$$

where $\Theta(x_1(t^-)) := J_y^{sw}(Rx_1(t^-))\Sigma(x_1(t^-)) \in \mathbb{R}^{1\times n}$ and $J_y^{sw}(q) := \partial y^{sw}/\partial q \in \mathbb{R}^{1\times n}$. The first condition is to ensure that the swing foot lifts from the ground at t^+. The second condition is to ensure that the impact results in no slipping [39]. The valid results are used to re-initialize the model for next step. Furthermore, the double support phase has been assumed to be instantaneous. If we define

$$\Omega := \{x = (x_1^T, x_2^T)^T \in S \,|\, \Theta(x_1)x_2 > 0\ ,\ \mu\left|\Gamma_2(x_1)x_2\right| - \left|\Gamma_1(x_1)x_2\right| > 0\}, \tag{20}$$

the hybrid model of the mechanical system can be given by

$$\begin{aligned} \dot{x} &= f(x) + g(x)\,u \qquad && x^- \notin S \\ x^+ &= \Delta(x^-) && x^- \in \Omega, \end{aligned} \tag{21}$$

where $x^-(t) := \lim_{\tau\to t^-} x(\tau)$. For $x^- \in S - \Omega$, this model is not valid. Also the validity conditions in (8) can not be expressed only as a function of x and they can be expressed as a function of (x,u).

3. Control system

Neural control of human locomotion is not yet fully understood, but there are many evidences suggesting that the main control of vertebrates is done by neural circuits called central pattern generators (CPG) in spinal cord which have been coupled with musculoskeletal system. These central pattern generators with reflexes can produce rhythmic movements such as walking, running and swimming.

A. Central pattern generator model

There are several mathematical models proposed for CPG. In this section, neural oscillator model proposed by Matsuoka has been used [24], [25]. In this model, each neural oscillator consists of two mutually inhibiting neurons (i.e. extensor neuron and flexor neuron). Each neuron is represented by the following nonlinear differential equations

$$\begin{aligned}\tau\dot{u}_{\{e,f\}i} &= -u_{\{e,f\}i} + w_{fe}y_{\{f,e\}i} - \beta v_{\{e,f\}i} + u_0 + Feed_{\{e,f\}i} + \sum_{j=1}^{n} w_{\{e,f\}ij}y_{\{e,f\}j} \\ \tau'\dot{v}_{\{e,f\}i} &= -v_{\{e,f\}i} + y_{\{e,f\}i} \\ y_{\{e,f\}i} &= \max(0, u_{\{e,f\}i}),\end{aligned} \tag{22}$$

where suffixes f and e mean flexor muscle and extensor muscle, respectively. Also suffix i means the i th oscillator. u_i is the inner state of i th neuron, y_i is the output of the i th neuron, v_i is a variable which represents the degree of self-inhibition effect of the i th neuron, u_0 is an external input from brain with a constant rate and $Feed_i$ is a feedback signal from the mechanical system which can be an angular position or an angular velocity. Moreover, τ and τ' are the time constants associated with u_i and v_i, respectively, β is a constant representing the degree of the self-inhibition influence on the inner state and w_{ij} is a connecting weight between the i th and j th neurons. Finally, the output of the neural oscillator is a linear combination of the extensor neuron inner state and the flexor neuron inner state

$$y_{NO,i} = -p_e u_{e,i} + p_f u_{f,i}. \tag{23}$$

The positive or negative value of $y_{NO,i}$ corresponds to activity of flexor or extensor muscle, respectively. The output of the neural oscillator can be used as a reference trajectory, joint torque and phase. In this chapter, it is used as a reference trajectory at joints. The studied robot (see Fig. 1) has four actuated joints (i.e. hip and knee joints of the legs). We assume that one neural oscillator has been used for generating reference trajectories at each of the actuated joints.

B. Tuning of the CPG parameters

The walking period is a very important factor since it much influences stability, maximum speed and energy consumption. The walking mechanism has its own natural frequency determined mainly by the length of the links of the legs. It appears that humans exploit the natural frequencies of their arms, swinging pendulums at comfortable frequencies equal to the natural frequencies [26]. Human arms can be thought of as masses connected by springs, whose frequency response makes the energy and the control required to move the arm vary with frequency [26]. Humans certainly learn to exploit the dynamics of their limbs for rhythmic tasks [42], [43]. Robotic examples of this idea include open-loop stable systems where the dynamics are exploited giving systems which require little or no active control for stable operation (e.g. PDW [11]). At the resonant frequency, the control need only inject a small amount of energy to maintain the vibration of the mass of the arm segment on the spring of the muscles and tendons. Extracting and using the natural frequency of the links of the robots is a desirable property of the robot controllers. According to these facts, we match the endogenous frequency of each neural oscillator with the resonant frequency of the corresponding link. On the other hand, when swinging or supporting motions of the legs are closer to the free motion, there will not be any additional acceleration and deceleration and the motion will be effective [44]. When no input is applied to the CPG, the frequency of it is called endogenous frequency. Endogenous frequency of the CPG is mainly determined by τ and τ'. In this section, we change the value of τ with constant value of τ/τ'. In this case, the endogenous frequency of CPG is proportional to $1/\tau$. It was pointed out that the proper value of the τ/τ' for stable oscillation is within $[0.1, 0.5]$ [42]. After tuning the time constants of the CPG, other parameters of CPG can be tuned by using the necessary conditions for free oscillation. These necessary conditions for free oscillation can be written as the following form [24], [25]

$$\begin{aligned} &i) \quad \beta > -w_{fe} - 1 \\ &ii) \quad w_{fe} < -(1 + \tau/\tau') \\ &iii) \quad u_0 > 0. \end{aligned} \tag{24}$$

Table I specifies the lengths, masses and inertias of each link of the robot studied in this chapter [3]. By these data and extracting and using resonant frequencies of the links, we match the endogenous frequency of the CPG with the resonant frequency of each link. In this case, τ is designed at $0.13\ (s)$ and $\tau' = 1.53\tau = 0.2\ (\mathrm{s})$ for all of the neural oscillators. According to conditions in (24), we tune β and w_{fe} to 2 and -2, respectively. Also u_0 is equal to 5. The amplitude of the output signal of the CPG is approximately proportional to u_0, p_e and p_f. The output parameters of the CPGs (i.e. p_e and p_f of oscillators at the knee and the hip joints) can be determined by the amplitude of the desired walking algorithm. Table II specifies the designed values of the output parameters of the oscillators at the knee and the hip joints of the robot.

	mass (kg)	length (m)	inertia (kgm^2)
Torso	12.00	0.625	1.33
Femur	6.80	0.40	0.47
Tibia	3.20	0.40	0.20

Table I. The parameters of the robot

	knee	hip
p_f	0.11	0.15
p_e	0.01	0.02

Table II. The output parameters of the cpg

4. Feedback network

It is well known in biology that the CPG network with feedback signals from body can coordinate the members of the body, but there is not yet a suitable biological model for feedback network. The control loop used in this section is shown in the Fig. 2 where $\tilde{\theta} := (q_1, q_2, q_3, q_4)^T$ encapsulates the actuated joint coordinates and there is not any feedback signal from the unactuated DOF (i.e. q_0). The feedback network in this control loop is for autonomous adaptation of the CPG network. In other hand, by using feedback network, the

CPG network (i.e. the higher level of the control system) can correct its outputs (i.e. reference trajectories) in various conditions of the robot.
In animals, the stretch reflexes act as feedback loop [44]. In this section, the feedback signals to the CPG neurons of the hip joints are the tonic stretch reflex as follows [22], [23]

$$\begin{aligned} Feed_{e,h} &= k_{tsr,h}(\theta_{hip} - \theta_{0,hip}) \\ Feed_{f,h} &= -k_{tsr,h}(\theta_{hip} - \theta_{0,hip}), \end{aligned} \tag{25}$$

where $k_{tsr,h}$ is a constant value and also $\theta_{0,hip}$ is the neutral point of this feedback loop at hip joints. We tune the $k_{tsr,h}$ and $\theta_{0,hip}$ to 1 and $0\ (rad)$, respectively.
One of important factors in control of walking is the coordination of the knee and the hip joints in each leg. For tuning the phase difference between the oscillators of the knee and the hip joints in each leg, we propose the following feedback structure which is applied only at oscillators of the knee joints

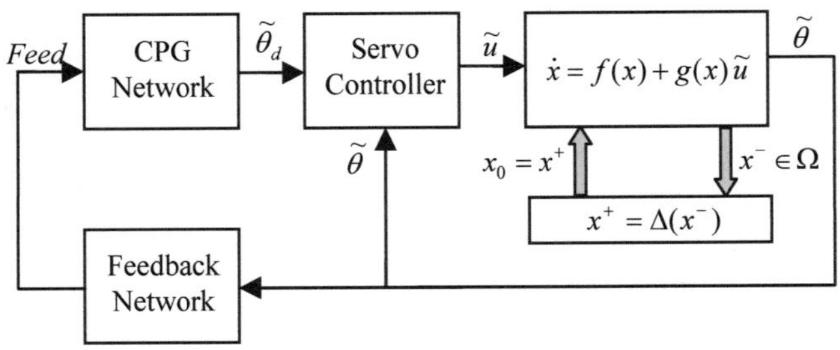

Fig. 2. The control loop used for the biped walker.

$$\begin{aligned} Feed_{e,k} &= k_{tsr,k}(\theta_{knee} - \theta_{0,knee})u(-\dot{\theta}_{hip}) + k_f(\theta_{hip} - \overline{\theta}_{0,hip})u(\dot{\theta}_{hip}) \\ Feed_{f,k} &= -k_{tsr,k}(\theta_{knee} - \theta_{0,knee})u(-\dot{\theta}_{hip}) - k_f(\theta_{hip} - \overline{\theta}_{0,hip})u(\dot{\theta}_{hip}), \end{aligned} \tag{26}$$

where $k_{tsr,h}$ and k_f are constant values , $\theta_{0,knee}$ is the neutral point of the tonic stretch reflex signal at knee joints and $u(.)$ is a unit step function. The first terms of feedback signals in (26) are the tonic stretch reflex terms. These terms are active in stance phase (i.e. $\dot{\theta}_{hip} < 0$). With these terms, we force the mechanical system to fix the stance knee at a certain angular position (i.e. $\theta_{0,knee}$) during the single support phase like the knee joints of the human being. We call $\theta_{0,knee}$ as the bias of the stance knee. In this section, we tune $k_{tsr,h}$ and $\theta_{0,knee}$ to 10 and $0.1\ (rad)$, respectively. The second terms in (26) are active in swinging phase (i.e. $\dot{\theta}_{hip} > 0$). These terms force the knee oscillator to increase its output at the beginning of swinging phase (i.e. $\theta_{hip} < \overline{\theta}_{0,hip}$). Also these terms force the knee oscillator to decrease its output at the end of swinging phase (i.e. $\theta_{hip} > \overline{\theta}_{0,hip}$). We tune k_f and $\overline{\theta}_{0,hip}$ to 4 and $0\ (rad)$, respectively.

5. Tuning of the weights in the CPG network

The coordination and the phase difference among the links of the biped robot in the discussed control loop are done by the synaptic weights of connections in the CPG network. There are two kinds of connections in the CPG network. One of them is the connections among the flexor neurons and the other one is the connections among the extensor neurons.

The neural oscillators in the CPG network can be relabeled as shown in the Fig. 3. According to this relabeling law,

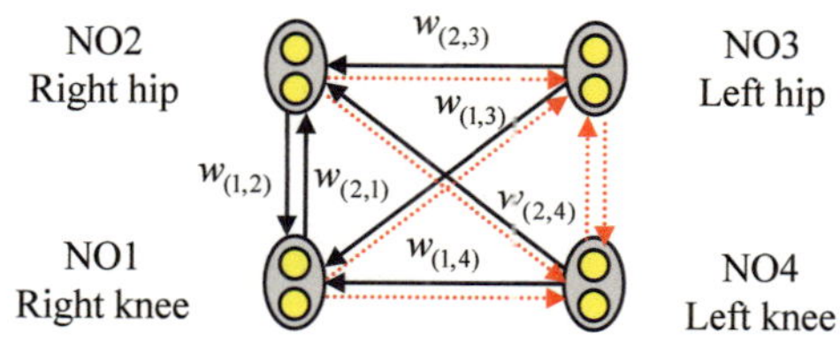

Fig. 3. The CPG network and the synaptic connections.

NO1, NO2, NO3 and NO4 correspond to the right knee, the right hip, the left hip and the left knee neural oscillators, respectively. We show the weight matrix among the flexor and extensor neurons by W_f and W_e, respectively. According to the symmetry between the right leg and the left leg, these matrixes can be written as the following form

$$W_{\{f,e\}} = \left[\begin{array}{cc|cc} 0 & w_{\{f,e\},(1,2)} & w_{\{f,e\},(1,3)} & w_{\{f,e\},(1,4)} \\ w_{\{f,e\},(2,1)} & 0 & w_{\{f,e\},(2,3)} & w_{\{f,e\},(2,4)} \\ \hline w_{\{f,e\},(2,4)} & w_{\{f,e\},(2,3)} & 0 & w_{\{f,e\},(2,1)} \\ w_{\{f,e\},(1,4)} & w_{\{f,e\},(1,3)} & w_{\{f,e\},(1,2)} & 0 \end{array}\right]. \tag{27}$$

This symmetry can be given by the following equations

$$\begin{aligned} w_{\{f,e\},(i,j)} &= w_{\{f,e\},(5-i,5-j)} && ;i,j = 1, \ \ldots \ ,4 \\ w_{\{f,e\},(i,i)} &= 0 && ;i = 1, \ \ldots \ ,4. \end{aligned} \tag{28}$$

In this chapter, we assume $W_f = W_e$. With this assumption and the symmetry between legs, there are six unknown weights which should be determined (bold lines in Fig. 3). For tuning the unknown weights of the CPG network, we should use a tool of the concept of stability for the biped robots. But the concept of stability and stability margin for biped robots is difficult to precisely define, especially for underactuated biped robots with point feet. Since the discussed robot in this chapter has point feet, the ZMP heuristic is not applicable for trajectory generation and verification of the dynamic feasibility of trajectories during execution. In addition, extracting the eigenvalues magnitude of the Poincaré return map may be sufficient for analyzing periodic bipedal walking but they are not sufficient for analyzing nonperiodic motions such as when walking over discontinuous rough terrain. Also, large disruptions from a limit cycle, such as when being pushed, cannot be analyzed using this technique. Some researchers [45] have suggested that angular momentum about the Center of Mass (CoM) should be minimized throughout a motion. As studied in [20], minimizing the angular momentum about the CoM is neither necessary nor sufficient condition for stable walking. According to these facts, for tuning the weights of the CPG network, we define the control problem of the underactuted biped walking as an optimization problem. By finding the optimal solution of the optimization problem, the unknown weights are determined. The total cost function of the optimization problem in this chapter is defined as a summation of sub cost functions and it can be given by

$$J(X) := a_1 J_1(X) + a_2 J_2(X) + a_3 J_3(X), \tag{29}$$

where

$$X := \left(w_{(1,2)}, w_{(1,3)}, w_{(1,3)}, w_{(2,1)}, w_{(2,3)}, w_{(2,4)} \right)^T \tag{30}$$

and $X \in \left[-0.5, 0.5\right]^6$. Also $a_i\ ; i = 1,2,3$ are the positive weights. The first sub cost function in (29) can be defined as a criterion of the difference between the distance travelled by the robot in the sagittal plane and the desired distance

$$J_1(X) := \frac{1}{D_m} \sum_{T_1 + \ldots + T_i \le t_f} sl(T_i), \tag{31}$$

where $sl(T_i)$ is the step length of the i th step, T_i is the time duration of the i th step and D_m is an upper bound of the traveled distance. Also, t_f is the duration of the simulation. This sub cost function is a good criterion of the stability.

The second sub cost function in (29) can be defined as the least value of the normalized height of the CoM of the mechanical system during simulation and it can be given by

$$J_2(X) := \min_{t \in [0\ \ t_f]} \frac{y_{cm}(t)}{y_{cm,\max}}, \tag{32}$$

where $y_{cm,\max}$ is the value of the height of the CoM where the vector q is equal to zero. Since the biped should maintain an erect posture during locomotion, this sub cost function is defined as a criterion of the erect body posture.

The regulation of the rate change of the angular momentum about the CoM is not a good indicator of whether a biped will fall but the reserve in angular momentum that can be utilized to help recover from push or other disturbance is important. We use the rate change of the angular momentum about the CoM for defining the third sub cost function. With

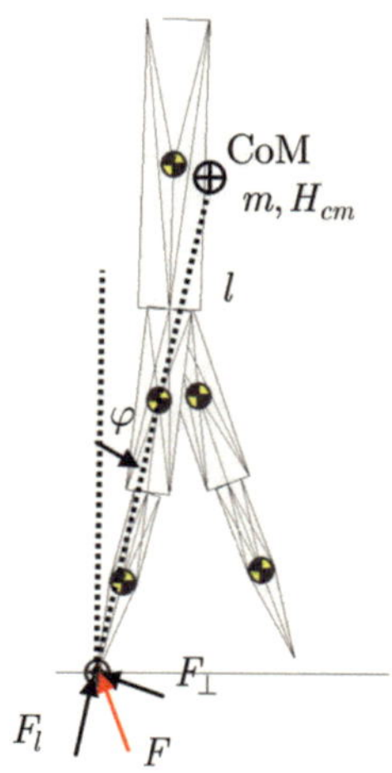

Fig. 4. The virtual inverted pendulum.

setting $x_{cm} = l\sin\varphi$ and $y_{cm} = l\cos\varphi$ in equation (5) where l is the distance from the stance leg end to the CoM and φ is the angle from the stance leg end to the CoM with vertical being zero (see Fig. 4), the equation (5) becomes

$$\begin{aligned} ml^2\ddot{\varphi} + 2ml\dot{l}\dot{\varphi} - mgl\sin\varphi &= -lF_{\perp} \\ m\ddot{l} - ml\dot{\varphi}^2 + mg\cos\varphi &= F_l, \end{aligned} \tag{33}$$

where $F_l := F_x^{ext,st}\sin\varphi + F_y^{ext,st}\cos\varphi$ and $F_{\perp} := -F_x^{ext,st}\cos\varphi + F_y^{ext,st}\sin\varphi$. Also, the total momentum about the stance leg end consists of the angular momentum of the CoM rotating the stance foot plus the angular momentum about the CoM

$$H_{tot} = ml^2\dot{\varphi} + H_{cm}, \tag{34}$$

where H_{tot} and H_{cm} are the angular momentums about the stance leg end and CoM, respectively. Also the net angular momentum rate change is equal to $\dot{H}_{tot} = mgx_{cm} = mgl\sin\varphi$ [3], [20]. With differentiating of equation (34) and setting $\dot{H}_{tot} = mgl\sin\varphi$ in it and comparing with equation (33), it can be shown that

$$\dot{H}_{cm} = lF_{\perp}. \tag{35}$$

Hence, the third sub cost function is defined as following

$$J_3(X) := \frac{1}{1 + \int_0^{t_f} |\ \dot{H}_{cm}(t)\ |^2\, dt} = \frac{1}{1 + \int_0^{t_f} |\ l(t)F_{\perp}(t)\ |^2\, dt}. \tag{36}$$

In this chapter, $a_1 = 4$, $a_2 = 1$ and $a_3 = 1$ and the control problem of the biped walking is defined as the optimal solution of the following optimization problem

$$\max_X\ J(X). \tag{37}$$

By using Genetic algorithm, the optimal solution can be determined. Genetic algorithm is one of the evolutionary algorithms based on the natural selection. In this section, the size of each generation in this algorithm is equal to 400, and at the end of each generation, 50% of chromosomes are preserved and the others are discarded. The roulette strategy is employed for selection and 100 selections are done by this strategy. With applying one-point crossover, 200 new chromosomes are produced. The mutation is done for all of the chromosomes with the probability of 10% except the elite chromosome which has the most fitness. Also, each parameter is expressed in 8 bits.

6. Simulation results

In this section, the simulation of a 5 link planar biped robot is done in MATLAB environment. Table I specifies the lengths, masses and inertias of each link of the robot. This is the model of RABBIT [3]. RABBIT has $50:1$ gear reducers between its motors and links. In this biped robot, the joint friction is modeled by viscous and static friction terms as described by $F(q,\dot{q}) := F_v\dot{q} + F_s\,\mathrm{sgn}(\dot{q})$. Joint PI controllers have been used as servo controllers. Because of the existence of the abrupt changes resulting from the impacts in the hybrid model, the servo controller does not include the derivative terms. We have designed $P_H = 30$, $P_K = 30$, $I_H = 10$ and $I_K = 10$ for the servo controllers at the hip and the knee joints. Also in optimization problem, we tune $D_m = 10\ (m)$ and $t_f = 10\ (s)$. By using Genetic algorithm, the optimal solution of the optimization problem in (37) is determined after 115 generations. The optimal solution of the optimization problem in (37) is equal to $X = \left(-0.063, 0.429, 0.172, 0.141, -0.109, -0.016\right)^T$.

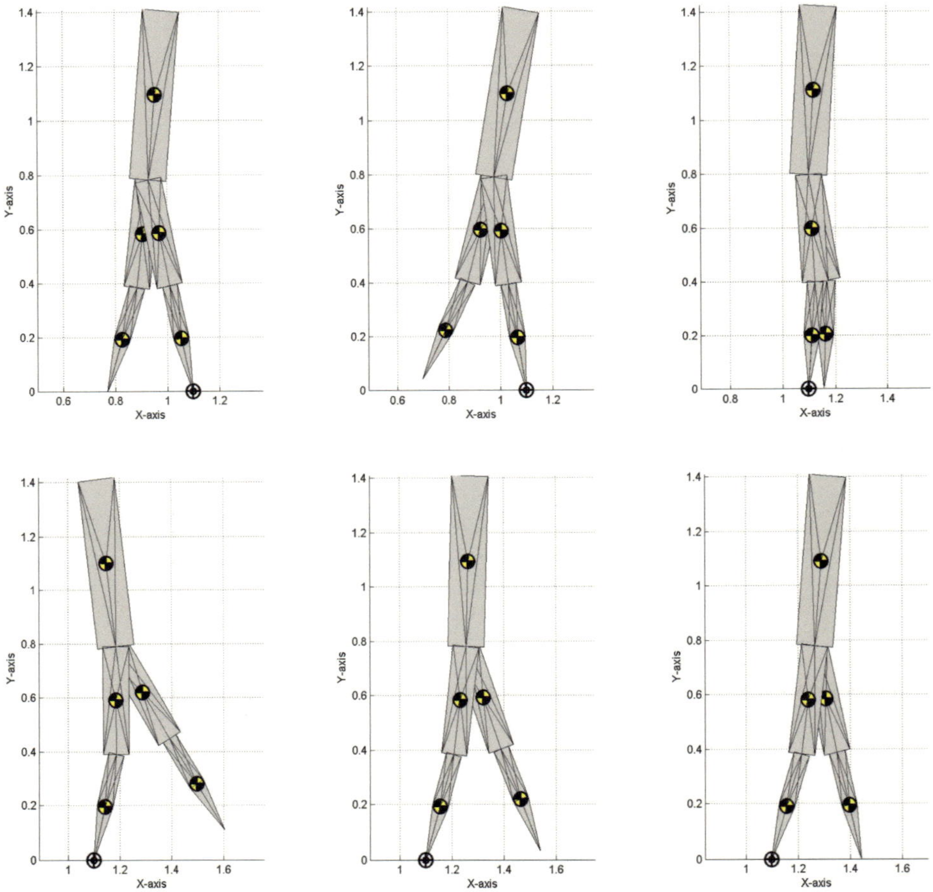

Fig. 5. The snapshots of one step for the biped robot with the best fitness.

The period of the neural oscillators in the biped robot with the best fitness is equal to $1.10\ (s)$. The time between consecutive impacts for this robot is equal to $T = 0.55\ (\mathrm{s})$. Also the step length during the walking (the distance between consecutive impacts) is equal to $sl = 0.33\ (\mathrm{m})$. The snapshots of one step for the best biped robot at limit cycle in this set of experiments are depicted in Fig. 5. In this picture, the left leg is taking a step forward. It can be seen that the swing leg performs a full swing and it allows sufficient ground clearance for the foot to be transferred to a new location. In Fig. 6, the CPG outputs and the joint angle positions of the leg joints during $10\ (s)$ are shown with dashed lines and solid lines, respectively. Figure 7 depicts the phase plot and the limit cycle of joint angle vs. velocity at the unactuated joint ($q_0 - \dot{q}_0$ plane) during $10\ (s)$. Also Fig. 8 depicts the limit cycles at the phase plots of the leg joints during $10\ (s)$.

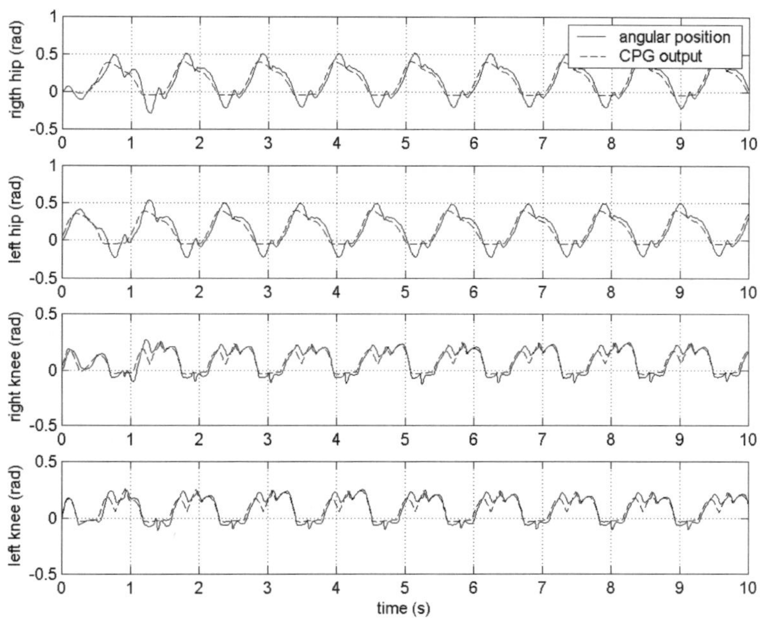

Fig. 6. The CPG outputs and the joint angle positionsof leg joints during $10\ (s)$.

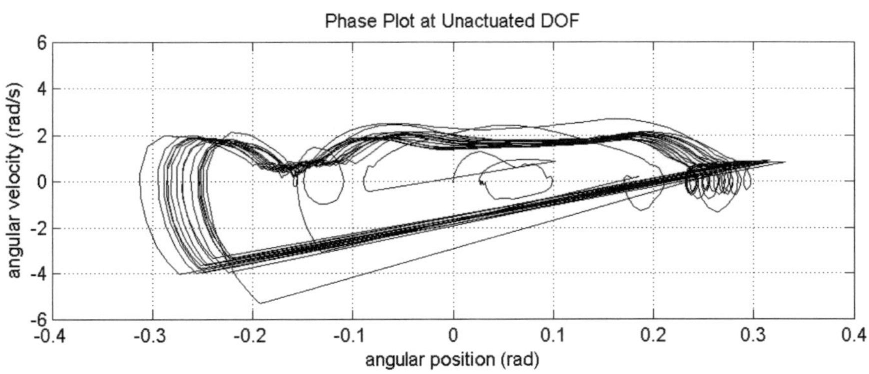

Fig. 7. The phase plot of joint angle vs. velocity at the unactuated joint ($q_0 - \dot{q}_0$ plane) during $10\ (s)$.

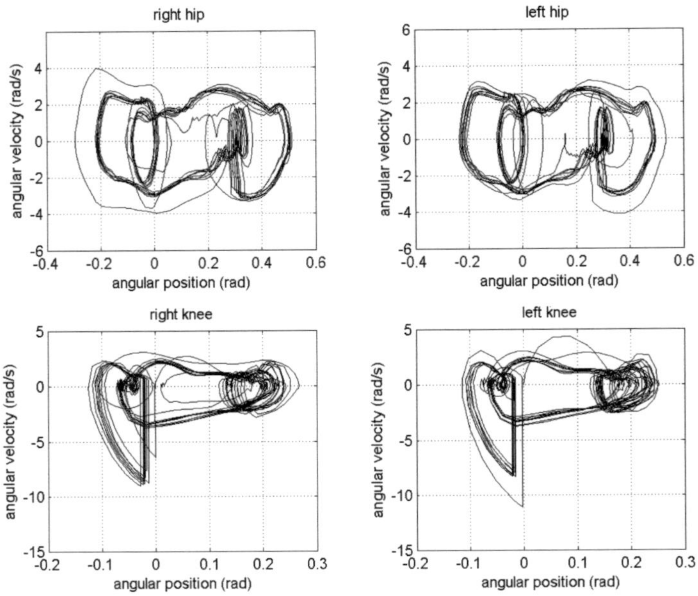

Fig. 8. The phase plots of joint angle vs. velocity at the leg joints during $10\ (s)$.

Control signals of the servo controllers during $10\ (s)$ are depicted in Fig. 9. The validity of the reduced single support phase model and impact model can be seen by plotting the ground reaction forces as plotted in Fig. 10.

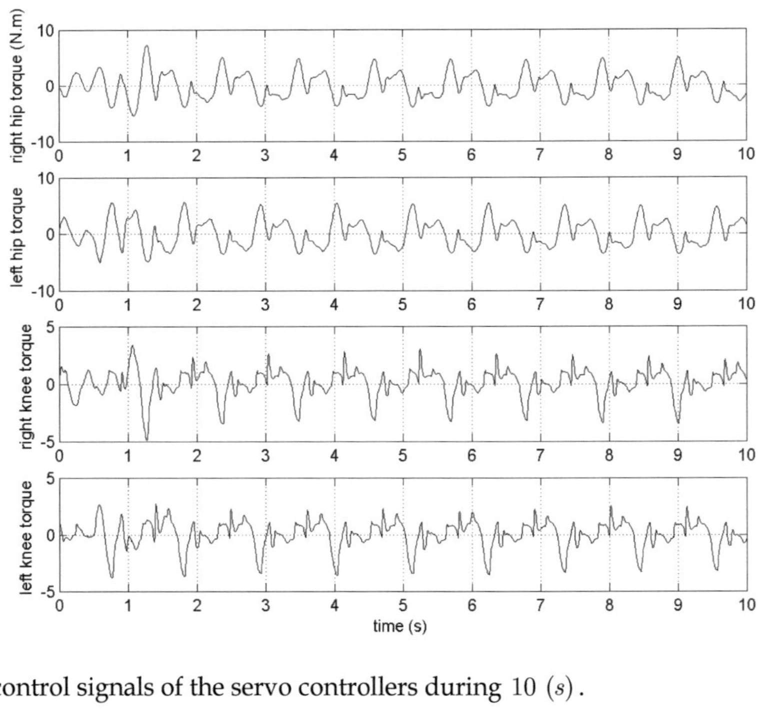

Fig. 9. The control signals of the servo controllers during $10\ (s)$.

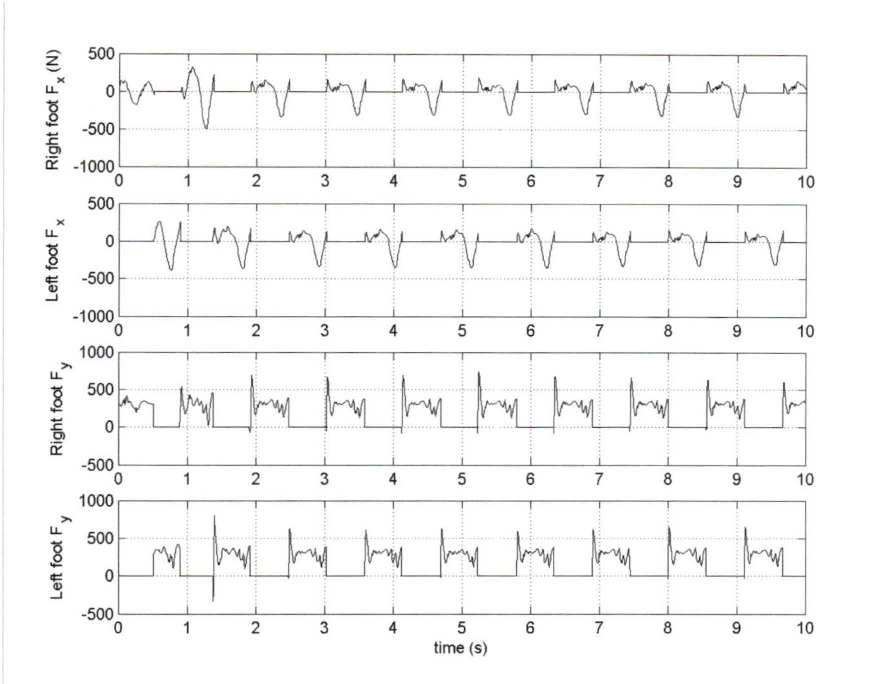

Fig. 10. The ground reaction forces at the leg ends during $10\ (s)$.

For evaluating the robustness of the limit cycle of the closed loop system, an external force as disturbance is applied to the body of the biped robot. We assume that the external force is applied at the center of mass of the torso and it can be given by $F_d(t) := F_d(u(t - t_d) - u(t - t_d - \Delta t_d))$ where F_d is the disturbance amplitude, t_d is the time when the disturbance is applied, Δt_d is the duration of the pulse and $u(.)$ is a unit step function. The stick figure of the robot for a pulse with amplitude $F_d = 25\ (N)$ and with pulse duration equal to $\Delta t_d = 0.5\ (s)$ which is applied at $t_d = 3\ (s)$ is shown in Fig. 11. This figure shows the robustness of the limit cycle due to disturbance. Also Fig. 12 shows the stable limit cycle at the unactuated joint. Figure 13 shows the maximum value of the positive and negative pulses vs. pulse duration which don't result in falling down.

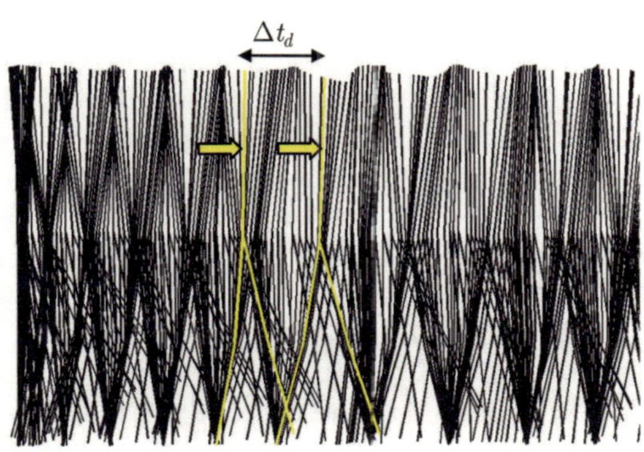

Fig. 11. Stick figure of the robot.

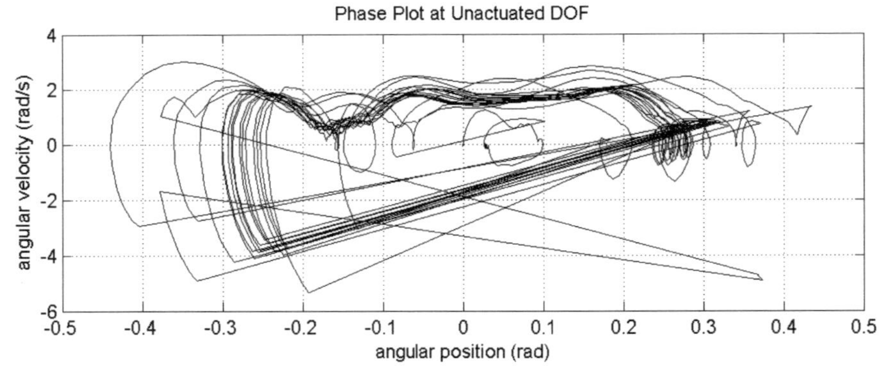

Fig. 12. The phase plot of joint angle vs. velocity at the unactuated joint.

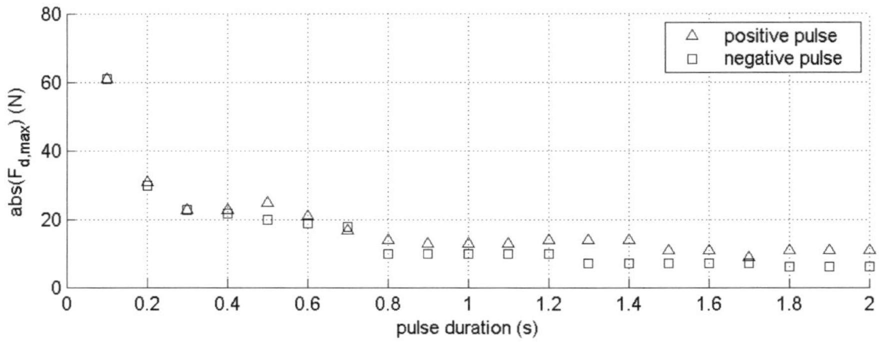

Fig. 13. Maximum amplitude of the pulse vs. pulse duration.

7. Conclusion

In this chapter, the hybrid model was used for modeling the underactuated biped walker. This model consisted of single support phase and the instantaneous impact phase. The double support phase was also assumed to be instantaneous. For controlling the robot in underactuated walking, a CPG network and a new feedback network were used. It is shown that the period of the CPG is the most important factor influencing the stability of the biped walker. Biological experiments show that humans exploit the natural frequencies of their arms, swinging pendulums at comfortable frequencies equal to the natural frequencies. Extracting and using the natural frequency of the links of the robots is a desirable property of the robot controller. According to this fact, we match the endogenous frequency of each neural oscillator with the resonant frequency of the corresponding link. In this way, swinging motion or supporting motion of legs is closer to free motion of the pendulum or the inverted pendulum in each case and the motion is more effective.

It is well known in biology that the CPG network with feedback signals from body can coordinate the members of the body, but there is not yet a suitable biological model for feedback network. In this chapter, we use tonic stretch reflex model as the feedback signal at the hip joints of the biped walker as studied before. But one of the most important factors in control of walking is the coordination or phase difference between the knee and the hip joints in each leg. We overcome this difficulty by introducing a new feedback structure for the knee joints oscillators. This new feedback structure forces the mechanical system to fix the stance knee at a constant value during the single support phase. Also, it forces the swing knee oscillator to increase its output at the beginning of swinging phase and to decrease its output at the end of swinging phase.

The coordination of the links of the biped robot is done by the weights of the connections in the CPG network. For tuning the synaptic weight matrix in CPG network, we define the control problem of the biped walker as an optimization problem. The total cost function in this problem is defined as a summation of the sub cost functions where each of them evaluates different criterions of walking such as distance travelled by the biped robot in the sagittal plane, the height of the CoM and the regulation of the angular momentum about the CoM. By using Genetic algorithm, this problem is solved and the synaptic weight matrix in CPG network for the biped walker with the best fitness is determined. Simulation results show that such a control loop can produce a stable and robust limit cycle in walking of the biped walker. Also these results show the ability of the proposed feedback network in correction of the CPG outputs. This chapter also shows that by using the resonant frequencies of the links, the number of unknown parameters in the CPG network is reduced and hence applying Genetic algorithm is easier.

8. References

[1] J. W. Grizzle, C. Moog, and C. Chevallereau, "Nonlinear control of mechanical systems with an unactuated cyclic variable," *IEEE Transactions on Automatic Control*, vol. 30, no. 5, pp. 559-576, May 2005.

[2] Y. Hurmuzlu, F. Genot, and B. Brogliato, "Modeling, stability and control of biped robots-a general framework," *Automatica*, vol. 40, pp. 1647-1664, 2004.

[3] C. Chevallereau, G. Abba, Y. Aoustin, F. Plestan, E.R. Westervelt, C. Canduas-de Wit, and J. W. Grizzle, "RABBIT: A testbed for advanced control theory," *IEEE Control Systems Magazine*, vol. 23, no. 5, pp. 57-79, October 2003.

[4] M. Vukobratovic and D. Juricic, "Contribution on the synthesis of biped gait," *IEEE Transactions on Biomedical Engineering*, vol. 16, no.1, pp. 1-6,1969.

[5] A. Takanishi, M. Ishida, Y. Yamazaki, and I. Kato, "The realization of dynamic walking robot WL-10RD," *Int. Conf. Advanced Robotics*, 1985, pp. 459-466.

[6] C. L. Shin, Y. Z. Li, S. Churng, T. T. Lee, and W. A. Cruver, "Trajectory synthesis and physical admissibility for a biped robot during the single support phase," *IEEE Int. Conf. Robotics and Automation*, 1990, pp. 1646-1652.

[7] K. Hirai, M. Hirose, Y. Haikawa, and T. Takenaka, "The development of honda humanoid robot," *IEEE Int. Conf. Robotics and Automation*, 1998, pp. 1321-1326.

[8] A. Dasgupta and Y. Nakamura, "Making feasible walking motion of humanoid robots from human motion capture data," *IEEE Int. Conf. Robotics and Automation*, 1999, pp. 1044-1049.

[9] A. Goswami, "Postural stability of biped robots and the foot rotation indicator (FRI) point," *International Journal of Robotic Research*, vol. 18, no. 6, pp. 523-533, June 1999.

[10] F. Plestan, J. W. Grizzle, E. R. Westervelt, and G. Abba, "Stable walking of a 7-DOF biped robot," *IEEE Transactions on Robotics and Automation*, vol. 19, no. 4, pp. 653-668, August 2003.

[11] T. McGeer, "Passive dynamic walking," *International Journal of Robotic Research*, vol. 9, no. 2, pp. 62-82, 1990.

[12] E. R. Westervelt, J. W. Grizzle, and D. E. Koditschek, "Hybrid zero dynamics of planar biped walkers," *IEEE Transactions on Automatic Control*, vol. 48, no. 1, pp. 42-56, January 2003.

[13] A. Isidori, *Nonlinear Control Systems: An Introduction*, 3rd ed. Berlin, Germany: Springer-Verlag, 1995.

[14] A. Isidori and C. Moog, "On the nonlinear equivalent of the notion of transmission zeros," in *Proc. IIASA Conf.: Modeling Adaptive Control*, C. Byrnes and A. Kurzhanski, Eds., Berlin, Germany, 1988, pp. 146-157.

[15] J. Guckenheimer and P. Holmes, *Nonlinear Oscillations, Dynamical Systems, and Bifurcations of Vector Fields*, corrected second printing ed., ser. Applied Mathematical Sciences. New York: Springer-Verlag, 1996, vol. 42.

[16] B. Morris and J. W. Grizzle, "A restricted Poincaré map for determining exponentially stable periodic orbits in systems with impulse effects: Application to bipedal robots," in *Proc. of IEEE 2005 Conference on Decision and Control*, 2005.

[17] S. Nersesov, V. Chellaboina, and W. Haddad, "A generalized of Poincaré's theorem to hybrid and impulsive dynamical systems," *Int. J. Hybrid Systems*, vol. 2, pp. 35-51, 2002.

[18] Y. Hurmuzlu, "Dynamics of bipedal gait - part 2: stability analysis of a planar five-link biped," *Journal of Applied Mechanics*, vol. 60, pp. 337-343, June 1993.

[19] J. H. Choi and J. W. Grizzle, "Feedback control of an underactuated planar bipedal robot with impulsive foot action," *Robotica*, vol. 23, pp. 567-580, September 2005.

[20] J. E. Pratt and R. Tedrake, "Velocity based stability margins for fast bipedal walking," http://www.ai.mit.edu/projects/leglab, 2007.

[21] S. Grillner, "Control of locomotion in bipeds, tetrapods and fish," *Handbook of Physiology II*, American Physiol. Society, Bethesda, MD, pp. 1179-1236, 1981.

[22] G. Taga, Y. Yamaguchi, and H. Shimizu, "Self-organized control of bipedal locomotion by neural oscillators," *Biolog. Cybern.*, vol. 65, pp. 147-159, 1991.

[23] G. Taga, "A model of the neuro-musculo-skeletal system for human locomotion II: real-time adaptability under various constraints," *Biolog. Cybern.*, vol. 73, pp. 113-121, 1995.

[24] K. Matsuoka, "Mechanism of frequency and pattern control in the neural rhythm generators," *Biolog. Cybern.*, vol. 56, pp. 345-353, 1987.

[25] K. Matsuoka, "Sustained oscillations generated by mutually inhibiting neurons with adaptation," *Biolog. Cybern.*, vol. 52, pp. 367-376, 1985.

[26] M. M. Williamson, "Neural control of rhythmic arm movements," *Neural Networks*, vol. 11, pp. 1379-1394, 1998.

[27] S. Miyakoshi, G. Taga, Y. Kuniyoshi, and A. Nagakubo, "Three dimensional bipedal stepping motion using neural oscillators-towards humanoid motion in the real word," *IROS98*, pp. 84-89, 1998.

[28] H. Kimura, Y. Fukuoka, Y. Hada, and K. Takase, "Three-dimensional adaptive dynamic walking of a quadruped rolling motion feedback to CPGs controlling pitching motion," *IEEE International Conference on Robotics and Automation*, 2002, pp. 2228-2233.

[29] H. Kimura and Y. Fukuoka, "Adaptive dynamic walking of the quadruped on irregular terrain -autonomous adaptation using neural system model," *IEEE International Conference on Robotics and Automation*, 2000, pp. 436-443.

[30] H. Ye, A. N. Michel, and L. Hou, "Stability theory for hybrid dynamical systems," *IEEE Trans. Automatic Control*, vol. 43, no. 4, pp. 461-474, Apr. 1998.

[31] B. Brogliato, "Some perspectives on the analysis and control of complementarity systems," *IEEE Transaction on Automatic Control*, vol. 48, no. 6, pp. 918-935, 2003.

[32] B. Brogliato, S. I. Niculescu, and M. Monteiro, "On the tracking control of a class of complementarity-slackness hybrid mechanical systems," *Systems and Control Letters*, vol. 39, pp. 255-266, 2000

[33] B. Brogliato, S. I. Niculescu, and P. Orhant, "On the control of finite dimensional mechanical systems with unilateral constraints," *IEEE Transactions on Automatic Control*, vol. 42, no. 2, pp. 200-215, 1997.

[34] H. Goldstein, *Classic Mechanics*, 2nd ed. Reading. MA: Addison Wesley, 1980.

[35] M. W. Spong and M. Vidyasagar, *Robot Dynamics and Control*, New York: Wiley, 1991.

[36] E. Dombre and W. Khalil, *Modeling, Identification and Control of Robots*, Paris: Hermes Sciences, 2002.

[37] R. M. Murray, Z. Li, and S. Sastry, *A Mathematical Introduction to Robotic Manipulation*, Boca Raton, FL: CRC Press, 1993.

[38] E. R. Westervelt and J. W. Grizzle, "Design of asymptotically stable walking for a 5-link planar biped walker via optimization," *2002 IEEE International Conference on Robotics and Automation, Wahington D.C.*, 2002, pp. 3117-3122.

[39] Y. Hurmuzlu, "Dynamics of bipedal gait-part I: objective functions and the contact event of a planar five-link biped," *Journal of Applied Mechanics*, vol. 60, pp. 331-336, June 1993.

[40] J. H. Choi and J. W. Grizzle, "Planar bipedal robot with impulsive foot action," *IEEE Conf. on Decision and Control*, Paradise Island, Bahamas, December 2004, pp. 296-302.

[41] Y. Hurmuzlu and D. Marghitu, "Rigid body collisions of planar kinematic chains with multiple contact points," *International Journal of Robotics Research*, vol. 13, no. 1, pp. 82-92, 1994.

[42] K. Schneider, R. F. Zernicke, R. A. Schmidt and T. J. Hart, "Changes in limb dynamics during the practice of rapid arm movements," *Journal of Biomechanics*, vol. 22, pp. 805-817, 1989.

[43] G. P. Bingham, R. C. Shmidt, and L. D. Rosenblum, "Hefting for a maximum distance throw: a smart perceptual mechanism," *Journal of Experimental Psychology: Human Perception and Performance*, vol. 15, no.3, pp. 507-528, 1989.

[44] H. Kimura, Y. Fukuoka, K. Konaga, Y. Hada, and K. Takase, "Towards 3D adaptive dynamic walking of a quadruped robot on irregular terrain by using neural system

model," *IEEE/RSJ International Conference on Intelligent Robots and Systems,* 2001, pp. 2312-2317.

[45] M. Abdullah and A. Goswami, "A biomechanicaly motivated two-phase strategy for biped upright balance control," *IEEE International Conference on Robotics and Automation,* 2005.